POLYNOMIAL COMPLETENESS
in
ALGEBRAIC SYSTEMS

POLYNOMIAL COMPLETENESS
in
ALGEBRAIC SYSTEMS

Kalle Kaarli
Alden F. Pixley

CRC Press
Taylor & Francis Group
Boca Raton London New York

CRC Press is an imprint of the
Taylor & Francis Group, an **informa** business
A CHAPMAN & HALL BOOK

CRC Press
Taylor & Francis Group
6000 Broken Sound Parkway NW, Suite 300
Boca Raton, FL 33487-2742

First issued in paperback 2019

ISBN-13: 978-1-58488-203-9 (hbk)
ISBN-13: 978-0-367-39833-0 (pbk)

**Visit the Taylor & Francis Web site at
http://www.taylorandfrancis.com**

**and the CRC Press Web site at
http://www.crcpress.com**

Library of Congress Cataloging-in-Publication Data

Kaarli, Kalle.
     Polynomial completeness in algebraic systems/Kalle Kaarli and Alden Pixley.
       p. cm.
     Includes bibliographical references and index.
     ISBN 1-58488-203-4 (alk. paper)
     1. Completeness theorem. 2. Polynomials. I. Pixley, Alden F. II. Title.

QA9.67 .K37 2000
511.3--dc21

00-034538

Contents

Preface

Historically, the theory of general algebraic systems—also known as universal algebra—was established as a discipline devoted to generalizing and systematizing results obtained in the study of specific algebraic structures. Among the latter a special role was played by Boolean algebras, a role which in many respects was somewhat comparable to the traditional organizing role of group theory in the broader mathematical arena. We have in mind here, for example, the theory and application of ultrafilters, their primary role in logic and especially in model theory, particularly the ultraproduct construction, Boolean valued models, sheaf constructions, and the theory of Boolean products. In this organizing role, Boolean algebras have provided important linkage between the sometimes disparate areas of algebra and analysis, set theory, mathematical logic, and computer science. In view of this special role it is not surprising that focusing on specific properties of Boolean algebras has led to the establishment of new directions in universal algebra. To mention just one, the famous Stone duality theorem, allowing one to apply topological methods in Boolean theory, has given rise to general duality theory which has spread far beyond the original limits of Boolean algebras. The purpose of the present monograph is to focus on and systematically extend another specific property of Boolean algebras: the property of *affine completeness*. We will develop this property in the context of a somewhat more general setting which we loosely refer to as *polynomial completeness*. It is remarkable how much of the flavor of Boolean algebras will persist in this vastly wider class of systems and, indeed, though we shall not discuss it in detail, work of other authors, most recently Brian Davey, David Clark, and Clifford Bergman, demonstrates that results of the generalized Stone duality theory extend to many of the structures which we shall discuss.

Why the name *affine completeness*, which sounds geometric? In fact, just as Boolean algebras and related systems, for example Heyting algebras, monadic algebras, etc., were invented to algebraize classical and related propositional logics, so also have various algebraic systems been devised which coordinatize various geometries; for example skew fields coordinatize the affine plane without Pappus' theorem. In 1970 Rudolf Wille, working in this same spirit, introduced the general concept of *congruence class geometries* wherein the appropriate coordinatizing algebraic systems turned out to be what he called *affine complete*. Thus our motivation for pursuing the study of affine completeness as a purely algebraic phenomenon is based on the somewhat remarkable fact that the processes of algebraizing both logic and geometry lead to the same common property.

Our starting point—both logically and historically—is the fact, again due to Marshall Stone, that the class of all Boolean algebras is generated by its smallest nontrivial member—the two element algebra—and that this algebra is *primal*, meaning that all functions on its universe are induced by terms, that is, are functions which are obtained by composition from the projections and the basic operations of the algebra. This fact is the mathematical basis of the algebraic theory of switching circuits and has a fundamental place in mathematical logic. The general study of primal algebras was initiated by Alfred Foster. In the 1950s and 1960s Foster (in part jointly with Alden Pixley) wrote a series of papers developing and expanding the theory of primal algebras. The development of this theory depended essentially on fundamental new work in the theory of varieties. Varieties, also called equational classes or primitive classes, are classes of algebraic systems which are defined by equations—the classes of groups, rings, and lattices are familiar examples. While their basic properties were discovered by Garrett Birkhoff in the early 1930s, in 1954 A. I. Mal'cev made a simple discovery—leading to a general theory—which enables one to make a direct connection between the defining equations of a variety and specific algebraic properties of the systems of the variety. The resulting theory of Mal'cev conditions flourished in the 1960s and 1970s and greatly enhanced the subsequent explosive growth of universal algebra.

Mal'cev's original discovery was a characterization of the algebraic property called *congruence permutability*, meaning that the least congruence relation of an algebra containing two given congruences, i.e., their join, is just their relation product, a property familiar from groups and rings where the join of normal subgroups and ideals is just their product and sum. Another important Mal'cev condition is *congruence distributivity*, meaning that the lattice of all congruence relations of an algebra is distributive, as is true in the case of the ring of integers. Since the ring of integers is both congruence permutable and distributive, the conjunction of these two properties, which is also a Mal'cev condition, is called *arithmeticity*. An important result of this early period was the observation that primal algebras can be characterized by ordinary general algebraic concepts: a finite algebra is primal if and only if it has no proper subalgebras, no nontrivial congruences and automorphisms and generates an arithmetical variety. This was probably the first significant application of the theory of Mal'cev conditions. Also, during this period a guiding belief of Foster—that whatever could be said about the variety of Boolean algebras could also be said in some corresponding form about the variety generated by any primal algebra—was justified in precise form in 1969 by T. K. Hu. Specifically, Hu showed that any primal algebra generates a variety which is equivalent as a category to the variety of Boolean algebras.

These early results encouraged these and other investigators to extend results about primality in several directions. Primality can be regarded as a kind of *completeness* of an algebra: an algebra is primal if it is maximally complete in term functions. However, one may require that not all, but only some (for example, just the congruence preserving functions), be term functions. Also one may replace term functions by polynomial functions, the class of functions, broader than the class of term functions, obtained by using constants as well as the basic operations in constructing new functions by composition. Further, it is also natural to study algebras in which certain functions (or partial functions) can be interpolated by term or polynomial functions on finite subsets of the algebra. All of these classes generalize primality and the properties defining it, and thus are examples of the general phenomenon, *polynomial completeness*, occuring in the title of the book.

For further orientation let us discuss in a little more detail some particular properties which generalize primality and whose study has generated a number of valuable ideas. A finite algebra is said to be *functionally complete* if all functions on its universe are induced by polynomials. Galois fields are the classical examples of such algebras and finite simple nonabelian groups are a more recently recognized class of examples. An interesting discovery of Wille's student, Heinrich Werner in 1970, was that the functional completeness of an algebra follows already from the polynomial representation of a particular single function, Pixley's so-called *(ternary) discriminator* function. It turned out that the role of the discriminator was not restricted to just Werner's discovery: this function was also used to characterize the important class of *quasiprimal* algebras, namely as finite algebras having the discriminator as a term function. These special functionally complete algebras may have quite different structures compared to primal algebras, but the behavior of the varieties generated by them is very similar to the variety generated by a primal algebra, and they constitute, at this level of generality, the most important generalization of the two element Boolean algebra. For example, quasiprimal generated varieties have played an important role in describing varieties with a decidable first order theory.

Since functionally complete algebras are always simple it is obviously of interest to extend this class to nonsimple algebras which have maximal sets of polynomial functions. Since polynomial functions of an algebra are always compatible with all congruence relations of the algebra it is natural to isolate for study algebras whose polynomial functions are precisely those which are compatible with the congruences. But this had already led Wille, in his study of geometries, to the principal topic of the present book: affine complete algebras. An algebra is *affine complete* if every congruence compatible function is induced by a polynomial. (Thus a finite algebra is functionally complete if and only if it is simple and affine complete.) Affine complete algebras are therefore the most important generalization of Boolean algebras in the present context of polynomial completeness and it is important to recall here that their connection with Boolean algebras was established quite early and, indeed, was first recognized by George Grätzer in 1962 (before many of the relevant universal algebraic tools were fully developed) when he proved that every Boolean algebra (that is, the entire variety of Boolean algebras) is affine complete!

Thus we are led to the study of both (1) affine complete algebras and (2) affine complete varieties, i.e., varieties all of whose members are affine complete. Unlike the case for functional completeness, there does not appear to be any interesting or significant way of characterizing all affine complete algebras. (That this may be so is suggested by the simple observation that *every* algebra is a reduct of an affine complete algebra.) What has been a more successful program than (1) is (1′): the description of the affine complete members of specific varieties. This program has been completed for, among others, the varieties of vector spaces over a given field, abelian groups, distributive lattices, semilattices, and both Stone and Kleene algebras. These results will be discussed extensively in Chapter 5 of this book.

But of course it is the entire variety of Boolean algebras which is affine complete, so that it is to the study (2) of affine complete varieties where we must go if we hope to capture much of the motivating Boolean theory. In particular, for affine complete varieties, the generating algebras will be the analogs of the two element Boolean algebra.

The study and description of affine complete varieties has indeed turned out to be a rich and fruitful topic and has involved and contributed to much of the contemporary theory of algebraic systems. Since Boolean algebras form an arithmetical variety of finite type the characterization of affine complete arithmetical varieties of finite type is of obvious interest and here our results are most satisfactory. In particular we shall show that such a variety is affine complete if and only if it is generated by a finite algebra having no proper subalgebras. Thus the generating algebra differs from the generator of the variety of Boolean algebras only in that it may have finite cardinality larger than 2 and it may fail to be either simple or rigid. Otherwise the analogy is strikingly close. Moreover, even in the case of nonarithmetical affine complete varieties we will still find striking persistence of much of the Boolean character. We believe that this, together with the importance of the new concepts developed along the way, more than justifies presentation of an account of the subject at this time. This is the focus of the first four chapters of the present book which, with the inclusion of Chapter 5 and other related material developed along the way, is intended to constitute a coherent, if not encyclopedic, introduction to the general topic of polynomial completeness.

The systematic study of affine complete varieties was initiated in the early 1980s by the present authors. Most of the subsequent development was done by Kalle Kaarli and, more recently, jointly by Kaarli and Ralph McKenzie.

We have written this book at a level which we hope might be readable by anyone with a minimal background in lattice theory, the elementary model theory of first order logic, and general algebraic systems (for example a student who has completed a first graduate course in the subject). In particular we believe that a reader of the very accessible book [Burris, Sankappanavar, 1981] should be quite well prepared. (This admirable text is now available free on the internet at the site www.thoralf.uwaterloo.ca.) An exception to these prerequisites may be encountered in some sections of Chapter 5 where some specialized knowledge of certain varieties may be required. Appropriate references are included when this seems likely.

With this background in mind, the first chapter, particularly Section 1.1, is a review of basic material and also is intended to fix terminology and notation. Section 1.2 also contains mostly background material (oriented towards our later requirements) except for Subsection 1.2.5 which contains material which has appeared previously only in journal articles.

Chapter 2 gathers together important material on equivalence lattices presented systematically together for the first time. Much of the material of Section 2.3 on compatible function lifting is new and is the basis for the proof (in Chapter 4) that every affine complete variety is congruence distributive, one of the most important facts about these varieties.

Chapter 3 starts with primal algebras and systematically develops their natural generalizations including affine completeness. We also introduce important related concepts of interpolation by polynomials. The importance of the roles played by congruence distributivity, congruence permutability, and near unanimity varieties is described in this chapter. There is a huge literature on primality and its generalizations, so beyond what we need for our central goal we only survey enough of the subject to put it into perspective in the larger context of polynomial completeness theory. We also include a brief overview of recent work on categorical equivalence which applies to many affine complete varieties and which generalizes Hu's basic result about the categorical equivalence of all primal varieties. It is hoped that this research will continue and that it may be possible to obtain a complete set of invariants, relative to categorical equivalence, for all affine complete varieties, at least of finite type.

Chapter 4 presents the main results, known to the present time, concerning affine complete varieties. Most important, it is shown that such varieties are always residually finite and congruence distributive. Also the three concepts, locally finite, finitely generated, and term equivalent to a variety of finite type, are shown to be equivalent. In these cases the variety is also shown to possess a near unanimity term. While we have a reasonable description of such varieties in the general case, as remarked above, a complete description in the arithmetical case is obtained and C. Bergman's recent characterization of categorical equivalence for this case is briefly discussed. Chapter 4 thus delineates the general territory and boundaries of the domain of affine complete varieties. We hope it will provide an incentive for others to join in the exploration of the more detailed aspects of the territory.

Chapter 5 describes the affine complete members in several special varieties as well as related interpolation properties. These results are of course of a quite different character than those of the earlier chapters since they are generally dependent on the methods common to the particular varieties (lattices, modules, etc.) being studied. This is also partly responsible for the length of the chapter.

Acknowledgments We are indebted to several mathematicians for their help and suggestions in preparing this book. These include especially Clifford Bergman, Ervin Fried, Miroslav Haviar, László Márki, Miroslav Ploščica, and Ivo Rosenberg, and Kaarli's student Vladimir Kutshmei.

We are also grateful for the support of our work provided by the Estonian Science Foundation (Grant number 3372).

Finally, we acknowledge, with thanks, permission from Birkhäuser Publishers, Ltd. to reproduce Figures 1.1 and 5.2 which were originally published in [Kaarli, McKenzie, 1997] and [Wille, 1977b] respectively.

Kalle Kaarli
University of Tartu
Tartu, Estonia

Alden Pixley
Harvey Mudd College
Claremont, California

March, 2000

Chapter 1

Algebras, Lattices, and Varieties

In Section 1.1 of this chapter we shall very briefly sketch some of the basic concepts about (universal) algebras, varieties of algebras, lattice theory, and a few ideas from the theory of clones, all presented with our subsequent requirements in mind. In Section 1.2 we develop material on congruence properties of algebras, again slanted towards our subsequent needs.

Most, but not all, of the general background can be found, sometimes in slightly different form, in the very readable introductory text of [Burris, Sankappanavar, 1981], or in the more advanced books [McKenzie, McNulty, Taylor, 1987] and [Grätzer, 1979]. A significant exception is Subsection 1.2.5 where most of the material has so far only appeared in journal articles. This new material is needed in Section 2.3 and the results developed there will be applied to prove a principal result of Chapter 4 (the congruence distributivity of all affine complete varieties). In Subsection 1.2.6 we briefly introduce the commutator for universal algebras; this material will not be needed until Chapter 5.

1.1 Algebras, languages, clones, varieties

1.1.1 Algebras

An **algebra**, also called a **universal algebra** or an **algebraic system,** is a system $\mathbf{A} = \langle A; F \rangle$ where A is a nonempty set (the **uni-**

verse of A) and F is a set of finitary operations on A, called the **basic operations** of **A**. The algebra is **trivial** if $|A| = 1$. In general by the cardinality of an algebra we mean the cardinal number $|A|$. By **finitary operations** we mean functions $f : A^m \to A$ where m (the **rank** or **arity** of f) is a nonnegative integer. If $m = 0$ we call f a **nullary** (necessarily constant) operation. The following are some common examples.

Binary systems Usually a binary operation is denoted by infix rather than the prefix notation indicated above. Thus $\times$ usually denotes a binary operation on a set; ordinarily $+$ is used if the intention is that the operation is commutative. A binary system $\mathbf{A} = \langle A; \times \rangle$ is called a **groupoid** if no special properties of the operation (i.e., axioms) are specified. (We are using the common convention of simply listing the basic operations in the set F without the enclosing braces $\{,\}$ if F is finite.) A groupoid is a **semigroup** if the operation satisfies the associative law. An **inverse semigroup** is one with the property that for each $x \in A$, there is a unique $y \in A$, called the inverse of x, such that $x \times y \times x = x$ and $y \times x \times y = y$. A semigroup is a **monoid** if in addition A contains an element e which is a two sided identity with respect to $\times$. Notice that if $\mathbf{A}$ is a monoid we can describe this by stipulating that the axiom $(\exists e)(\forall x)(e \times x = x \times e = x)$ is satisfied by the algebra. Alternatively it is possible to consider a monoid to be an algebra $\mathbf{A} = \langle A; \times, e \rangle$, where e is a nullary operation, satisfying the axiom $(\forall x)(e \times x = x \times e = x)$, (and $\times$ is associative). The importance of the difference lies primarily in the fact that in the latter case the axioms are universally quantified equations while in the former case an existential quantifier is required. Similarly, groups are commonly described either as semigroups satisfying the usual existentially quantified axioms asserting the existence of inverses and an identity or, alternatively, as algebras having as basic operations the binary operation $\times$ and also a unary inverse operation $^{-1}$, and possibly also a nullary identity operation e, in either of which cases the axioms can be taken to be universally quantified equations. Inverse semigroups can also be taken with a unary inverse operation.

Rings Usually rings are described most economically as algebras $\mathbf{R} = \langle R; +, \cdot \rangle$ subject to familiar axioms which may or may not require the existence of an identity element (also often called a unity). One way of describing rings, using only (universally quantified) equational axioms, is as algebras $\langle R; +, -, 0, \cdot \rangle$ where $-$ is the unary inverse relative to the commutative $+$, 0 is the nullary identity relative to $+$, and $\cdot$ is the binary multiplication making $\langle R; \cdot \rangle$ a semigroup. This semigroup is a reduct of the original ring. In general algebra **A** is a **reduct** of algebra **B** if they have the same universe and the basic operations of **A** form a subset of those of **B**.

Some authors also consider algebras with some **partial operations** among their basic operations and call the resulting structures **partial algebras**. Partial operations are operations whose domain is a proper subset of A^k. For example, division can be taken as a basic partial operation in any field. The other operations are often then called "total" operations. A **finite partial operation** is one for which the domain is a finite subset of A^k. For example, the Lagrange Interpolation Theorem tells us that in any field a finite partial operation can be *interpolated* by a unique polynomial of minimal degree.

Sometimes, when either common usage or the context make it seem more appropriate, we shall use the word *function* instead of *operation* even if the latter word might be preferable in terms of the preceding definition. For example sometimes the word "operation" is reserved for basic operations (partial or total) and "function" (partial or total) just as an alternative to operation (partial or total). The context will always make the meaning clear.

While we shall frequently wish to consider partial operations we shall never, in this book, encounter partial algebras. This will make many results easier to state and, as discussed in the next section, presents no problems.

Lattices Recall that lattices are ordered sets in which each pair of elements a, b has both a least upper bound $(a \vee b)$, and a greatest lower bound $(a \wedge b)$, their *join* and *meet* respectively. An important fact is that by taking $\vee$ and $\wedge$ as binary operations, a lattice can be taken as an algebra $\mathbf{L} = \langle L; \vee, \wedge \rangle$ satisfying the familiar dual equational axioms which assert that the operations are each idempotent, associative, commutative, and together satisfy the dual absorption axioms. Conversely, if we start with a lattice as an algebra described in this way we can define an order relation $\leq$ by taking $a \leq b$ to mean either $a \wedge b = a$ or equivalently $a \vee b = b$. Thus the two ways of considering lattices are interchangeable. If a lattice has a largest element it is denoted by 1 and if it has a least element it is denoted by 0. Lattices with both 0 and 1 are called **bounded lattices**.

Distributive lattices are those which satisfy either (and therefore both) of the dual distributive laws of $\vee$ with respect to $\wedge$ or $\wedge$ with respect to $\vee$. **Modular lattices** are those in which the distributive laws hold among three elements in those cases where two of the three are comparable with respect to the order relation. This fact is expressible by means of either of the two equivalent dual modular laws. It is important to observe that considered as algebras (with or without 0 or 1) the classes of lattices, distributive lattices, and modular lattices can all be axiomatized by means of equations.

It is a familiar fact that in distributive lattices, complements, when they exist, are unique. Therefore, in a **Boolean lattice**, which is defined to be a complemented distributive lattice, the complement of each element can be denoted as the value of a unary operation on the lattice. If we take complementation as one of the basic operations the resulting algebra is called a **Boolean algebra**. Hence the difference between a Boolean algebra and a Boolean lattice depends only on whether complementation is taken as a basic operation or not. Boolean algebras are then equationally axiomatizable while Boolean lattices are not.

In a lattice with 0 a **pseudocomplement** of an element a is an element a^* with the properties: $a \wedge a^* = 0$, and $a \wedge x = 0$ implies $x \leq a^*$; thus an element can have at most one pseudocomplement. A **pseudocomplemented lattice** is one in which each element has a pseudocomplement. Finally, a lattice **L** is **complete** if every subset X of L (including the empty set) has both a meet $\bigwedge X$ and join $\bigvee X$. Since join and meet in a complete lattice are *infinitary* operations, complete lattices are not special kinds of algebras in our sense. An element a of a complete lattice is **compact** if for all $X \subseteq L$, if $a \leq \bigvee X$, then $a \leq \bigvee X_0$ for some finite subset $X_0 \subseteq X$. A complete lattice is **algebraic** if each element is the join of a set of compact elements. An important fact is that every distributive algebraic lattice is pseudocomplemented. In particular we shall use the fact that each finite distributive lattice is pseudocomplemented.

Modules Since a module **M** over a ring **R** involves multiplication by ring elements as well as the abelian group operation on module elements, modules do not seem to fit our definition of algebras. But by the simple device of introducing, for each ring element r a unary operation f_r, defined for each $m \in M$ by $f_r(m) = rm$, we can consider (in this case) left **R**-modules, for fixed **R**, as algebras

$\mathbf{M} = \langle M; +, \{f_r\}_{r \in R} \rangle$ subject to the usual module axioms. If we also take the module subtraction as a basic operation the axioms may be taken to be universal equations.

1.1.2 Languages and structures

In addition to the simple direct way of thinking of algebras, just described, in order to generally discuss collections of algebras having "the same operations" and hence such concepts as direct products or homomorphisms, we really need to think of algebras from a logical standpoint, that is as structures in the context of model theory. Thus we sketch a little of this background.

A **first order language** L consists of a set O of operation symbols, a set R (disjoint from O) of relation symbols, and a function r which assigns to each member s of $O \cup R$ a nonnegative integer $r(s)$ called the *rank* (or *arity*) of s. Thus a first order language is a system $L = \langle O; R; r \rangle$. If R is empty L is an **algebraic language** while if O is empty L is a **relational language**. In this book we shall be almost entirely concerned with algebraic languages, but it is worthwhile here to recall a few basic facts about more general first order languages. An L-**structure** is a system

$$\mathbf{A} = \langle A; f^{\mathbf{A}}(f \in O); s^{\mathbf{A}}(s \in R) \rangle$$

which consists of a nonempty set A, an $r(f)$-ary operation $f^{\mathbf{A}}$ on A for each $f \in O$, and an $r(s)$-ary relation $s^{\mathbf{A}}$ on A for each $s \in R$. While an $r(f)$-ary operation on A is a function with domain $A^{r(f)}$ and range in A, an $r(s)$-ary relation on A is a nonempty subset of $A^{r(s)}$. The empty set is not assigned an arity. Notice that operations are therefore necessarily "total" operations (having domain equal to $A^{r(f)}$). L-structures are also often called **models** of L. The models of L are called **algebras** if L is algebraic and are called **relational structures** if L is a relational language. The **type** of a structure $\mathbf{A}$ is its language. Thus a structure is of finite type if it has only a finite number of operations and relations. The structure $\mathbf{A}$ is **finite** if its **universe** A is a finite set. If $\mathbf{A}$ is a structure of finite type with operations $f_1, \ldots, f_n$ and relations $s_1, \ldots, s_m$ we usually write $\mathbf{A} = \langle A; f_1, \ldots, f_n; s_1, \ldots, s_m \rangle$. Occasionally there are cases where we want to consider structures which do not immediately fit into our description of them as models of some first order language. As noted above, a case in point occurs when we want to study algebras having some partial operations. But at least as far as model theory is concerned this is usually not a problem since operations are just relations with special first order properties; hence we can just consider partial operations to be relations subject to these properties. In this way an algebra with partial operations becomes a structure which is no longer just an algebra.

Whether we think of algebras in the simplest way as just sets with operations or as models of first order languages is sometimes not very important and certainly for any algebra in the former sense we can always index the operations by an appropriate set and introduce an appropriate first order language. Thus, in particular we can refer to the type of an algebra in either case; and for this reason we often think of the type as simply the set of its operations (taken as symbols, together with their arities). To distinguish between algebras in these two senses it is common to refer to algebras as models of first order languages as **indexed** algebras, and in the other sense as **nonindexed** algebras. If we refer to say "algebras **A** and **B** of the same type" it is implicit that we consider them to be indexed algebras. Just as is common in most of classical group and ring theory, if the context permits, we often just use an operation symbol, f, instead of the corresponding operations $f^{\mathbf{A}}, f^{\mathbf{B}}$, etc., which the formalism would require.

We shall assume that the reader is familiar with the general model theoretic concepts of **satisfaction and truth** in the context of first order languages. For our purposes these are adequately discussed in [Burris, Sankappanavar, 1981]; we review here the basic notation and results we shall need in the sequel. First recall that for a first order language $L = \langle O; R; r \rangle$, the **terms** and the **formulas** of L are certain recursively defined strings of symbols over an alphabet consisting of the union of the sets of operation symbols O, relation symbols R, a countable set of variables $\{x_1, x_2, \ldots\}$, the punctuation symbols: left parenthesis $\{(\}$, right parenthesis $\{)\}$ and comma $\{,\}$, and the logical symbols $\{\&, \lor, \neg, \rightarrow, \leftrightarrow, \forall, \exists, \approx\}$.

The simplest terms are the variables and the nullary operation symbols. All other terms have the form $f(t_1, \ldots, t_{r(f)})$ where f is an operation symbol and $t_1, \ldots, t_{r(f)}$ are terms. A term is called n-ary if no more than n distinct variables occur in it.

The simplest formulas are of form $t_1 \approx t_2$ or $s(t_1, \ldots, t_{r(s)})$ where

$t_1, \ldots, t_{r(s)}$ are terms and s is a relation symbol. They are called **atomic formulas**. All other formulas are obtained from atomic formulas by means of logical symbols. A formula ϕ is a **sentence** if each variable of ϕ is in the scope of some quantifier, i.e., if ϕ has no free variables. The concepts of satisfaction and truth of formulas are then defined recursively. For a sentence ϕ of L, and an L-structure **A**, we write $\mathbf{A} \models \phi$ (**A** models ϕ) to mean that ϕ is true in **A**; more generally, if K is a class of L-structures and Γ is a set of sentences of L,

$$K \models \Gamma \quad \text{means that} \quad \mathbf{A} \models \phi \quad \text{for all} \quad \mathbf{A} \in K \quad \text{and} \quad \phi \in \Gamma.$$

In this case we say that K is a class of models of Γ. (Hence models of L, defined earlier as another name for L-structures, may be thought of as just models of the empty set of sentences.) The symbol $Mod(\Gamma)$ will denote the class of all models of Γ. By the cardinality of the language L, denoted by $|L|$, we mean the cardinality of the set $O \cup R$ of operation symbols and relation symbols. Two basic theorems from model theory which we shall require are the following:

Theorem 1.1.1 (Compactness Theorem) *A set Γ of sentences has a model iff every finite subset of Γ has a model.*

Since satisfiability and truth are equivalent for sentences, sometimes the Compactness Theorem is stated briefly as:

A set of sentences is satisfiable iff it is finitely satisfiable.

Theorem 1.1.2 (Löwenheim-Skolem Theorem) *If a set Γ of sentences of language L has an infinite model then it has models of every infinite cardinality $\kappa \geq |L|$.*

More on these theorems together with simple applications can be found in [Burris, Sankappanavar, 1981] and [Grätzer, 1979].

Basic Algebraic Concepts Let **A** and **B** be algebras of the same type. Then **B** is a **subalgebra** of **A** if $B \subseteq A$ and each basic operation of **B** is the restriction of the corresponding operation of **A**, i.e., for each operation symbol f of the type, $f^{\mathbf{B}}$ is the restriction to B of $f^{\mathbf{A}}$. We write $\mathbf{B} < \mathbf{A}$ to indicate that **B** is a subalgebra of **A**. If B is a proper subset of A we say that **B** is a **proper** subalgebra of

A. A **subuniverse** of **A** is a subset $B \subseteq A$ which is closed under each of the basic operations of **A**. We denote the collection of subuniverses of **A** by Sub **A**.

Note that if **B** < **A** then B is a subuniverse of A. However, note also that the empty set may be a subuniverse of **A** and in this single case it cannot be the universe of any subalgebra of **A**. On the other hand, if the type contains nullary operations, and only in this case, then every subuniverse is nonempty. Since we allow the empty set to be a subuniverse perhaps we should also allow an algebra to have an empty universe and this is commonly done in category theory (where a category often has an initial object). Nonetheless we shall follow the usual convention for algebra as described above despite the seeming inconsistency; the main reason for excluding the "empty algebra" is that this convention allows us to state many definitions and theorems without making special exception for the empty algebra.

For illustration let **G** be a group with the usual multiplication × as the only operation. Then a subuniverse of **G** can be empty and, even when not, need not be the universe of a subgroup. If **G** has in addition the unary inverse operation $^{-1}$, then every nonempty subuniverse is the universe of a subgroup. If we further include the nullary identity operation e then each subuniverse is the universe of a subgroup. This suggests how we can, by appropriately choosing the basic operations, usually resolve conflicts between our notion of subalgebra and the standard terminology of classical algebra. But, for example, there is no really practical way in which the subfields of a field can be construed as subalgebras without resorting to partial algebras.

Allowing the empty set to be a subuniverse also implies that arbitrary collections of subuniverses are always closed under the taking of intersections. From this it follows that the collection of all subuniverses of an algebra becomes a complete lattice which we denote by **Sub A**. In particular for each subset X of the universe A of algebra **A**, let

$$\mathrm{Sg}^{\mathbf{A}}(X) = \bigcap \{ B : X \subseteq B \text{ and } B \text{ is a subuniverse of } \mathbf{A} \}$$

(dropping the superscript if the context is clear), and call this the subuniverse of **A** generated by X. Thus Sg (applied to the union of a collection of subuniverses) is the join in the subuniverse lattice. Of course $\mathrm{Sg}(X)$ can also be obtained by successively applying the basic operations to the members of X until a "closed set" is obtained. If $X \subseteq A$ and $\mathrm{Sg}(X)=A$ we say that X generates $\mathbf{A}$ (or X is a set of generators for $\mathbf{A}$). An algebra is **finitely generated** if it has a finite generating set. The lattice **Sub A** is an algebraic lattice whose compact elements are its finitely generated subuniverses. The operator $\mathrm{Sg}^{\mathbf{A}}$ which acts on subsets of A is an example of an *algebraic closure operator*. (See [McKenzie, McNulty, Taylor, 1987].)

If $\mathbf{A}$ and $\mathbf{B}$ are algebras of the same type a map $h : A \rightarrow B$ is a homomorphism of $\mathbf{A}$ into $\mathbf{B}$ provided h commutes with each of the basic operations. This means that for each operation symbol f of the type and $x_1, \ldots, x_{r(f)} \in A$,

$$f^{\mathbf{B}}(h(x_1), \ldots, h(x_{r(f)})) = h(f^{\mathbf{A}}(x_1, \ldots, x_{r(f)})).$$

For algebras $\mathbf{A}$ and $\mathbf{B}$ the notation $h : \mathbf{A} \rightarrow \mathbf{B}$ indicates that h is a homomorphism of $\mathbf{A}$ into $\mathbf{B}$. If h is an isomorphism (i.e., establishes a 1-1 correspondence between A and B) the same notation is used. The set of all automorphisms and the group of automorphisms of $\mathbf{A}$ are denoted by Aut $\mathbf{A}$ and **Aut A** respectively.

Given any family of algebras $\mathbf{A}_i$, $i \in I$, of the same type, the **direct product** $\mathbf{A} = \prod(\mathbf{A}_i : i \in I)$ of the algebras $\mathbf{A}_i$ is the algebra with universe $A = \prod(A_i : i \in I)$ and operations defined componentwise. The $\mathbf{A}_i$ are called **factors** of the direct product. The elements of $\prod(A_i : i \in I)$ are choice functions, i.e., the functions $a : I \rightarrow \bigcup(A_i : i \in I)$ such that $a(i) \in A_i$ for every $i \in I$. Instead of $a(i)$ we often write a_i and also $a = (a_i)_{i \in I}$. Thus, for every operation symbol f, we have

$$f^{\prod(\mathbf{A}_i : i \in I)}(a^1, \ldots, a^{r(f)})_i = f^{\mathbf{A}_i}(a_i^1, \ldots, a_i^{r(f)})$$

where $a^1, \ldots, a^{r(f)}$ are elements of $\prod(A_i : i \in I)$. As usual, the direct product of finitely many algebras $\mathbf{A}_1, \ldots, \mathbf{A}_n$ is denoted by $\mathbf{A}_1 \times \ldots \times \mathbf{A}_n$. If all of the factors of a direct product are the same algebra $\mathbf{B}$ then $\mathbf{A}$ is a **direct power** of $\mathbf{B}$. If a subalgebra $\mathbf{S}$ of a direct power of $\mathbf{B}$ contains the diagonal Δ consisting of all constants then $\mathbf{S}$ is a **diagonal subalgebra** and S is a **diagonal subuniverse**. For example, if $\mathbf{B}$ has a nullary operation and has no proper subalgebras then every subuniverse of a direct power of $\mathbf{B}$ contains the diagonal. If the direct power has only finitely many factors the diagonal is represented as $\Delta = \{(x, \ldots, x) : x \in B\}$.

If $A = \prod(A_i : i \in I)$ then for every $i \in I$ there is a projection mapping $p_i : A \to A_i$ defined via $p_i(a) = a_i$. It is easy to see that in the case of the direct product of algebras all projection maps are homomorphisms.

A **congruence relation** on **A** is an equivalence relation θ on A which has the **substitution property** with respect to each basic operation. This means that in addition to being an equivalence relation θ is also a subuniverse of the direct product $\mathbf{A} \times \mathbf{A}$. For any congruence θ of **A** the **quotient algebra** $\mathbf{A}/\theta$ has elements a/θ, the θ-congruence classes (or θ-blocks), and operations

$$f^{\mathbf{A}/\theta}(x_1/\theta, \ldots, x_{r(f)}/\theta) = f^{\mathbf{A}}(x_1, \ldots, x_{r(f)})/\theta$$

for all $f \in O$ and $x_1, \ldots, x_{r(f)} \in A$. Given any homomorphism h on **A**, $\ker h = \{(x, y) : h(x) = h(y)\}$ is a uniquely determined congruence on **A**; conversely any congruence relation θ on **A** is the kernel of the quotient homomorphism $h_\theta : \mathbf{A} \to \mathbf{A}/\theta$ defined by $h_\theta(x) = x/\theta$.

In groups, rings, and many other classical algebras each congruence is uniquely determined by any one of its blocks. This phenomenon does not occur in general, for example in lattices.

The set Con **A** of all congruences of **A** is a subset of the set Eqv A of all equivalence relations on A. Eqv A is ordered by set inclusion, with the least equivalence, equality on A ($\{(x, x) : x \in A\}$), denoted by 0_A and the greatest equivalence, diversity on A ($A \times A$), denoted by 1_A. If θ and ϕ are in Eqv A and $\theta \subseteq \phi$ this means that for each $x \in A$, $x/\theta \subseteq x/\phi$. For this reason inclusion between equivalences is frequently called **refinement** and $\theta \subseteq \phi$ is usually denoted by writing $\theta \leq \phi$. Relative to refinement the equivalence relations on A form a lattice. This lattice is the algebra $\mathbf{Eqv}\, A = \langle \text{Eqv}\, A; \wedge, \vee \rangle$, where for $\theta, \phi \in$ Eqv A, the greatest lower bound, $\theta \wedge \phi$, is the intersection $\theta \cap \phi$ and the least upper bound, $\theta \vee \phi$, is the transitive closure of the union of θ and ϕ, that is

$$\theta \vee \phi = \theta \cup \phi \cup (\theta \circ \phi) \cup (\theta \circ \phi \circ \theta) \cup \cdots$$

where $\circ$ is the relation product. In fact, since the intersection of an *arbitrary* family of equivalence relations on A is an equivalence relation on A, the members of Eqv A form a complete lattice.

An important observation is that the intersection of an arbitrary family of equivalence relations having the substitution property again has the substitution property and that also if θ and ϕ are equivalence relations having the substitution property then $\theta \circ \phi$ also has the substitution property. From this it follows that the set of all congruences of $\mathbf{A}$, Con $\mathbf{A}$, is a subuniverse of the lattice **Eqv** A, i.e., the meet and join of congruences is computed the same as is their meet and join as equivalence relations. The **congruence lattice Con A** is the corresponding complete sublattice (subalgebra) of **Eqv** A. Since **Con A** is complete, it follows that for any set X of elements of $A \times A$ the intersection of all congruences containing X always exists. We denote it by $\mathrm{Cg}^{\mathbf{A}}(X)$ (dropping the superscript when the context is clear) and call it the congruence **generated** by X. For $X = \{(a,b)\} \subseteq A \times A$ we write $\mathrm{Cg}(a,b)$ and call this a **principal** congruence of $\mathbf{A}$. The lattice **Con A** is algebraic with compact elements $\mathrm{Cg}(X)$ where X is a finite subset of $A \times A$. Like Sg, the operator Cg is an algebraic closure operator.

If elements a and b are related by a congruence relation θ, we express this fact by any of the equivalent notations,

$$(a,b) \in \theta, \quad a\theta b, \quad a \equiv b(\theta), \quad a \equiv b \,(\mathrm{mod}\,\theta), \quad a \equiv_\theta b,$$

depending on the particular context.

An algebra $\mathbf{A}$ is **simple** if $|A| > 1$ and if 0_A and 1_A are the only congruences of $\mathbf{A}$. More generally, if $\mathbf{A}$ has at least one congruence relation larger than 0_A (and hence $|A| > 1$) and if the meet of all congruences which are larger than 0_A is a congruence μ, which is also larger than 0_A, then $\mathbf{A}$ is called **subdirectly irreducible** which we often abbreviate as SI. The congruence $\mu > 0_A$ is called the **monolith** of $\mathbf{A}$. Obviously a simple algebra is necessarily subdirectly irreducible. The importance of subdirectly irreducible algebras is due to G. Birkhoff: If the algebras $\mathbf{A}_i$, $i \in I$, are all of the same type, then an algebra $\mathbf{A}$ is a **subdirect product** of the $\mathbf{A}_i$ if $\mathbf{A}$ is a subalgebra of the direct product of the $\mathbf{A}_i$, i.e., $\mathbf{A} < \Pi(\mathbf{A}_i : i \in I)$ and in addition each of the projection maps $p_i : A \to A_i$ is onto A_i. The $\mathbf{A}_i$ are called **subdirect factors** of $\mathbf{A}$. If $\mathbf{A}$ is a subdirect product, the kernels of the projections have meet 0_A. More generally if $f : \mathbf{A} \to \Pi(\mathbf{A}_i : i \in I)$ is any isomorphism embedding $\mathbf{A}$ onto a subdirect product, then the kernels of the maps $p_i f$ have meet 0_A. The isomorphism f is a **representation** of $\mathbf{A}$ as a subdirect product. On the other hand if θ_i, $i \in I$, is a family of congruences of $\mathbf{A}$, and $\bigwedge(\theta_i : i \in I) = 0_A$, then $\mathbf{A}$ is naturally isomorphic to a subdirect product of the algebras $\mathbf{A}/\theta_i$. Hence the representations of an algebra $\mathbf{A}$ as a subdirect product correspond to the families of congruences of $\mathbf{A}$ having meet 0_A. In such a representation, the quotient algebra $\mathbf{A}/\theta_i$ is subdirectly irreducible just in case θ_i is completely meet irreducible, i.e., is not the meet of any family of congruences, all strictly larger than θ_i. One of G. Birkhoff's fundamental theorems is the following:

Theorem 1.1.3 *Every nontrivial algebra is isomorphic to a subdirect product of subdirectly irreducible algebras.*

Thus we can think of the representation of an algebra as a subdirect product of subdirectly irreducibles as loosely analogous to the representation of an integer as the product of primes and although this representation is very important and useful it does have significant limitations. In the first place, the proof of the theorem is not constructive and in many cases we have little understanding of the nature of the subdirect factors. For example, the subdirectly irreducible groups constitute a proper class. Also the representation is far from being unique. For example, let $q_1, q_2, \ldots$ be any infinite sequence of distinct positive integers and for each integer n let n_i be the residue of n modulo q_i, $i = 1, 2, \ldots$. Then the mapping

$$n \mapsto (n_1, n_2, \ldots)$$

is an isomorphism of the ring of integers $\mathbf{Z}$ onto a subdirect product of the residue class rings $\mathbf{Z}/(q_i)$. On the other hand in this book we shall be primarily concerned with the study of congruence distributive varieties, to be discussed later; in this setting the analogy with the prime factorization of the integers is much stronger.

Finally we recall the basic facts concerning the construction of ultraproducts of algebras. Let I be some arbitrary fixed index set. By an **ultrafilter** on I we mean an ultrafilter in the power set Boolean algebra of all subsets of I. (In any lattice, filters are defined dually to the concept of ideals, i.e., a **filter** is a meet closed set which contains all elements above any of its members. Ultrafilters are **maximal filters** or **maximal dual ideals**.) That every filter (ideal) is contained in an ultrafilter (maximal ideal) is an axiom of set theory which is known to be strictly weaker than the axiom of choice. Because of duality and the fact that in a Boolean algebra prime ideals are necessarily maximal, this axiom is commonly called the **prime ideal**

theorem. A subset F of a Boolean algebra has the **finite inter-section property** if each pair of its members has a nonzero meet. Since each such subset can be embedded in a filter the prime ideal theorem is often taken to be the statement: *any subset of a Boolean algebra having the finite intersection property can be embedded in an ultrafilter.*

For each $i \in I$ let $\mathbf{A}_i$ be an L-algebra. For a given ultrafilter U on I, define the relation θ_U on $A = \Pi(A_i : i \in I)$ by

$$(a, b) \in \theta_U \quad \Longleftrightarrow \quad \{i \in I : a_i = b_i\} \subset U.$$

From the fact that U is a filter it follows that the relation θ_U is a congruence relation on the direct product $\Pi(\mathbf{A}_i : i \in I)$. We think of the members of an ultrafilter U as being "large" subsets of I and accordingly say that $a = b$ *almost everywhere* to mean that the pair (a, b) is in the congruence θ_U. The quotient algebra $\mathbf{A}/\theta_U$ is called an **ultraproduct** of the $\mathbf{A}_i$. If all of the algebras $\mathbf{A}_i$ are isomorphic, then $\mathbf{A}/\theta_U$ is called an **ultrapower** of $\mathbf{A}_i$. The basic fact about ultraproducts is Łoś's Theorem, which asserts that any first order L-formula ϕ is satisfied in an ultraproduct by a particular assignment of values to the free occurrences of its variables if and only if the assignment satisfies ϕ almost everywhere. (More precisely, if $x_1, \ldots, x_k$ are the variables occurring freely in ϕ and are assigned the values $a^1/\theta_U, \ldots, a^k/\theta_u$ from $\mathbf{A}/\theta_U$, then ϕ is satisfied by this assignment in $\mathbf{A}/\theta_U$ iff the set of indices $i \in I$ for which ϕ is satisfied by $a_i^1, \ldots, a_i^k$ in $\mathbf{A}_i$ is a member of U.) In particular, if ϕ has no free occurrences of variables, i.e., is a sentence, then ϕ is true in an ultraproduct iff it is true almost everywhere. This is the form in which we shall need Łoś's Theorem. We state it formally as follows:

Theorem 1.1.4 *If ϕ is an L-sentence and U is an ultrafilter on I, then*

$$\Pi(\mathbf{A}_i : i \in I)/\theta_U \models \phi \quad \Longleftrightarrow \quad \{i \in I : \mathbf{A}_i \models \phi\} \in U.$$

1.1.3 Clones and related topics

If A is a nonempty set, for positive integers i and n, $i \leq n$, the n-ary **projection operations** $p_i^n : A^n \to A$ are defined by

$$p_i^n(x_1, \ldots, x_n) = x_i \quad \text{for all} \quad x_1, \ldots, x_n \in A.$$

A **clone** on A is a set of operations on A which contains all of the projection operations on A and is closed under composition of operations.

If $\mathbf{A}=\langle A; F\rangle$ is an algebra we associate three important clones with $\mathbf{A}$. First, the **clone of term operations** (or term functions), which we denote by $\mathrm{Clo}\,\mathbf{A}$, is the clone which is generated by the "basic" operations F of $\mathbf{A}$. Thus t is a term operation of $\mathbf{A}$ iff t is a composition of basic operations and projections. It is easy but important to observe that the term operations come from terms which justifies their name. Indeed, if $t(x_1,\ldots,x_n)$ is any term of the language of algebra $\mathbf{A}$ then there is a unique way to compute $t(a_1,\ldots,a_n)\in A$. In this way each term which contains n distinct variables uniquely determines an n-ary operation $t^{\mathbf{A}}$ on A. Normally, if no confusion occurs, we drop the superscript and denote $t^{\mathbf{A}}$ by t. Notice that we can write

$$\mathrm{Sg}^{\mathbf{A}}(X) = \{t(a_1,\ldots,a_n) : a_1,\ldots,a_n \in X,\ t \text{ a term}\}$$

for the subuniverse of $\mathbf{A}$ generated by $X \subseteq A$.

Second, the **clone of polynomial operations** (or polynomial functions) is the clone generated by $\mathrm{Clo}\,\mathbf{A}$ together with all of the constant 0-ary operations on A. This clone is denoted by $\mathrm{Pol}\,\mathbf{A}$. Obviously $\mathrm{Clo}\,\mathbf{A} \subseteq \mathrm{Pol}\,\mathbf{A}$. The n-ary members of $\mathrm{Clo}\,\mathbf{A}$ (respectively $\mathrm{Pol}\,\mathbf{A}$) are denoted by $\mathrm{Clo}_n\,\mathbf{A}$ (respectively $\mathrm{Pol}_n\,\mathbf{A}$). Thus we have the following important but simple observation that if $p \in \mathrm{Pol}_n\,\mathbf{A}$ then for some integer $m \geq 0$ there is a term operation $t \in \mathrm{Clo}_{n+m}\,\mathbf{A}$ and $\mathbf{a} = (a_1,\ldots,a_m) \in A^m$ such that $p(\mathbf{x}) = t(\mathbf{x},\mathbf{a})$ for all elements $\mathbf{x} \in A^n$. (We shall always denote finite ordered tuples by boldface lower case letters.) For an algebra $\mathbf{A}$ we often denote by $\mathbf{A}^+$ the algebra obtained from $\mathbf{A}$ by adding new nullary operations, one for each element of A. With this notation we have $\mathrm{Clo}\,\mathbf{A}^+ = \mathrm{Pol}\,\mathbf{A}$.

Third is the **clone of congruence compatible operations**, also called the **clone of compatible operations** or the **clone of congruence preserving operations**. It is the clone consisting of all finitary operations f on A, with the property that each congruence relation of $\mathbf{A}$ has the substitution property with respect to f. It is easy to see that this set of operations is a clone; it is denoted by $\mathrm{Comp}\,\mathbf{A}$. Obviously $f \in \mathrm{Comp}\,\mathbf{A}$ iff the algebra obtained from $\mathbf{A}$ by adding f as a new operation has the same congruence lattice as $\mathbf{A}$.

In order to describe what is often the most useful way of characterizing membership in Comp $\mathbf{A}$, for $\mathbf{a}, \mathbf{b} \in A^m$ we let

$$\mathrm{Cg}(\mathbf{a}, \mathbf{b}) = \mathrm{Cg}(a_1, b_1) \vee \cdots \vee \mathrm{Cg}(a_m, b_m).$$

The right side of this equality is just $\mathrm{Cg}(\{(a_1, b_1), \ldots, (a_m, b_m)\})$. Now if f is an m-ary operation on algebra $\mathbf{A}$, then f is defined on $A \times A$ componentwise, i.e., by

$$f((a_1, b_1), \ldots, (a_m, b_m)) = (f(\mathbf{a}), f(\mathbf{b})).$$

Hence we have the convenient description

$$f \in \mathrm{Comp}\,\mathbf{A} \quad \Longleftrightarrow \quad (\forall \mathbf{a}, \mathbf{b} \in A^m)[(f(\mathbf{a}), f(\mathbf{b})) \in \mathrm{Cg}(\mathbf{a}, \mathbf{b})].$$

The collection of all clones on a given set A is clearly ordered by set inclusion. Relative to this order there is a largest element $O(A)$, the clone of all finitary operations on A, and a least clone $o(A)$, the clone consisting of just the various finitary projection operations p_j^n on A. Since the intersection of any set of clones on A is a clone on A, it follows that the clones on A form a complete lattice under set inclusion. If $\mathbf{A}$ is an algebra then we obviously have

$$o(A) \subseteq \mathrm{Clo}\,\mathbf{A} \subseteq \mathrm{Pol}\,\mathbf{A} \subseteq \mathrm{Comp}\,\mathbf{A} \subseteq O(A) \qquad (1.1)$$

and for each n

$$o_n(A) \subseteq \mathrm{Clo}_n\,\mathbf{A} \subseteq \mathrm{Pol}_n\,\mathbf{A} \subseteq \mathrm{Comp}_n\,\mathbf{A} \subseteq O_n(A) \qquad (1.2)$$

(except for the first inclusion when $n = 0$, since there are no nullary projections).

Some of the most important algebras studied in this book are obtained as in the definition below, by collapsing various of the inclusions occurring in (1.1) and (1.2), and in this sense are specific examples of what we have referred to as *polynomial completeness*.

Before giving this definition we recall that in Section 1.1.2 we noted how the presence or absence of nullary basic operations altered the lattice of subuniverses. Now we encounter another instance of the same kind of phenomenon. For example, traditionally, an algebra $\mathbf{A}$ has, rather loosely, been defined to be primal if "all of the operations on A are term operations" so that it should be possible to define primality by the equality $\mathrm{Clo}\,\mathbf{A} = O(A)$. Formally the latter would imply $\text{Clo}_0 \mathbf{A} = O_0(A)$ meaning that every nullary operation on a primal algebra is a nullary term operation. However, an algebra has nullary term operations iff it has at least one nullary basic operation. Thus, the type of every primal algebra should contain at least one nullary operation symbol. This, however, has not been a traditional intention of the definition of primality. If f is any nullary operation then for any $n \geq 1$ there is an n-ary operation g which is thus necessarily different from f but which has the same constant value as f, and in this sense g *represents* f. The traditional intention of the definition of primality was that all operations in $O(A)$ are so *representable* by term functions. Thus it is not strictly correct to define $\mathbf{A}$ to be primal iff $\text{Clo}\,\mathbf{A}=O(A)$. These remarks also apply to the concept of congruence primality as well and account for the technical form of the definitions below. In later definitions to come, with this understanding in mind, we shall often refer to Clo $\mathbf{A}$ *representing* certain functions in this sense. See [McKenzie, McNulty, Taylor, 1987] for a more precise discussion of this issue. For the present we have the following:

Definition An algebra $\mathbf{A}$ is

  (i) **primal** if $\text{Clo}_n \mathbf{A} = O_n(A)$ for $n > 0$;

  (ii) **functionally complete** if $\text{Pol}\,\mathbf{A} = O(A)$;

  (iii) **congruence primal** if $\text{Clo}_n \mathbf{A} = \text{Comp}_n \mathbf{A}$ for $n > 0$;

  (iv) **affine complete** if $\text{Pol}\,\mathbf{A} = \text{Comp}\,\mathbf{A}$.

For primal and functionally complete algebras (and often in the case of congruence primal algebras), these definitions will be restricted to finite algebras. This does not mean that the infinite versions do not exist; they do, but the theory associated with them is not nearly so satisfactory, particularly in the context of varieties, which is our primary interest. Why this is so will become more apparent as we proceed. At this point, however, we should at least notice that an infinite primal or functionally complete algebra must necessarily have uncountable type.

To avoid the technicalities concerning nullary operations (and also nullary relations), for the remainder of this section we shall assume that none of the algebras considered has nullary operations.

We also redefine $O(A)$ to be the set of all at least unary operations on the nonempty set A. Further, we let $R(A)$ denote the set of all at least unary relations r on A:

$$r \in R(A) \quad \Longleftrightarrow \quad r \subseteq A^n \text{ for some positive integer } n.$$

For subsets $R \subseteq R(A)$ and $F \subseteq O(A)$, we define

$$clo(R) = \{f \in O(A) : (\forall r \in R) \; f \text{ preserves } r\},$$
$$inv(F) = \{r \in R(A) : (\forall f \in F) \; f \text{ preserves } r\}.$$

If $r \subseteq A^n$ and $f \in O(A)$ is m-ary, then the expression "*f preserves r*" means precisely that r is a subuniverse of the direct power algebra $\langle A; f \rangle^n$, the algebra with universe A^n and the operation f applied componentwise. The empty relation (which has no arity) is preserved by every nonnullary operation (and by no nullary operation). We also commonly use the phrases "*f* is compatible with *r*", "*r* supports *f*", and "*r* is closed under *f*" to mean the same thing as "*f* preserves *r*". If $R \subseteq R(A)$ then "*f* preserves *R*" means that f preserves r for all $r \in R$ and likewise for $F \subseteq O(A)$.

The statement "*f* preserves *r*" can generally be best visualized by forming an $m \times n$ matrix M with rows which are n-tuples of r and with columns which are m-tuples in A^m. Let $f(M)$ be the n-tuple obtained by applying f to each column of M. With these conventions then the statement "*f* preserves *r*" means that for every such matrix M, we have $f(M) \in r$.

In particular if θ is an equivalence relation on A then the matrices are $m \times 2$. Thus θ is a congruence relation of algebra $\langle A; F \rangle$ just in case f preserves θ for all $f \in F$. Notice also that if $\mathbf{A}$ is an algebra and M is $m \times 2$ where the columns are arbitrary elements $\mathbf{a}, \mathbf{b} \in A^m$, then $Cg(\mathbf{a}, \mathbf{b})$, as defined above, is just the congruence of $\mathbf{A}$ generated by the rows of M.

While we shall not be concerned with clones of partial operations on a set, for future use we do need to explain the meaning of the statement "*f* preserves *r*" when f is a partial operation on A; this means precisely: for all matrices M *whose columns lie in the domain of f* and with rows in r we have $f(M) \in r$.

The relation "*f* preserves *r*" between operations and relations on A is fundamental in the sense that most of the basic concepts of algebra can be defined by means of this relation. For example, the congruences of the algebra $\langle A; F \rangle$ are just the members of $inv(F) \cap$ Eqv A. Likewise the subuniverses, automorphisms, and endomorphisms of an algebra can be obtained by restricting $inv(F)$ to special types of relations. The definitions above of primal, functionally complete, affine complete, and congruence primal algebras can also be reduced to statements involving the preservation relation. The same will be true of the other types of "polynomially complete" algebras to be defined later in the course of our development.

The compositions $clo \circ inv$ and $inv \circ clo$ are closure operators on the subsets of $O(A)$ and of $R(A)$ respectively. The clones on A are then the nonempty closed sets in $O(A)$ while the nonempty closed sets in $R(A)$ are called **coclones**. The operators "clo" and "inv" are examples of **polarities** and induce a **Galois connection** between the closed subsets (clones) of $O(A)$ and the closed subsets (coclones) of $R(A)$. For a general discussion of Galois connections between closed set systems see [McKenzie, McNulty, Taylor, 1987]. In our particular context this means that for subsets $R, R_1, R_2 \subseteq R(A)$ and $F, F_1, F_2 \subseteq O(A)$, the following inclusions hold:

$$R \subseteq inv(clo(R)), \qquad R_1 \subseteq R_2 \Rightarrow clo(R_1) \supseteq clo(R_2),$$
$$F \subseteq clo(inv(F)), \qquad F_1 \subseteq F_2 \Rightarrow inv(F_1) \supseteq inv(F_2),$$
$$clo(inv(clo(R))) = clo(R), \qquad inv(clo(inv(F))) = inv(F).$$

From this Galois connection it follows that the coclones on A, as well as the clones, also form a lattice and the maps clo and inv are dual isomorphisms between the two. An important remark is that $F \subseteq O(A)$ is a clone iff $F = clo(R)$ for some $R \subseteq R(A)$; alternatively, iff $F = clo(inv(F))$ (and dually for coclones). Note that if $\mathbf{A} = \langle A; F \rangle$, then Comp $\mathbf{A} = clo(\text{Con } \mathbf{A})$ and Clo $\mathbf{A} = clo(inv(F))$.

We now review a few important facts about clones on a finite set. From the observations above it follows that in the lattice of all clones on A the compact elements are those generated by some finite set of operations in $O(A)$. Hence each clone is the join of compact elements so that the set of all clones on A is an algebraic lattice.

If A has more than two elements then it is known that the lattice of all clones on A has the cardinality of the continuum and its structure, although it is the topic of much research, is still only poorly understood. It follows from the uncountability of the lattice, for example, that many clones are not finitely generated; how this actually occurs is also not very clear. The most important knowledge we have about this lattice is the celebrated result of I. Rosenberg ([Rosenberg, 1970]) which precisely classifies the maximal elements of this lattice, which are finite in number, into 6 classes which are determined by 6 types of relations in $R(A)$. For future reference we list these as follows:

First, they include the following three types of binary relations: (1) partial orders on A with both a least and greatest element, (2) permutations of A all of whose cycles have the same prime order, and (3) nontrivial equivalence relations on A (i.e., excluding 0_A and 1_A).

Next, a 4-ary relation r on A is called *affine* if there exists an abelian group operation $+$ on A such that

$$(a, b, c, d) \in r \iff a + b = c + d.$$

The affine relation is called *prime* if $\langle A; + \rangle$ is an elementary abelian p-group. The relations of type (4) are all prime affine relations.

An h-ary relation r on A ($h \geq 1$) is *totally symmetric* if for any permutation α on $\{1, \ldots, h\}$,

$$(a_1, \ldots, a_h) \in r \iff (a_{\alpha(1)}, \ldots, a_{\alpha(h)}) \in r.$$

It is *totally reflexive* if $(a_1, \ldots, a_h) \in r$ whenever $a_i = a_j$ for some $1 \leq i \neq j \leq h$. The *center* of r is the set of all $a \in A$ such that for any $a_2, \ldots, a_h \in A$, $(a, a_2, \ldots, a_h) \in r$. The relation r is *central* if it is both totally symmetric and totally reflexive and has a center which is a nonempty proper subset of A. The relations of Rosenberg's type (5) are all central relations.

Finally let $2 < k \leq |A|$ and let $m \geq 1$. A family $T = \{\theta_1, \ldots, \theta_m\}$ of equivalence relations on A is *k-regular* if each θ_i has index k and the intersection of any m blocks chosen from among the members of T is nonempty. The relation r *determined by T* consists of all $(a_1, \ldots, a_k) \in A_k$ with the property that for each $1 \leq j \leq m$ at least two elements among $a_1, \ldots, a_k$ are equivalent modulo θ_j. It is easy to see that an r determined by T is both totally reflexive and totally symmetric. The Rosenberg relations of type (6) are those which are determined by a k-regular family of equivalence relations.

Now we can precisely state Rosenberg's theorem.

Theorem 1.1.5 *A clone on a finite set A is maximal in the lattice of all clones on A iff it has the form $clo(\{r\})$ for some relation r of one of the above 6 types.*

Rosenberg's original proof has been somewhat simplified by R. Quackenbush ([Quackenbush, 1979]). We will return to the discussion of Rosenberg's theorem and its implications in Chapter 3.

If $|A| = 2$ the situation is very different: E. L. Post ([Post, 1941]) showed that there are just countably many clones on A and completely classified them; the maximal elements of this lattice are precisely 5 in number and are very easily described by the resulting specialization of Rosenberg's theorem to the case $|A| = 2$. Moreover, each clone on A is finitely generated and thus has the form Clo **A** where the (two element) algebra **A** has only finitely many operations.

Equivalence lattices and their clones For a set A, any sublattice **L** of **Eqv** A is called an **equivalence lattice** (on A). If both 0_A and 1_A are in **L** then **L** is called a 0-1 equivalence lattice. For such an **L** then $F = clo(L)$, the clone of all **L-compatible operations** (or **L-compatible functions**) is the set of all operations on A which preserve all equivalence relations in L. As noted above, the congruence lattice of the algebra **A** $= \langle A; F \rangle$ certainly contains **L** but it is important to recognize that it is, in general, larger (i.e., $\langle A; clo(L) \rangle$ is not usually congruence primal). The following four element example illustrating this is taken from [Jónsson, 1972]:

Let $A = \{a, b, c, d\}$ and define equivalence relations on A by

$$\theta_1 = \{\{a, b\}, \{c\}, \{d\}\}; \quad \theta_2 = \{\{a, c\}, \{b, d\}\}; \quad \theta_3 = \{\{a, d\}, \{b, c\}\}.$$

Each pair of these relations has meet 0_A and join 1_A. Thus let **L** be the equivalence lattice on A with $L = \{0_A, 1_A, \theta_1, \theta_2, \theta_3\}$. Then let $\phi = \{\{a\}, \{b\}, \{c, d\}\}$. Assume $f : A^m \to A$ preserves L but not ϕ. Then for some $(a_1, \ldots, a_i, \ldots, a_m) \in A^m$ and $b \in A$, with $(a_i, b) \in \phi$ we must have $(f(a_1, \ldots, a_i, \ldots, a_m), f(a_1, \ldots, b, \ldots, a_m)) \notin \phi$. Hence the unary operation $g(x) = f(a_1, \ldots, x, \ldots, a_m)$ preserves L but not ϕ. Hence if there is any operation with the property that it is in $clo(L)$ but does not preserve ϕ, there must be a unary operation with this property.

But, as the reader can easily verify by a short computation, the only unary operation on A which preserves L, aside from the identity and the constant operations, is the operation g defined by

$$g(a) = b, \quad g(b) = a, \quad g(c) = d, \quad g(d) = c,$$

and this also preserves ϕ. It follows from this that no algebra with universe A can have **L** as its congruence lattice. In general then, not every 0-1 equivalence lattice is the congruence lattice of some algebra.

On the other hand, in [Quackenbush, Wolk, 1971] it is shown that if **L** is a finite 0-1 *distributive* equivalence lattice on A then, in fact, **Con A**=**L** for suitable **A**. In their proof of this fact they show that it is possible to find, for each equivalence relation θ on A which is not in L, a unary operation f_θ on A which preserves each element of L but not θ. Thus F can be taken to be the set of all such f_θ. Hence in this case $\langle A; clo(L) \rangle$ is congruence primal.

While not every equivalence lattice is the congruence lattice of an algebra it should be noted here that the celebrated Grätzer-Schmidt Theorem (see, e.g., [Grätzer, 1979]) asserts that every algebraic lattice is *isomorphic* to the congruence lattice of some algebra. In particular every finite lattice is isomorphic to a congruence lattice.

Coclone algebras Just as we defined clones to be sets of functions which a) contain the projections, and b) are closed under composition, it is to be expected that coclones can also be described as sets of relations satisfying closure conditions analogous to a) and b) and indeed this is the case. A discussion of this can be found in both [Rosenberg, 1979] and [Pöschel, Kalužnin, 1979]. A very readable elementary account is in the book [Pippenger, 1997]. At this point we simply present one way to describe coclones by operations; it is useful since if A is finite it makes the set $R(A)$ of all finitary relations into an algebra in a straightforward way. (An equivalent way, equally straightforward, using different operations appears in [Pöschel, Kalužnin, 1979].) We use the description presented in [C. Bergman, 1998] which is a slight variation of that appearing in [Rosenberg, 1979]. This will be important for us only in Section 3.5.

Let r and s be members of $R(A)$, say $r \subseteq A^m$ and $s \subseteq A^n$. We define the following operations:

$$\zeta(r) = \{(x_2, x_3, \ldots, x_m, x_1) : (x_1, x_2, \ldots, x_m) \in r\}$$
$$\text{(rotation)},$$

$$\tau(r) \;=\; \{(x_2, x_1, x_3, \ldots, x_m) : (x_1, x_2, x_3, \ldots, x_m) \in r\}$$
$$\text{(transposition)},$$
$$\nu(r) \;=\; \{(x_1, \ldots, x_m, y) : (x_1, \ldots, x_m) \in r,\; y \in A\}$$
$$\text{(cylindrification)},$$
$$\pi(r) \;=\; \{(x_1, x_2, \ldots, x_{m-1}) : (\exists x_m)\,(x_1, \ldots, x_m) \in r\}$$
$$\text{(projection)},$$
$$\delta \;=\; \{(x, x) : x \in A\} \quad \text{(diagonal)},$$
$$r \sqcap s \;=\; \{(x_1, x_2, \ldots, x_k) : (x_1, \ldots, x_m) \in r, (x_1, \ldots, x_n) \in s\}$$
$$k = \max(m, n), \quad \text{(intersection)}.$$

If $m = 1$, each of the unary operations ζ, τ, and π is taken to be the identity function. Notice that for $m = n$ the operation $\sqcap = \cap$ while for $m > n$, $\sqcap$ can be composed from ordinary set intersection and the $(m-n)$-fold iteration of ν: specifically, in this case $r \sqcap s = r \cap \nu^{m-n}(s)$.

With these operations the system $\mathbf{R} = \langle R(A); \zeta, \tau, \nu, \pi, \sqcap, \delta \rangle$ is then an algebra. Since δ is a nullary operation of $\mathbf{R}$, this is a one-time violation of our general restriction for this section, but a consequence of this is that the empty set is not a subuniverse. The basic fact which we wish to observe is the following theorem. The general result is due to D. Geiger ([Geiger, 1968]). The proof for the instructive special case $|A| = 2$ appears in [Pippenger, 1997].

Theorem 1.1.6 *If A is finite, then a nonempty subset $R \subseteq R(A)$ is a coclone on A iff R is a subuniverse of the algebra $\mathbf{R}$.*

Since δ is nullary it generates the least coclone on A, often denoted by D_A. In the literature $\mathbf{R}$ and its subalgebras have been known by a variety of different names, including "subdirect closure systems" in [Rosenberg, 1979], and "relation algebras" in the book [Pöschel, Kalužnin, 1979]. These names are all somewhat unsatisfactory. (See [C. Bergman, 1998] where they are called "Krasner algebras" despite the fact that this name has been used to describe different structures.) We shall call them **coclone algebras** since this seems to adequately capture their significance.

Finally, it is worth noting that it is possible to change our original "closed under composition" definition of clones and to redefine them as sets of functions closed with respect to operations which are close to being dual to the operations on relations defined above. This approach, which goes back to A. I. Mal'cev, is discussed in both [Pippenger, 1997] and [Pöschel, Kalužnin, 1979].

We shall use these meager facts about coclone algebras only in Section 3.5 where they will play a significant role.

Term equivalence and weak isomorphism Now let us drop our temporary requirement that algebras have no nullary basic operations. If the two algebras $\mathbf{A}$ and $\mathbf{B}$ have the same universe and $\mathrm{Clo}_n\,\mathbf{A} = \mathrm{Clo}_n\,\mathbf{B}$ for all positive integers n, we say that $\mathbf{A}$ and $\mathbf{B}$ are **term equivalent**. The intent of this definition is to ensure that the term functions of each are representable by the term functions of the other. (If neither has nullary operations we can require just $\mathrm{Clo}\,\mathbf{A} = \mathrm{Clo}\,\mathbf{B}$.) If they have precisely the same polynomial operations, i.e., $\mathrm{Pol}\,\mathbf{A} = \mathrm{Pol}\,\mathbf{B}$, they are called **polynomially equivalent**. Thus term equivalence implies polynomial equivalence but not conversely. The interdefinability of each Boolean algebra and a corresponding Boolean ring is an example of term equivalence.

Algebras $\mathbf{A}$ and $\mathbf{B}$ are called **weakly isomorphic** if $\mathbf{A}$ is isomorphic to an algebra $\mathbf{C}$ which is term equivalent to $\mathbf{B}$.

1.1.4 Varieties

We quickly review basic properties and our notation for varieties of algebras: classes of algebras defined by (universally quantified) equations. Thus the examples of semigroups, groups, rings, modules, etc., from Section 1.1.1 are varieties if the operations are appropriately chosen. In this discussion it is thus important that we always consider indexed algebras.

Suppose L is a given first order algebraic language. An **equation** or **identity** (an L-equation or L-identity) is a first order sentence of the form $(\forall \mathbf{x})(t \approx s)$ where t and s are terms whose variables occur in $\mathbf{x}$. For brevity we ordinarily omit the universal quantifier and just write an equation as $t \approx s$. By $Eq(K)$ we denote the set of all equations true in every algebra in K. As before, if Γ is a set of L-equations, $Mod(\Gamma)$ denotes the class of all models of Γ.

The mappings Eq and Mod are polarities and therefore induce a Galois connection between the closed sets Γ of L-equations and the closed classes K of L-algebras. To emphasize this we can restate the definitions formally as follows:

$$Eq(K) = \{t \approx s : (\forall \mathbf{A} \in K)(\mathbf{A} \models t \approx s)\},$$
$$Mod(\Gamma) = \{\mathbf{A} : (\forall t \approx s \in \Gamma)(\mathbf{A} \models t \approx s)\}.$$

An **algebraic theory**, also called an **equational theory**, of type L is a set of equations of the form $Eq(K)$ for some class K of algebras. A **variety** of type L is a class of algebras of the form $Mod(\Gamma)$ for some set Γ of equations. It follows that a set Θ of equations is an algebraic theory iff $\Theta = Eq(Mod(\Theta))$ and a class K of algebras is a variety iff $K = Mod(Eq(K))$. The variety generated by a class K, the least variety containing K, denoted by $V(K)$, is then given by $V(K) = Mod(Eq(K))$. The Galois connection then establishes a 1-1 correspondence, a dual isomorphism, between the lattice of equational theories and the lattice of varieties of type L. Let K be a class of algebras of a given type and let $\mathbf{F}$ be an algebra of that type. Let X be a set of generators of $\mathbf{F}$. Then $\mathbf{F}$ is called **free** for K (or K-free) with free generators X provided every mapping of X to a set of generators of an algebra $\mathbf{A}$ in K can be uniquely extended to a homomorphism of $\mathbf{F}$ onto $\mathbf{A}$. An important fact, made explicit by G. Birkhoff, is that for any class K of algebras, when an algebra free for K exists and is in K then it is uniquely determined within K, up to isomorphism, by the class K and the cardinality $\kappa = |X|$ of any set of free generators. When this occurs we denote this algebra by $\mathbf{F}_K(X)$ or $\mathbf{F}_K(\kappa)$. We conveniently think of $\mathbf{F}_K(X)$ as consisting of equivalence classes of terms in the generators X where two terms t and s are in the same class provided $\mathbf{A} \models t \approx s$ for every $\mathbf{A}$ in K, i.e., provided $t \approx s \in Eq(K)$. When we wish to emphasize this representation we often call $\mathbf{F}_K(\kappa)$ the K-*free term algebra with* κ *free generators.*

Further, if we consider the family of all mappings of X into members of K it is easy to see that $\mathbf{F}_K(X)$ is isomorphic to a subalgebra of a direct product of copies of the algebras $\mathbf{A}$ in K, one copy of $\mathbf{A}$ for each mapping of X into $\mathbf{A}$. Therefore if the given class K is closed under the taking of direct products and subalgebras of its members, in particular if K is a variety, then all algebras free for K are contained in K.

For any class K of L-algebras, $I(K)$, $H(K)$, $S(K)$, and $P(K)$ denote, respectively, the classes of all isomorphic copies, of all homomorphic images, of all subalgebras, and of all direct products, of algebras in K. The symbols $P_s(K)$ and $P_U(K)$ denote respectively the classes of all subdirect products and of all ultraproducts of members of K. The operators I, H, S, P, P_s, P_U are usually called **class operators**. We denote the class of all finite direct products of K by $P_f(K)$. The finite members of a class K are likewise denoted by K_f and likewise, by K_{fg}, the class of all finitely generated algebras in the class K. Using these conventions we can summarize the basic facts about varieties which we shall require. All are essentially due to G. Birkhoff and are established in detail in [Burris, Sankappanavar, 1981]. (We use ω to denote the cardinality of the set of natural numbers.)

Theorem 1.1.7 *If K is a class of algebras of the same type, then*

(i) $V(K) = HSP(K)$;

(ii) K *is a variety iff* $K = HSP(K)$;

(iii) *If K is a variety then it is generated by a single algebra, namely by the free algebra having countably many free generators, that is* $K = HSP(\mathbf{F}_K(\omega))$;

(iv) *If K is a variety then K is generated by the class K_{SI} of subdirectly irreducible algebras in K, and in fact* $K = IP_s(K_{SI})$;

(v) *For a nonnegative integer n, if $t(x_1,\ldots,x_n)$ and $s(x_1,\ldots,x_n)$ are terms whose variables are contained in the set $\{x_1,\ldots,x_n\}$, then the identity $t \approx s$ is in $Eq(K)$ iff in $\mathbf{F}_K(n)$,*

$$t(a_1,\ldots,a_n) = s(a_1,\ldots,a_n),$$

where $a_1,\ldots,a_n$ are free generators of $\mathbf{F}_K(n)$.

A variety K is said to be **finitely generated** if $K = V(\mathbf{A})$ for some finite algebra $\mathbf{A}$. An algebra is **locally finite** if each of its finitely generated subalgebras is finite. Thus a finite algebra is locally finite but not conversely. A variety is **locally finite** if each of its members is locally finite. A criterion is that all finitely generated free algebras in the variety be finite. Because of this it is clear that as in the case of single algebras, if a variety is finitely generated it is locally finite, but not conversely. The variety of *Monadic algebras* is an example. (See [McKenzie, McNulty, Taylor, 1987], p 226, for this example.)

Because of part (iv) of Theorem 1.1.7, a consequence of Theorem 1.1.3, we are particularly interested in the size of K_{SI}, the class of (isomorphism types of) subdirectly irreducibles (SIs) in a variety K and the sizes of the SIs themselves. In general, for a nontrivial variety the number of isomorphism types of the SIs can be as small as 1, as for example in the case of distributive lattices, or, at the other extreme, K_{SI} can be a proper class; for example this is true in the variety of all groups, as mentioned earlier.

If every SI algebra in a variety is simple then the variety is called **semisimple**. The variety of distributive lattices is semisimple.

For a variety K and a cardinal κ we say that K is **residually less than** κ provided each SI in K has cardinality less than κ. In case K is residually less than ω we say that K is **residually finite**. Other terminology which occurs in the literature: K is **residually small** means that K is residually less than κ for some cardinal κ; otherwise K is **residually large**.

In Section 1.1.3 we defined term equivalence of two algebras $\mathbf{A}$ and $\mathbf{B}$ to mean that $A = B$ and $\mathrm{Clo}_n \mathbf{A} = \mathrm{Clo}_n \mathbf{B}$ for $n > 0$. Then $\mathbf{A}$ and $\mathbf{B}$ were defined to be weakly isomorphic if $\mathbf{A}$ is isomorphic to an algebra term equivalent to $\mathbf{B}$. It is natural to extend this concept to varieties: Suppose L_1 and L_2 are algebraic languages and V_1 is an L_1-variety and V_2 an L_2-variety. The varieties V_1 and V_2 are **term equivalent** provided the free algebras $\mathbf{F}_{V_1}(\omega)$ and $\mathbf{F}_{V_2}(\omega)$ are weakly isomorphic. This concept occurs most often when we have some single variety V (with language L) and decide for reasons of convenience to change the basic operations to some other, more complex terms, in such a way that the process is reversible. The best known example is the variety of Boolean algebras versus the variety of Boolean rings. Whether we take the ring operations $\{+, \times, 0, 1\}$, or the Boolean operations $\{\vee, '\}$, as basic is usually a matter of taste, since each can be represented as terms in the other. For a more formal and complete discussion of term equivalence of varieties the reader is referred to [McKenzie, McNulty, Taylor, 1987].

Exercises

1. For a variety V of L-algebras show that if $\kappa \neq 0$, then $\mathbf{F}_V(\kappa)$ exists iff V has nontrivial members or $\kappa = 1$. $\mathbf{F}_V(0)$ exists iff L contains a nullary operation.

2. If the class K consists of a single algebra $\mathbf{A}$, we denote the variety generated by $\mathbf{A}$ by $V(\mathbf{A})$ and the free algebras for $V(\mathbf{A})$ by $\mathbf{F_A}(\kappa)$. Show that if it exists then

$$\mathbf{F_A}(\kappa) \in IS(\mathbf{A}^{|A|^\kappa}).$$

3. Show that for an integer $m \geq 0$ an m-ary operation f on an algebra $\mathbf{A}$ is a term operation of $\mathbf{A}$ iff f preserves all subuniverses of $\mathbf{A}^{|A|^m}$.

   Show that $\mathbf{F_A}(m)$ is isomorphic to the algebra of all m-ary term operations on $\mathbf{A}$.

4. Prove:

   a) If $\mathbf{A}$ is the p element ring of integers modulo p (prime), then $\mathbf{F_A}(1) \cong \mathbf{A}^{p-1}$.

   b) If $\mathbf{A}$ is the cyclic group of order n, then $\mathbf{F_A}(1) \cong \mathbf{A}$.

5. Prove that a finite algebra $\mathbf{A}$ is primal if $\mathbf{F_A}(n) \cong \mathbf{A}^{|A|^n}$ for each nonnegative integer n.

6. Prove that the two element chain is the only subdirectly irreducible distributive lattice.

7. Show that an algebra is simple iff $\mathrm{Comp}\,\mathbf{A} = O(A)$.

8. Show that every subdirectly irreducible modular lattice of finite height is simple.

1.2   Congruence properties

1.2.1   Mal'cev conditions

The theory of Mal'cev conditions provides a way of establishing a connection between the algebraic theory (set of equations) defining a given variety and certain algebraic properties of the variety. In this section we shall summarize some basic Mal'cev conditions which will be most important in our subsequent work. We shall not discuss the general theory here; W. Taylor ([Taylor, 1973]) developed this theory and first provided a characterization of those algebraic properties definable by Mal'cev conditions.

We begin with the classic theorem of A. I. Mal'cev which initiated the general theory. Recall that the join $\theta \vee \phi$ of two congruence relations θ and ϕ of an algebra is the transitive closure of their union, which means that the join is the union of all of their finite relation products, as described in Section 1.1.2. In certain important cases the join is given in the simplest possible way as simply the relation product, $\theta \vee \phi = \theta \circ \phi$; it is easy to see that this occurs iff $\theta \circ \phi = \phi \circ \theta$ and we say that θ and ϕ **permute**. If all pairs of congruence relations of an algebra permute we say that the algebra is **congruence permutable**. If all algebras of a variety are congruence permutable we say the variety is congruence permutable. (We often abbreviate "congruence permutable" as CP.) Mal'cev's theorem ([Mal'cev, 1954]) is the following.

Theorem 1.2.1 *A variety V is congruence permutable iff there is a term $m(x, y, z)$ in the language of V such that*

$$m(x, y, y) \approx x, \quad m(x, x, y) \approx y$$

are equations of V.

PROOF Notice that to show that the existence of such a term implies CP it is sufficient to show for all $\mathbf{A} \in V$, $\theta, \phi \in \mathrm{Con}\,\mathbf{A}$, that $\theta \circ \phi \leq \phi \circ \theta$. Hence suppose $\mathbf{A} \in V$, $\theta, \phi \in \mathrm{Con}\,\mathbf{A}$ and $(x, z) \in \theta \circ \phi$. Then for some element $y \in A$, $(x, y) \in \theta$ and $(y, z) \in \phi$. Let m be a term satisfying the equations of the theorem. Then by the substitution property,

$$x = m(x, z, z) \equiv_\phi m(x, y, z) \equiv_\theta m(x, x, z) = z$$

so that $(x, z) \in \phi \circ \theta$.

For later reference we make the following technical observation: the proof above can be construed to show that for a single algebra $\mathbf{A}$, to prove that $\mathbf{A}$ is CP, it suffices to have only the following information: for each triple of elements $a, b, c \in A$ there is a ternary partial operation m_{abc} on A, whose domain contains

$$\{(a, a, c), (a, b, c), (a, c, c)\},$$

and which is compatible with all congruences of $\mathbf{A}$, where defined, and which also satisfies the equalities

$$m_{abc}(a, a, c) = c \quad \text{and} \quad m_{abc}(a, c, c) = a.$$

To prove the converse implication, let $\mathbf{F}$ be the free algebra in V with free generators x, y, z. Then $\mathbf{F}$ is CP. Consider the principal congruences $\mathrm{Cg}(x, y)$ and $\mathrm{Cg}(y, z)$ of $\mathbf{F}$. Since

$$(x, z) \in \mathrm{Cg}(x, y) \circ \mathrm{Cg}(y, z),$$

by permutability it follows that also

$$(x, z) \in \mathrm{Cg}(y, z) \circ \mathrm{Cg}(x, y).$$

Hence for some term $m(x, y, z)$ (element of F) we have

$$(x, m(x, y, z)) \in \mathrm{Cg}(y, z) \quad \text{and} \quad (m(x, y, z), z) \in \mathrm{Cg}(x, y).$$

Let g be the endomorphism of $\mathbf{F}$ defined by the assignments $g(x) = x$, $g(y) = g(z) = z$. Since $(x, m(x, y, z)) \in \ker g$ we have

$$x = g(x) = g(m(x, y, z)) = m(g(x), g(y), g(z)) = m(x, z, z)$$

and since this equality holds on the generators x, z it follows that $x \approx m(x, z, z)$ is an identity of V (by Theorem 1.1.7). The other identity follows likewise. $\qquad\bullet$

The most familiar congruence permutable variety is the variety of groups. If we take group product and inverse ($\cdot$ and $^{-1}$) as basic operations of this variety then $m(x, y, z) = x \cdot y^{-1} \cdot z$ illustrates the theorem.

Another special property of congruence lattices is the **distributive law**

$$\theta \wedge (\phi \vee \psi) = (\theta \wedge \phi) \vee (\theta \wedge \psi)$$

(which is of course equivalent to the dual equality or either of the usual dual inequalities). If $\mathbf{A}$ is an algebra in which $\mathbf{L} = \mathrm{Con}\,\mathbf{A}$ is distributive then $\mathbf{A}$ is **congruence distributive** (abbreviated CD). If $\mathbf{A}$ is both congruence distributive and congruence permutable the algebra is called **arithmetical**. This property may be expressed by either the equality

$$\theta \wedge (\phi \circ \psi) = (\theta \wedge \psi) \circ (\theta \wedge \phi),$$

or the inequality

$$\theta \wedge (\phi \circ \psi) \leq (\theta \wedge \psi) \circ (\theta \wedge \phi)$$

(for all $\theta, \phi, \psi \in L$). To see this observe that the distributivity together with permutability certainly implies the equality, which in turn implies the inequality. Then, supposing the inequality, substitute $\phi \vee \psi$ for θ in the inequality and infer $\phi \circ \psi \leq \psi \circ \phi$, and thus permutability. Hence $\circ = \vee$ so that the inequality becomes $\theta \wedge (\phi \vee \psi) \leq (\theta \wedge \psi) \vee (\theta \wedge \phi)$ which implies the distributive law.

A variety is congruence distributive or arithmetical provided each of its members has the corresponding property.

Arithmetical algebras are the natural generalization to universal algebra of arithmetical rings which are in turn a generalization of Dedekind domains. (Dedekind domains are integral domains in which each ideal is the product of prime ideals.) These domains are discussed in the book of [Zariski, Samuel, 1958] where it is shown that their lattices of ideals are distributive. It is further shown that this property is, in turn, precisely equivalent to the Chinese remainder condition, i.e., for ideals $I_1, \ldots, I_n$ and elements $a_1, \ldots, a_n$ of the ring **R**, the finite system of congruences

$$x \equiv a_i \pmod{I_i}, \quad i = 1, \ldots, n,$$

is solvable iff the compatibility conditions $x_i \equiv x_j \pmod{(I_i + I_j)}$ are met for all $1 \leq i < j \leq n$. The classical Chinese remainder theorem for the integers is a special case. In [Fuchs, 1949], L. Fuchs introduced the terminology **arithmetical rings** to designate not necessarily commutative (but associative) rings whose lattice of two sided ideals is distributive. Arithmetical algebras were introduced in [Pixley, 1972a] as a further natural generalization, observing that all rings are already congruence permutable. We will discuss an important further abstraction to lattices of equivalence relations in Chapter 2 where we will observe that the proof given in [Zariski, Samuel, 1958] of the equivalence of the Chinese remainder theorem to distributivity of the ideal lattice carries over almost directly to the most abstract level of equivalence lattices.

The following theorem ([Pixley, 1963]) gives the Mal'cev conditions for arithmeticity of varieties. Notice that the conditions are the same as for congruence permutability except that one more equation $p(x, y, x) \approx x$ is required.

Theorem 1.2.2 *A variety V is arithmetical iff there exists a term*

$p(x, y, z)$ *of the type of V such that*

$$p(x, y, y) \approx x, \quad p(x, y, x) \approx x, \quad p(x, x, z) \approx z$$

are equations of V.

PROOF Let $\mathbf{A} \in V$; $\theta, \phi, \psi \in \mathrm{Con}\,\mathbf{A}$, and let $(x, z) \in \theta \wedge (\phi \circ \psi)$. Then $(x, z) \in \theta$ and for some $y \in A$, $(x, y) \in \phi$ and $(y, z) \in \psi$. Let p be a term satisfying the equations of the theorem. Then, by the substitution property,

$$x = p(x, y, x) \equiv_\theta p(x, y, z) \equiv_\theta p(z, y, z) = z, \quad \text{and}$$

$$x = p(x, y, y) \equiv_\psi p(x, y, z) \equiv_\phi p(y, y, z) = z,$$

so that $(x, z) \in (\theta \wedge \psi) \circ (\theta \wedge \phi)$. By the remarks above it follows that V is arithmetical.

As in the corresponding part of the proof of Theorem 1.2.1 the same kind of technical remark can be made here, i.e., to prove arithmeticity of a single algebra $\mathbf{A}$ it is sufficient to have certain special $\mathrm{Con}\,\mathbf{A}$-compatible partial operations. The reader can easily formulate the condition, which will be useful later.

For the converse, let $\mathbf{F}$ be the free algebra in V with free generators x, y, z. Then $\mathbf{F}$ is arithmetical. Consider the principal congruences $\mathrm{Cg}(x, z)$, $\mathrm{Cg}(x, y)$, and $\mathrm{Cg}(y, z)$. Since

$$(x, z) \in \mathrm{Cg}(x, z) \wedge (\mathrm{Cg}(x, y) \circ \mathrm{Cg}(y, z)),$$

from arithmeticity it follows that also

$$(x, z) \in (\mathrm{Cg}(x, z) \wedge \mathrm{Cg}(y, z)) \circ (\mathrm{Cg}(x, z) \wedge \mathrm{Cg}(x, y)).$$

Hence for some term $p(x, y, z)$ (element of $\mathbf{F}$) we have

$$\begin{aligned}(x, p(x, y, z)) &\in \mathrm{Cg}(x, z) \wedge \mathrm{Cg}(y, z), \\ (p(x, y, z), z) &\in \mathrm{Cg}(x, z) \wedge \mathrm{Cg}(x, y).\end{aligned}$$

Hence we obtain

$$\begin{aligned}(x, p(x, y, z)) &\in \mathrm{Cg}(x, z), \\ (x, p(x, y, z)) &\in \mathrm{Cg}(y, z), \\ (p(x, y, z), z) &\in \mathrm{Cg}(x, y),\end{aligned}$$

and from here we proceed as in the proof of Theorem 1.2.1.   •

The next theorem presents the Mal'cev conditions characterizing congruence distributivity and is due to B. Jónsson ([Jónsson, 1967]). They are more complex than those for congruence permutability or arithmeticity: instead of a single term, in general we now require many. This is because the join of a pair of congruence relations may, in general, involve relation products involving an arbitrarily large number of factors. The Mal'cev conditions described in Theorems 1.2.1 and 1.2.2, which characterize a property of a variety by means of identities involving only a single term, are examples of what are usually called **strong** Mal'cev conditions to distinguish them from those of the more general type of Theorem 1.2.3 below. For the proof the reader is referred to [McKenzie, McNulty, Taylor, 1987]; the proof given there is somewhat simpler than the original.

Theorem 1.2.3 *A variety V is congruence distributive iff for some integer $n \geq 2$ and ternary terms $d_0, \ldots, d_n$, of the type of V, the following are equations of V:*

$$
\begin{aligned}
d_0(x, y, z) &\approx x, \\
d_i(x, y, x) &\approx x \quad \text{for all} \ \ 0 \leq i \leq n, \\
d_i(x, x, z) &\approx d_{i+1}(x, x, z) \quad \text{for all even} \ \ i < n, \\
d_i(x, z, z) &\approx d_{i+1}(x, z, z) \quad \text{for all odd} \ \ i < n, \\
d_n(x, y, z) &\approx z.
\end{aligned}
$$

The lattice term $u(x, y, z) = (x \vee y) \wedge (y \vee z) \wedge (z \vee x)$ (and also its dual) is a ternary **majority** operation on any lattice, since

$$u(x, y, x) \approx x, \quad u(x, x, z) \approx x, \quad u(x, z, z) \approx z$$

are identities of any lattice. The simplest example of Theorem 1.2.3 is obtained by taking V to be the variety of lattices, $n = 2$, and $d_1(x, y, z) = u(x, y, z)$. We should also note that the conditions of the theorem for a given n imply the conditions for any larger n. In this sense the existence of a ternary majority term is the strongest form of the theorem and any variety having a ternary majority term is CD.

Operations (terms) $d_0, \ldots, d_n$ satisfying the conditions of Theorem 1.2.3 are called **Jónsson operations (functions)** (terms).

Operations or terms $m(x, y, z)$ or $p(x, y, z)$ satisfying Theorem 1.2.1 or Theorem 1.2.2 are called **Mal'cev** or **Pixley operations (functions)** or terms respectively.

It is easy to verify that for any Mal'cev, majority, and Pixley operations, m, u, and p respectively, $u(x, m(x, y, z), z)$ is a Pixley operation and $p(x, p(x, y, z), z)$ is a majority operation. Thus one sees that the existence of a Pixley term for a variety is equivalent to the existence of both a majority and a Mal'cev term. In particular, in the presence of congruence permutability all of the conditions of Theorem 1.2.3 collapse to the existence of a majority term.

In the variety of Boolean algebras we can take

$$p(x, y, z) = u(x, y', z)$$

(where u is a lattice majority term and $'$ is Boolean complementation) to show via Theorem 1.2.2 that the variety is arithmetical.

Another important example of a Pixley term is for the variety generated by the Galois field $GF(s^k)$ (s prime) with operations $+$ and $\cdot$ and

$$p(x, y, z) = z + (x + (s - 1)z)(x + (s - 1)y)^{s^k - 1}.$$

More generally, let V be the variety generated by a finite set of Galois fields. Let $s_1, \ldots, s_n$ be the distinct prime characteristics of the fields and let $k_1, \ldots, k_n$ be the least integers such that for each $i = 1, \ldots, n$, each of the fields of characteristic s_i is embedded in $GF(s_i^{k_i})$. (Thus k_i is the least common multiple of the exponents associated with the characteristic s_i.) Also let $p_i(x, y, z)$ be the term constructed as above for $GF(s_i^{k_i})$ and let a_i be a solution of the numerical congruence

$$s_1 \cdots s_{i-1} s_{i+1} \cdots s_n x \equiv 1 \,(\mathrm{mod}\, s_i).$$

Then

$$p(x, y, z) = \sum_{i=1}^{n} s_1 \cdots s_{i-1} s_{i+1} \cdots s_n a_i p_i(x, y, z)$$

is a Pixley term for V. This observation proves the "if" direction of the following theorem of Michler and Wille ([Michler, Wille, 1970]):

A variety of rings is arithmetical iff it is generated by a finite number of finite fields.

The proof of other direction of the theorem apparently requires deeper but well known facts from ring theory.

An important special Pixley operation is the ternary **discriminator** operation (function) t ([Pixley, 1971]). Let S be any nonempty set and define $t : S^3 \to S$ by

$$t(x, y, z) = \begin{cases} z & \text{if } x = y, \\ x & \text{if } x \neq y. \end{cases}$$

A closely related operation is the 4-ary **switching operation (function)** (also called the **normal transform**) n defined on any set by

$$d(x, y, z, w) = \begin{cases} z & \text{if } x = y, \\ w & \text{if } x \neq y. \end{cases}$$

On any set these two operations are interdefinable, specifically

$$d(x, y, z, w) = t(t(x, y, z), t(x, y, w), w) \text{ and } t(x, y, z) = d(x, y, z, x)$$

so the choice is often only a matter of convenience.

The discriminator and the normal transform are both examples of what is called a *pattern function*, in the sense that the value of the function is always one of its arguments, which one depending on the pattern of equalities holding among the arguments. Since we shall use them several times in later chapters we define them more precisely here. Following R. Quackenbush ([Quackenbush, 1974]) we say that two n-tuples $(a_1, \ldots, a_n), (b_1, \ldots, b_n) \in A^n$ have the same *pattern* if for all $1 \leq i < j \leq n$, $a_i = a_j$ iff $b_i = b_j$. An n-ary operation f on A is a **pattern function** if

1. $f(a_1, \ldots, a_n) \in \{a_1, \ldots, a_n\}$ for all n-tuples of A, and

2. if $f(a_1, \ldots, a_n) = a_i$ and the n-tuples $(a_1, \ldots, a_n), (b_1, \ldots, b_n)$ have the same pattern, then $f(b_1, \ldots, b_n) = b_i$.

Obviously projections are pattern functions, the trivial ones; notice that a pattern function must be at least ternary to avoid being a projection.

An important observation is that if an algebra **A** has a discriminator term (or polynomial), then the algebra is simple: indeed if θ

is any congruence relation on the algebra and $(x, y) \in \theta$ with $x \neq y$, then for any element $z \in A$,

$$z = t(x, x, z) \equiv_\theta t(x, y, z) = x,$$

which shows that $\theta = 1_A$.

In the preceding examples the Pixley operations for Boolean algebras and varieties of arithmetical rings are the discriminator on the two element Boolean algebra and $GF(p^k)$ respectively. Since all operations on a primal algebra are, by definition, term operations, a simple but important observation is that, just as in these cases, every primal algebra has a discriminator term operation. This elementary fact was the starting point for much of the content of this book.

A variety V is a **discriminator variety** if there is a ternary term t of V which is the discriminator on each subdirectly irreducible member of V. Thus a discriminator variety V is arithmetical and the remark above shows that the variety is also semisimple (that is, each SI is simple).

A finite algebra having a discriminator term, and hence any finite subdirectly irreducible member of a discriminator variety, is called **quasiprimal**. Obviously each nontrivial subalgebra of a quasiprimal algebra is also quasiprimal. In Chapter 3 we shall see that a finitely generated discriminator variety is generated by finitely many quasiprimal algebras and that each member of such a variety is isomorphic to a subdirect product of quasiprimal algebras. Such a variety, not surprisingly, has much in common with the variety generated by a primal algebra and, in particular, with the variety of Boolean algebras. The book [Werner, 1978] is recommended for many more examples of quasiprimal algebras; it emphasizes the model theory, in particular the decidability of the first order theory, of finitely generated discriminator varieties. Less is known about infinitely generated discriminator varieties.

Finally, notice that in the Mal'cev conditions appearing in Theorems 1.2.1, 1.2.3, and 1.2.2, only three variable equations occur. This has the following corollary.

Corollary 1.2.4 *A variety V is CD, CP, or arithmetical iff the free algebra $\mathbf{F}_V(3)$ has the corresponding property.*

Exercises

1. Prove that a bounded distributive lattice is complemented iff it is CP.

2. Prove that an arbitrary distributive lattice is relatively complemented iff it is CP.

3. Let V be a discriminator variety with algebraic language L and let $\phi = (\forall \mathbf{x})\psi$ be a sentence where ψ contains no quantifiers. Show that one can effectively find a single identity ϵ defined in L such that for each subdirectly irreducible member $\mathbf{A}$ of V, $\mathbf{A} \models \phi \leftrightarrow \epsilon$. ([McKenzie, 1975])

4. Let $\mathbf{A}$ be an algebra of finite type which is the direct product of finitely many quasiprimal algebras with the same discriminator term. Exhibit a finite set of identities Σ of $\mathbf{A}$ such that every identity of $\mathbf{A}$ is a consequence of Σ (i.e., show that $\mathbf{A}$ has a finite equational base). ([Fried, Pixley, 1979]) (An important theorem of K. Baker (Advances in Math. **24**, 207-243, 1977) states that any finite algebra in a CD variety of finite type has a finite equational base.)

5. Let V be an arithmetical variety of finite type whose equational theory Σ is finitely based.

   a) Show that, in fact, Σ is 1-based.

   b) If V is not arithmetical but is CD by virtue of Σ containing a majority term (instead of a Pixley term) then show that Σ is 2-based, but not generally 1-based. (R. Padmanabhan and R. Quackenbush, *Equational theories of algebras with distributive congruence lattices*, Proc. Amer. Math. Soc. **41**, 373-377, 1974)

1.2.2   Definable principal congruences

A basic lemma due to A. I. Mal'cev (usually called Mal'cev's Lemma) gives a precise description of the principal congruences of a given algebra. Specifically, if $\mathbf{A}$ is an algebra of type L and a, b, c, d are elements of A, then Mal'cev's Lemma asserts that

$$(a, b) \in \mathrm{Cg}^{\mathbf{A}}(c, d) \quad \Longleftrightarrow \quad \mathbf{A} \models \phi(a, b, c, d), \qquad (1.3)$$

where $\phi(x,y,u,v)$ is a certain existentially quantified conjunction of L-equations which contains the free variables x,y,u,v. The formula $\phi(x,y,u,v)$ is called a **principal congruence formula**.

While in general the formula $\phi(x,y,u,v)$ depends on the choice of $a,b,c,d \in A$, there are important cases where the dependence is not so strict. The most important case is the following: A variety V has **definable principal congruences**, DPC for short, if there is a finite set of principal congruence formulas such that for each $\mathbf{A} \in V$ and $a,b,c,d \in A$, (1.3) holds for some formula of the set.

The variety of all commutative rings with identity element is the simplest common example of this phenomenon. In this variety a principal congruence is determined by the congruence block (ideal) containing the zero element of the ring. Hence we have

$$(a,b) \in \mathrm{Cg}(c,d) \quad \Longleftrightarrow \quad (\exists x)[a - b = x(c - d)].$$

Thus this variety has DPC and in fact the single principal congruence formula

$$\phi(x,y,u,v) = (\exists z)[x - y \approx z(u - v)]$$

defines principal congruences throughout this variety. The variety of all distributive lattices also has DPC. Indeed it is not difficult to prove that in any distributive lattice

$$(a,b) \in \mathrm{Cg}(c,d) \quad \Longleftrightarrow \quad (a \vee c \vee d = b \vee c \vee d)\&(a \wedge c \wedge d = b \wedge c \wedge d).$$

It can be seen that this description of principal congruences also applies to the variety of Boolean algebras, but in this case we also have the formally simpler formula

$$(a,b) \in \mathrm{Cg}(c,d) \quad \Longleftrightarrow \quad a \vee (c \oplus d) = b \vee (c \oplus d)$$

as well. Here $\oplus$ is the Boolean symmetric difference operation:

$$x \oplus y = (x \wedge y') \vee (x' \wedge y).$$

A primary reason for the importance of DPC for a variety V is that the condition implies that the class V_{SI} of subdirectly irreducible members of V is first order definable. This is so since an algebra $\mathbf{A}$ is SI iff the sentence

$$(\exists x,y)[(x \not\approx y)\&(\forall u,v)(u \not\approx v \rightarrow (x,y) \in \mathrm{Cg}^{\mathbf{A}}(u,v))]$$

is true in **A**. When this is the case then the primary tools of model theory, including the Compactness and Löwenheim-Skolem Theorems, and the ultraproduct construction, can be applied to V_{SI}. Because every variety is generated by its subdirectly irreducible members, in the presence of DPC we can generally draw strong conclusions about the residual character of V. For example if V has DPC and arbitrarily large finite SI members, then taking their ultraproduct we infer that V contains an infinite SI member and hence (using the Löwenheim-Skolem Theorem), that V_{SI} contains algebras of arbitrarily large infinite cardinality.

A variety V has **equationally definable principal congruences** (EDPC for short) if there is a single formula ϕ which is a quantifier free conjunction of finitely many equations

$$t_i(x, y, u, v) \approx s_i(x, y, u, v)$$

and for which (1.3) holds throughout V. As exhibited above, the varieties of distributive lattices and Boolean algebras have EDPC. It can be shown that the existential quantifier occurring in the formula above describing principal congruences in commutative rings with identity cannot be removed. Hence EDPC is a strictly stronger condition than DPC. The most remarkable consequence of EDPC for a variety is that it implies that the variety is congruence distributive ([Köhler, Pigozzi 1980]). Many of the important varieties having EDPC are also congruence permutable and hence arithmetical. Since arithmetical algebras and varieties play a large role in subsequent chapters, we shall now devote some discussion to EDPC in arithmetical varieties. To begin, the following theorem describes precisely when an arithmetical variety has EDPC. The proof, which is rather technical, can be found as a special case of results presented in [Fried, Kiss, 1983].

Theorem 1.2.5 *A variety V is arithmetical and has EDPC iff there is a 4-ary term $d(u, v, x, y)$ of V such that*

$$d(u, u, x, y) \approx x \quad and \quad d(u, v, x, x) \approx x$$

are identities of V and such that for all $\mathbf{A} \in V$ and $u, v, x, y \in A$,

$$(x, y) \in Cg^{\mathbf{A}}(u, v) \quad \Longleftrightarrow \quad d(u, v, x, y) = y. \tag{1.4}$$

Notice that the conditions of the theorem do imply that V is arithmetical, for the conditions are easily seen to imply that the term

$$t(u, v, x) = d(u, v, x, u)$$

is a Pixley term for V. Moreover this definition of $t(u, v, x)$ is the same as that used to define the ternary discriminator in terms of the 4-ary switching operation defined in Section 1.2.1. This suggests that arithmetical varieties with EDPC can be characterized in terms of an appropriate Pixley term for the variety. We do not know if this can always be done, but the following theorem (extracted from a result in [McKenzie, 1975]) describes an interesting and important way in which it sometimes can be done. One reason for its importance is that it applies to discriminator varieties (see Theorem 1.2.10, below). Also, in Section 2.2.4 we shall see that in any arithmetical variety principal congruence formulas can always be taken to be of the form

$$\phi(x, y, u, v) = (\exists \mathbf{z})[t(u, v, x, \mathbf{z}) \approx t(u, v, y, \mathbf{z})]$$

where the term $t(u, v, x, \mathbf{z})$ induces a Pixley operation on some finite subset containing u, v, x, y; thus the equation in (1.5) of the theorem is just the special case where no quantifier is needed.

Theorem 1.2.6 *Let V be an arithmetical variety and let $t(x, y, z)$ be a Pixley term for V. Then V has EDPC with principal congruences defined for all $\mathbf{A} \in V$ by*

$$(x, y) \in \mathrm{Cg}^{\mathbf{A}}(u, v) \quad \Longleftrightarrow \quad t(u, v, x) = t(u, v, y) \qquad (1.5)$$

if and only if for any m and each m-ary basic operation f of V,

$$t(u, v, f(x_1, \ldots, x_m)) \approx t(u, v, f(t(u, v, x_1), \ldots, t(u, v, x_m))) \quad (1.6)$$

is an equation of V.

Because the conditions (1.6) are equations the theorem, in effect, asserts that for a given Pixley term t witnessing the arithmeticity of V, the class of all algebras in V having EDPC via (1.5) forms a subvariety V_t of V. Since the equations (1.6) contain the basic operations f of V, the varieties of the form V_t are not generally a Mal'cev definable class, though there is an obvious resemblance.

PROOF First observe that for any algebra $\mathbf{A}$ of V and $u, v \in A$,

$$R(u, v) = \{(x, y) : t(u, v, x) = t(u, v, y)\}$$

is an equivalence relation. By the identities $t(x, y, x) \approx x \approx t(x, y, y)$ it follows that $(u, v) \in R(u, v)$. Next, if θ is any congruence containing (u, v) then, using the third identity characterizing arithmeticity, we have that $(x, y) \in R(u, v)$ implies

$$x = t(u, u, x) \equiv_\theta t(u, v, x) = t(u, v, y) \equiv_\theta t(u, u, y) = y.$$

Thus $R(u, v) \leq \theta$ and from this it follows that

$$R(u, v) \leq \mathrm{Cg}^{\mathbf{A}}(u, v).$$

Using (1.6) we easily check that if $(x_i, y_i) \in R(u, v)$, $i = 1, \ldots, m$, then

$$t(u, v, f(x_1, \ldots, x_m)) = t(u, v, f(y_1, \ldots, y_m))$$

so $(f(x_1, \ldots, x_m), f(y_1, \ldots, y_m)) \in R(u, v)$; thus $R(u, v)$ is a congruence. This shows that satisfaction of (1.6) implies (1.5). Conversely, if (1.5) holds then $R(u, v) = \mathrm{Cg}^{\mathbf{A}}(u, v)$. It follows that each

$$x_i = t(u, u, x_i) \equiv t(u, v, x_i) \ (\mathrm{Cg}^{\mathbf{A}}(u, v))$$

and so

$$f(x_1, \ldots, x_m) \equiv f(t(u, v, x_1), \ldots, t(u, v, x_m)) \ (\mathrm{Cg}^{\mathbf{A}}(u, v))$$

by the substitution property for f. Then by the definition of $R(u, v)$ we finally have (1.6). •

Notice that the proof of the above theorem shows that in any arithmetical variety the relation $R(u, v)$ is always an equivalence relation which contains (u, v) and $R(u, v) \leq \mathrm{Cg}^{\mathbf{A}}(u, v)$. Further, $R(u, v)$ is a congruence iff the equality holds. Let us call an arithmetical variety a **principal arithmetical variety** when it has a Pixley term $t(x, y, z)$ for which $R(u, v)$ is always a congruence, and hence when V has EDPC via this term and (1.5). Thus V is a principal arithmetical variety iff V has a Pixley term satisfying (1.6). Notice that V may be principal arithmetical relative to many different Pixley terms and for any arithmetical variety with Pixley term t the subvariety V_t is principal arithmetical with respect to that t. While Theorem 1.2.6 does not claim that principal arithmetical varieties exhaust the class of arithmetical varieties having EDPC, it does give us the following sharper form of the question raised above: can arithmetical varieties with EDPC be characterized in terms of an appropriate Pixley term?

Problem 1.2.7 *If V is an arithmetical variety having EDPC, is there a Pixley term t defining principal congruences in V by (1.5), i.e., for which $V_t = V$?*

The problem, alternatively, asks if from a given 4-ary operation d satisfying the conditions of Theorem 1.2.5 for V there can be composed a Pixley term satisfying (1.5) (or (1.6))? The obvious first choice would be $t(u,v,x) = d(u,v,x,u)$ as remarked above, but it is not clear that this works. One case where it does work is in discriminator varieties.

Theorem 1.2.8 *A variety V is a discriminator variety iff V is a semisimple principal arithmetical variety.*

PROOF If V is a discriminator variety, then as noted at the end of Section 1.2.1, V is semisimple. Also if t is the discriminator on each SI then obviously equations (1.6) hold on all SIs and hence in V. Therefore (1.5) holds in V. Conversely, if V is semisimple, then for any simple member **A** of V (1.5) implies that for $u \neq v$, any pair $(x,y) \in \mathrm{Cg}(u,v) = 1_A$; so $t(u,v,x) = t(u,v,u) = u$, and hence t is the discriminator.   ●

Thus if V is a discriminator variety with d a 4-ary term which induces the switching operation on each SI member of V, then certainly the identities appearing in Theorem 1.2.5 hold in V and also the right to left implication of (1.4) holds. Now the term

$$t(u,v,x) = d(u,v,x,u)$$

induces the discriminator on each SI; hence, by Theorem 1.2.8, if $(x,y) \in \mathrm{Cg}^{\mathbf{A}}(u,v)$ for $\mathbf{A} \in V$ then $d(u,v,x,u) = d(u,v,y,u)$. Now let u',v',x',y' be the projections of u,v,x,y in any SI factor of **A**. Then $d(u',v',x',u') = d(u',v',y',u')$ so $(x',y') \in \mathrm{Cg}(u',v')$. Then $d(u',v',x',y') = y'$ (from the definition of the switching operation). Hence $d(u,v,x,y) = y$ holds in **A**, i.e., the left to right direction of

(1.4) holds. Therefore the conditions of Theorem 1.2.5 hold.

Another connection between arithmetical varieties with EDPC and principal arithmetical varieties is the following:

Theorem 1.2.9 *If V is an arithmetical variety with Pixley term t and if V has EDPC with the 4-ary term d satisfying the conditions of Theorem 1.2.5, then V_t consists of the class of all models of the identity*

$$d(d(u,v,x,y),y,t(u,v,x),t(u,v,y)) \approx t(u,v,y). \qquad (1.7)$$

The theorem thus provides another way of defining V_t as a sub-variety of an arithmetical variety having EDPC. Problem 1.2.7 can be therefore restated as: is there always a way of defining a Pixley operation t in terms of d in such a way that the identity (1.7) holds?

PROOF In view of the formula (1.4), the identity (1.7) is equivalent to the condition that for every elements u,v,x,y of each $\mathbf{A} \in V$,

$$(t(u,v,x),t(u,v,y)) \in \mathrm{Cg}(d(u,v,x,y),y). \qquad (1.8)$$

If $\mathbf{A} \in V_t$ then by Theorems 1.2.6 and 1.2.5 we have the following equivalences:

$$t(u,v,x) = t(u,v,y) \quad \Leftrightarrow \quad (x,y) \in \mathrm{Cg}(u,v) \quad \Leftrightarrow \quad d(u,v,x,y) = y,$$

for arbitrary $u,v,x,y \in A$. Applying this to the quotient algebra $\mathbf{A}/\phi$ where $\phi = \mathrm{Cg}(d(u,v,x,y),y)$ we get that (1.8) is satisfied in $\mathbf{A}$; hence $\mathbf{A}$ models (1.7).

For the converse, suppose that the algebra $\mathbf{A} \in V$ satisfies (1.8) and $(x,y) \in \mathrm{Cg}(u,v)$. By Theorem 1.2.5 then $d(u,v,x,y) = y$ which implies $t(u,v,x) = t(u,v,y)$. Thus $\mathbf{A} \in V_t$.   •

Finally we have another interesting characterization of discriminator varieties due to R. McKenzie ([McKenzie, 1975]):

Theorem 1.2.10 *A variety is a discriminator variety with discriminator term $t(x,y,z)$ if and only if it is a principal arithmetical variety relative to the Pixley term $t(x,y,z)$ and*

$$t(x,t(x,y,z),y) \approx y \qquad (1.9)$$

is an identity of the variety.

In effect the theorem asserts that each discriminator variety is a semisimple subvariety V_{disc} (consisting of all models of (1.9)) of some principal arithmetical variety. In general then, for any arithmetical variety V and any given Pixley term t, we have the chain of subvarieties $V_{disc} \subseteq V_t \subseteq V$. As is indicated in the proof of Theorem 1.2.8, t induces the discriminator on any simple member of V_t. Since it is well known that every nontrivial variety contains a simple algebra, it follows that in every case V_t is trivial iff V_{disc} is trivial.

PROOF If the variety V has a term $t(x, y, z)$ which induces the discriminator on each SI member of V then it is easy to check that equations (1.6) and (1.9) are identities of each SI and hence of V.

Conversely, if V is a principal arithmetical variety relative to $t(x, y, z)$ and satisfies (1.9), let $\mathbf{A} \in V$ be SI. Then by definition there are elements $a, b \in A$, $a \neq b$, such that $(a, b) \in Cg^{\mathbf{A}}(c, d)$ whenever $c \neq d$. Hence for any $x \in A$, if $t(a, b, x) \neq a$ it follows that $(a, b) \in Cg^{\mathbf{A}}(a, t(a, b, x))$, i.e.,

$$t(a, t(a, b, x), a) = t(a, t(a, b, x), b);$$

but the left side of this equation equals a, by arithmeticity, and the right side equals b by (1.9). This contradiction shows that $t(a, b, x) = a$ for all x and this means that $Cg^{\mathbf{A}}(a, b) = 1_A$, which in turn implies that $Cg^{\mathbf{A}}(x, y) = 1_A$ whenever x and y are distinct in A. Consequently $t(x, y, z) = t(x, y, x) = x$ if $x \neq y$. Since arithmeticity implies that $t(x, y, z) = z$ if $x = y$, it follows that $t^{\mathbf{A}}$ is the discriminator on A. •

Exercises

1. Verify the formulas given at the beginning of the section for principal congruences in distributive lattices and Boolean algebras.

2. Show that every finitely generated arithmetical variety in which the congruence lattice of each SI is a chain has DPC. ([Pixley, 1984])

1.2.3   Near unanimity operations and congruence distributivity

In Section 1.2.1 we observed that the existence of a Pixley term for a variety is equivalent to the existence of both a majority and a Mal'cev term, and that in the presence of a Mal'cev term the existence of any Jónsson terms implies the existence of a majority term. Near unanimity operations generalize ternary majority operations.

Definition For $n \geq 2$, an $(n+1)$-ary **near unanimity** operation is an $(n+1)$-ary operation u satisfying

$$u(x, \ldots, x, y, x, \ldots, x) = x$$

where the single occurrence of y can occur in any of the $n+1$ argument positions. Thus a ternary majority operation is just a 3-ary near unanimity operation.

If $u(x_0, \ldots, x_n)$ is an $(n+1)$-ary near unanimity operation then obviously

$$u(u(x_0, \ldots, x_n), x_2, \ldots, x_{n+1})$$

is an $(n+2)$-ary near unanimity operation. Thus for $n = 2, 3, \ldots$ the properties of having $(n+1)$-ary near unanimity operations on a given set are (at least formally) successively weaker. That they are actually strictly successively weaker is an easy consequence of a more complete discussion of their properties in Section 3.2. A **near unanimity term** for an algebra is a term which induces a near unanimity operation. For the present we want only the following important result.

Theorem 1.2.11 *Any variety having a near unanimity term is congruence distributive.*

This fact is due to A. Mitschke ([Mitschke, 1978]) who showed (by construction) that if V has an $(n+1)$-ary near unanimity term then it has Jónsson terms $d_0, \ldots, d_{2(n-2)}$ (see Theorem 1.2.3). The following lemma gives an alternative proof (due to E. Fried) of the theorem, not by constructing explicit Jónsson terms, but by directly demonstrating congruence distributivity.

Lemma 1.2.12 *If an algebra* **A** *has an $(n+1)$-ary congruence compatible near unanimity operation ($n \geq 2$), then* **A** *is congruence distributive.*

PROOF For $\theta, \phi, \psi \in \text{Con } \mathbf{A}$ let $(a, b) \in \theta \wedge (\phi \vee \psi)$ so that $(a, b) \in \theta$ and also $(a, b) \in \phi \vee \psi$, so that from the definition of $\vee$ we can choose some elements $x_0 = a, x_1, \ldots, x_k = b \in A$ such that $(x_i, x_{i+1}) \in \phi \cup \psi$ for all $i < k$. Now, if $k < n$ then we may take $x_{k+1} = \cdots = x_n = b$. If, however, $k > n$ then by the above remark $\mathbf{A}$ also has a $(k+1)$-ary congruence compatible near unanimity operation. Thus, without loss of generality $k = n$.

Let u be an $(n+1)$-ary near unanimity operation in Comp $\mathbf{A}$. Then for each i we have

$$(u(a, \ldots, a, x_i, b, \ldots, b), u(a, \ldots, a, x_{i+1}, b, \ldots, b)) \in \phi \cup \psi,$$

and also

$$
\begin{aligned}
u(a, \ldots, a, x_i, b, \ldots, b) &\equiv_\theta & u(a, \ldots, a, x_i, a, \ldots, a) \\
&= & u(a, \ldots, a, x_{i+1}, a, \ldots, a) \\
&\equiv_\theta & u(a, \ldots, a, x_{i+1}, b, \ldots, b).
\end{aligned}
$$

Then for each i we have

$$(u(a, \ldots, a, x_i, b, \ldots, b), u(a, \ldots, a, x_{i+1}, b, \ldots, b)) \in \theta \wedge (\phi \cup \psi).$$

But $\theta \wedge (\phi \cup \psi) = (\theta \wedge \phi) \cup (\theta \wedge \psi) \subseteq (\theta \wedge \phi) \vee (\theta \wedge \psi)$. Consequently, by transitivity we have

$$(u(a, \ldots, \underline{a}, b, \ldots, b), u(a, \ldots, a, \underline{b}, \ldots, b)) \in (\theta \wedge \phi) \vee (\theta \wedge \psi)$$

where for each argument position (underlined), the right side is obtained from the left by replacing the last a by a b. Hence we obtain

$$a = u(a, \ldots, a, b) \equiv u(a, \ldots, a, b, b) \equiv \cdots \equiv u(a, b, \ldots, b) = b$$

modulo $(\theta \wedge \phi) \vee (\theta \wedge \psi)$. Hence

$$\theta \wedge (\phi \vee \psi) \leq (\theta \wedge \phi) \vee (\theta \wedge \psi)$$

so **Con A** is distributive. $\bullet$

1.2.4   Congruence permutability and rectangular subalgebras

For algebras **A** and **B** let S be a subuniverse of the direct product **A** × **B**. We say that S is **rectangular** if

$$(x,y),(x,v),(u,v) \in S \quad \Longrightarrow \quad (u,y) \in S,$$

i.e., if three vertices of a rectangle are in S then so is the fourth. These special kinds of subuniverses are very close to being congruences. They are characteristic of congruence permutable varieties.

Theorem 1.2.13   (a) *A subuniverse S of a direct square* **A** × **A** *is a congruence relation of* **A** *iff S is both rectangular and diagonal (i.e., contains* $\Delta = \{(x,x) : x \in A\}$*).*

(b) *A variety V is congruence permutable iff each subuniverse of the direct product of each pair of algebras in V is rectangular.*

(c) *In a congruence permutable variety V a subuniverse S of a direct square* **A** × **A** *is a congruence of* **A** *iff S is diagonal.*

PROOF   To prove part (a) first let S be both rectangular and diagonal. Then for $(x,y) \in S$ we also have $(x,x),(y,y) \in S$, so rectangularity implies $(y,x) \in S$. Thus S is a symmetric relation in $A \times A$; it is transitive since

$$(x,y),(y,z) \in S \quad \Longrightarrow \quad (x,y),(z,y),(z,z) \in S \quad \Longrightarrow \quad (x,z) \in S,$$

by symmetry and rectangularity. Conversely, if S is a congruence relation of **A**, then S is diagonal and if $(x,y),(x,v),(u,v) \in S$, then symmetry and transitivity imply $(u,y) \in S$.

To prove part (b) first suppose V is CP and by Theorem 1.2.1 let $m(x,y,z)$ be a Mal'cev term for V. Then if S is a subuniverse of a direct product of two algebras in V and $(x,y),(x,v),(u,v) \in S$, then applying m to these elements we have

$$m((x,y),(x,v),(u,v)) = (m(x,x,u), m(y,v,v)) = (u,y)$$

so that S is rectangular. Conversely, if all subuniverses of products in V are rectangular, let **F** be freely generated in V with free generators x, y. Then the subalgebra of **F** × **F** generated by the three pairs $(x, x), (x, y), (y, y)$ is rectangular and hence contains the element (y, x). This means that there is a term m such that

$$m((x, x), (x, y), (y, y)) = (y, x)$$

which, in turn, means that $m(x, x, y) = y$ and $m(x, y, y) = x$. By Theorem 1.1.7 it follows that these are equations of V and, by Theorem 1.2.1, that V is CP.

Part (c) of the theorem is immediate from the first two parts. ●

Historically, rectangular subalgebras were first considered in a slightly different way. For algebras **A** and **B**, with subalgebras $\mathbf{A}_1 < \mathbf{A}, \mathbf{A}_2 < \mathbf{B}, \phi_i \in \mathrm{Con}\,\mathbf{A}_i, i = 1, 2$, suppose

$$\sigma : \mathbf{A}_1/\phi_1 \rightarrow \mathbf{A}_2/\phi_2$$

is an isomorphism. Then

$$S = \{(a_1, a_2) \in A_1 \times A_2 : \sigma(a_1/\phi_1) = a_2/\phi_2\}$$

is a subuniverse of $\mathbf{A} \times \mathbf{B}$. Since it is the union of the elements (blocks $a_1/\phi_1 \times a_2/\phi_2$) of the graph of σ, we shall call the subalgebra **S** a **graph subalgebra** of $\mathbf{A} \times \mathbf{B}$ (defined by σ).

Theorem 1.2.14 *A subalgebra of* $\mathbf{A} \times \mathbf{B}$ *is a graph subalgebra iff it is rectangular.*

PROOF For a graph subalgebra **S** let $(x, y), (x, v), (u, v) \in S$. Then

$$\sigma(x/\phi_1) = y/\phi_2, \quad \sigma(x/\phi_1) = v/\phi_2, \quad \sigma(u/\phi_1) = v/\phi_2.$$

The last two equations imply $x/\phi_1 = u/\phi_1$ and from this and the first equation we obtain $\sigma(u/\phi_1) = y/\phi_2$ so $(u, y) \in S$. Hence **S** is rectangular.

Conversely, if **S** is rectangular with first and second projections $\mathbf{A}_1 < \mathbf{A}, \mathbf{A}_2 < \mathbf{B}$, then for $a, b \in A_1$ define ϕ_1 by

$$(a, b) \in \phi_1 \quad \Longleftrightarrow \quad \exists y \in A_2 \text{ with } (a, y), (b, y) \in S;$$

ϕ_1 is reflexive and symmetric and has the substitution property directly from the definition. Finally, if $(x, y), (y, z) \in \phi_1$, then so is the triple $(y, x), (y, z), (z, z)$, so that by rectangularity so is (z, x), verifying transitivity. Hence ϕ_1 is a congruence of $\mathbf{A}_1$. Analogously define $\phi_2 \in \text{Con }\mathbf{A}_2$ by

$$(a, b) \in \phi_2 \quad \Longleftrightarrow \quad \exists x \in A_1 \text{ with } (x, a), (x, b) \in S.$$

By the definitions of ϕ_i it is immediate that for $(a_1, a_2) \in S$, the map $\sigma(a_1/\phi_1) = a_2/\phi_2$ defines an isomorphism and S is contained in the graph subalgebra defined by σ. To show that $\mathbf{S}$ equals this graph subalgebra, suppose $(a, b) \in S$ and $u \in A_1, v \in A_2$ are such that $(u, v) \in a/\phi_1 \times b/\phi_2$. We must show that $(u, v) \in S$. But we have $(u, a) \in \phi_1$ and $(v, b) \in \phi_2$ so that for some $x \in A_1, y \in A_2$,

$$(x, v), (x, b) \in S \text{ and } (u, y), (a, y) \in S.$$

Then rectangularity applied to $(x, b), (a, b), (a, y) \in S$ implies that $(x, y) \in S$, and then applied again to $(u, y), (x, y), (x, v) \in S$ implies $(u, v) \in S$. Hence $\mathbf{S}$ is the graph subalgebra defined by σ.   •

   If $\mathbf{S}$ is any subalgebra of $\mathbf{A} \times \mathbf{B}$ with first and second projections $\mathbf{A}_1$ and $\mathbf{A}_2$ then $\mathbf{S}$ is certainly contained in the rectangular subalgebra $\mathbf{A}_1 \times \mathbf{A}_2$ (the product of its projections). Also the intersection of all rectangular subalgebras containing $\mathbf{S}$ is obviously rectangular; since it is the least rectangular subalgebra of $\mathbf{A} \times \mathbf{B}$ containing $\mathbf{S}$, we call it the **rectangular hull** of $\mathbf{S}$. Thus the rectangular hull always exists (and a variety is CP iff each subalgebra of a product equals its rectangular hull). We can also construct the rectangular hull of $\mathbf{S}$ as a graph subalgebra: define $\phi_i \in \text{Con }\mathbf{A}_i$ by

$$\phi_1 = \bigvee\{\text{Cg}(a, b) : \exists y \in A_2 \text{ with } (a, y) \text{ and } (b, y) \text{ in } S\},$$

$$\phi_2 = \bigvee\{\text{Cg}(c, d) : \exists x \in A_1 \text{ with } (x, c) \text{ and } (x, d) \text{ in } S\}.$$

For each $a_1/\phi_1 \in A_1/\phi_1$ choose $a_2 \in A_2$ such that $(a_1, a_2) \in S$ and define $\sigma(a_1/\phi_1) = a_2/\phi_2$. From the definitions of ϕ_1 and ϕ_2 it is easy to check that $\sigma : A_1/\phi_1 \to A_2/\phi_2$ is an isomorphism and that the graph subalgebra defined by σ is the rectangular hull of $\mathbf{S}$. These observations will be very useful later in our development.

   The preceding two theorems together prove the well known fact (due essentially to I. Fleischer [Fleischer, 1955]) that a variety is CP iff all subalgebras of direct products of pairs of algebras in the variety are graph subalgebras. (The latter property has been known for groups for over a hundred years.)

A final fact about subdirect products in CP varieties is the following: recall that an algebra $\mathbf{A}$ is isomorphic to a direct product of a finite set of (quotient) algebras iff there is a finite set of congruences $\theta_1, \ldots, \theta_n$ of $\mathbf{A}$ such that

a) $\theta_1 \wedge \cdots \wedge \theta_n = 0_A$, and

b) for each $2 \leq i \leq n$, $\theta_1 \wedge \cdots \wedge \theta_{i-1}$ and θ_i permute and have join 1_A.

In this case $\mathbf{A}$ is isomorphic to the direct product of the quotient algebras $\mathbf{A}/\theta_i$. From this fact we have the following observation.

Lemma 1.2.15 *If* $\mathbf{A}$ *is an algebra in a CP variety and is isomorphic to an irredundant subdirect product of a finite set of simple algebras, then* $\mathbf{A}$ *is actually isomorphic to the direct product of the members of the set.*

"Irredundant" here means that $\mathbf{A}$ is not isomorphic to a subdirect product of a proper subset of the given factors.

PROOF Because of the subdirect representation we have congruences $\theta_1 \wedge \cdots \wedge \theta_n = 0$. Since each factor is simple, each θ_i is maximal and by permutability we thus have for each $i \geq 2$, $(\theta_1 \wedge \cdots \wedge \theta_{i-1}) \vee \theta_i = \theta_i$ or 1_A. The first possibility leads to redundancy, so each of these joins is 1_A and the product is therefore direct. ●

1.2.5 Congruence distributivity and skew congruences

If $\mathbf{A}_i, 1 \leq i \leq n$, are algebras of the same type and $\theta_i \in \mathrm{Con}\, \mathbf{A}_i$, then the **product congruence** $\theta_1 \times \cdots \times \theta_n$ on $\mathbf{A}_1 \times \cdots \times \mathbf{A}_n$ is defined by

$$((a_1, \ldots, a_n), (b_1, \ldots, b_n)) \in \theta_1 \times \cdots \times \theta_n \text{ iff } (a_i, b_i) \in \theta_i \text{ for } 1 \leq i \leq n.$$

If an algebra $\mathbf{A}$ is embedded in $\mathbf{B} \times \mathbf{C}$ as a subdirect product and the congruence $\rho \in \mathrm{Con}\, \mathbf{A}$ cannot be represented as a restriction of some product congruence of $\mathbf{B} \times \mathbf{C}$ then ρ is said to be a **skew congruence** of $\mathbf{A}$. Of course the property of being a skew congruence depends on the given subdirect decomposition.

It has long been known that the absence of skew congruences in finite *direct* products is a Mal'cev condition ([Fraser, Horn, 1970]), and that this condition is implied by congruence distributivity. The results of this section are mostly from [Kaarli, McKenzie, 1997]. First we establish Theorem 1.2.20 which implies that the absence of skew congruences in finite *subdirect* products is also a Mal'cev condition and is, in fact, precisely equivalent to congruence distributivity. Apart from the general interest of this theorem, and more important for our subsequent treatment of affine complete varieties, the section is primarily devoted to recent more technical results from [Kaarli, McKenzie, 1997] concerning skew congruences and congruence distributivity.

The following easy lemma from [Burris, Sankappanavar, 1981] gives a lattice-theoretic criterion for a congruence to be nonskew.

Lemma 1.2.16 *Let* $\mathbf{A}$ *be a subdirect product in* $\mathbf{A}_1 \times \cdots \times \mathbf{A}_n$ *and let* θ_i *be the kernel of projection* $\mathbf{A} \to \mathbf{A}_i$, $i = 1, \ldots, n$. *Then the following are true:*

(i) *A congruence* ρ *of* $\mathbf{A}$ *is nonskew with respect to this subdirect decomposition iff* $(\theta_1 \vee \rho) \wedge \cdots \wedge (\theta_n \vee \rho) = \rho$.

(ii) *If* $\mathbf{A}/\rho$ *is SI then* ρ *is nonskew with respect to this subdirect decomposition iff* $\theta_i \leq \rho$ *for some* $i \in \{1, \ldots, n\}$.

Next we prove another easy lemma which will be helpful if we have to show that a given equivalence lattice is distributive.

Lemma 1.2.17 *Let* $\mathbf{L}$ *be a lattice such that every element of* L *is a meet of completely meet irreducible elements of* L. *The following conditions are equivalent:*

(1) $\mathbf{L}$ *is distributive;*

(2) *for every* $\rho, \mu \in L$ *where* ρ *is completely meet irreducible and* μ *covers* ρ *there exists a lattice homomorphism* $f : \mathbf{L} \to \mathbf{D}_2$ *which separates* ρ *and* μ;

(3) *for every* $\rho, \beta, \gamma \in L$ *where* ρ *is completely meet irreducible, if* $\beta \wedge \gamma \leq \rho$ *then either* $\beta \leq \rho$ *or* $\gamma \leq \rho$.

PROOF The implication (1)$\Rightarrow$(2) is an easy consequence of the fact that $\mathbf{D}_2$ is the only SI distributive lattice.

(2) $\Rightarrow$ (3). Let μ be the unique cover of ρ in $\mathbf{L}$ and $f : \mathbf{L} \to \mathbf{D_2}$ be a lattice homomorphism such that $f(\rho) = 0$, $f(\mu) = 1$. Assume that $\beta \wedge \gamma \leq \rho$ but $\beta \not\leq \rho$ and $\gamma \not\leq \rho$. Then $\mu \leq (\beta \vee \rho) \wedge (\gamma \vee \rho)$ which implies

$$
\begin{aligned}
1 = f(\mu) &\leq f((\beta \vee \rho) \wedge (\gamma \vee \rho)) \\
&= (f(\beta) \vee f(\rho)) \wedge (f(\gamma) \vee f(\rho)) \\
&= (f(\beta) \wedge f(\gamma)) \vee f(\rho) \\
&= f((\beta \wedge \gamma) \vee \rho) \\
&= f(\rho) = 0,
\end{aligned}
$$

a contradiction.

(3)$\Rightarrow$(1). Assume (3) and suppose that $\mathbf{L}$ is not distributive. Then there exist $\alpha, \beta, \gamma \in L$ such that

$$
\alpha \vee (\beta \wedge \gamma) = \sigma < \tau = (\alpha \vee \beta) \wedge (\alpha \vee \gamma).
$$

Since σ is a meet of completely meet irreducible elements of $\mathbf{L}$, there exists a completely meet irreducible element $\rho \in L$ such that $\sigma \leq \rho$ but $\tau \not\leq \rho$. Let μ be the unique cover of ρ in $\mathbf{L}$ and $f : \mathbf{L} \to \mathbf{D_2}$ be a lattice homomorphism which separates ρ and μ, i.e., $f(\rho) = 0$, $f(\mu) = 1$. Then obviously $f(\sigma) = 0$, but $\mu \leq \rho \vee \tau$ implies

$$
1 = f(\mu) \leq f(\rho \vee \tau) = f(\rho) \vee f(\tau) = f(\tau).
$$

Hence $f(\tau) = 1$ and we see that f separates σ and τ also. Then, however,

$$
f(\alpha) \vee (f(\beta) \wedge f(\gamma)) < (f(\alpha) \vee f(\beta)) \wedge (f(\alpha) \vee f(\gamma))
$$

which contradicts the distributivity of $\mathbf{D_2}$. This proves the implication (3)$\Rightarrow$(1) and hence the lemma. $\bullet$

We wish to consider the relationships between several lattice-theoretic properties of an algebra. These are defined as follows:

(1) $\mathbf{A}$ is CD, i.e., $\mathbf{Con\,A}$ is a distributive lattice.

(2) $\mathbf{A}$ has no skew congruence for subdirect decompositions into finitely many factors, i.e., whenever $\{\rho, \theta_1, \ldots, \theta_k\} \subseteq \mathbf{Con\,A}$ and $\theta_1 \wedge \cdots \wedge \theta_k = 0$ then $(\theta_1 \vee \rho) \wedge \cdots \wedge (\theta_k \vee \rho) = \rho$.

(3) **A** has no skew congruence for subdirect decompositions into two factors, i.e., whenever $\{\rho, \theta_1, \theta_2\} \subseteq \mathrm{Con}\ \mathbf{A}$ and $\theta_1 \wedge \theta_2 = 0$ then $(\theta_1 \vee \rho) \wedge (\theta_2 \vee \rho) = \rho$.

(4) If $\mathbf{A}/\rho$ is SI and $\beta \wedge \gamma = 0$ (with $\rho, \beta, \gamma \in \mathrm{Con}\ \mathbf{A}$) then $\rho \geq \beta$ or $\rho \geq \gamma$.

(J) If **A** is a subdirect product of algebras $\mathbf{A}_i$, $i \in I$, and ρ is a congruence of **A** such that $\mathbf{A}/\rho$ is SI, then there exists an ultrafilter U on I such that $\theta_U \leq \rho$.

($\mathrm{J_{SI}}$) This is (J) for subdirect products with subdirectly irreducible factors.

Our basic observation below will be that if K is a class of algebras closed under the formation of homomorphic images then all six properties above are equivalent for K, in the sense that if K satisfies one, it satisfies all. It is important to note however, as we shall, that for a single algebra, there are actually three inequivalent properties among those defined above.

B. Jónsson [Jónsson, 1967] proved the very important result that (1) implies (J) for any algebra. Every SI algebra satisfies (2), (3) and (4) and hence these properties cannot imply (1). Also, (3) does not imply (2), as the following example shows.

Example 1.2.18 *An algebra which has no skew congruences with respect to subdirect decompositions into two factors but has skew congruences with respect to subdirect decompositions into three factors.*

Let **A** be any algebra whose congruence lattice is isomorphic to the lattice pictured in Figure 1.1. (There exists such an algebra by the Grätzer-Schmidt Theorem.) The completely meet irreducible congruence ρ does not dominate any of $\alpha_1, \alpha_2, \alpha_3$ yet $\alpha_1 \wedge \alpha_2 \wedge \alpha_3 = 0_A$. Nevertheless, it can be checked that whenever $\beta \wedge \gamma = 0_A$ in $\mathrm{Con}\,\mathbf{A}$ and σ is any one of the four completely meet irreducible elements of $\mathrm{Con}\,\mathbf{A}$, then $\sigma \geq \beta$ or $\sigma \geq \gamma$. Now our claims about existence of skew congruences follow from Lemma 1.2.16.

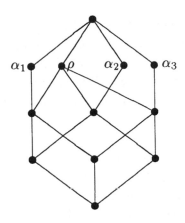

Figure 1.1

Lemma 1.2.19 *For any algebra* **A** *we have*

$$(1) \implies (2) \iff (J) \iff (J_{SI}) \implies (3) \iff (4).$$

PROOF The implications $(1) \Rightarrow (2) \Rightarrow (3) \Leftrightarrow (4)$ are obvious from the preceding discussions, and $(J) \Rightarrow (J_{SI})$ is trivial. The proof that (2) implies (J) is due to B. Jónsson.

To prove that $(J_{SI}) \Rightarrow (2)$, assume that (2) fails. Then **A** actually has a congruence ρ such that $\mathbf{A}/\rho$ is SI, and for some congruences $\theta_1, \ldots, \theta_k$, where $\theta_1 \wedge \cdots \wedge \theta_k = 0$, the inclusion $\theta_i \leq \rho$ holds for no i. Each of the congruences θ_i is the meet of some completely meet irreducible congruences, and in fact there must exist, for some set J presented as a union of subsets J_i, $i \in \{1, \ldots, k\}$, completely meet irreducible congruences ρ_j, $j \in J$, such that

$$\theta_i = \bigwedge_{j \in J_i} \rho_j \quad \text{for each } i \in \{1, \ldots, k\}.$$

Setting $\mathbf{A}_j = \mathbf{A}/\rho_j$, we have that **A** is a subdirect product of the SI algebras $\mathbf{A}_j$. If U is any ultrafilter on J, then some one of the sets J_i belongs to U, implying that $\theta_i = \bigwedge_{j \in J_i} \rho_j \leq \theta_U$ and so $\theta_U \leq \rho$ must be false. Thus we have shown that when (2) fails then (J_{SI}) fails also. ●

Theorem 1.2.20 *Let K be a class of algebras closed with respect to homomorphic images. If K has any one of the properties* (1), (2), (3), (4), (J), (J_{SI}) *then it has all of them.*

Therefore the absence of skew congruences for finite subdirect products is a Mal'cev definable property which is equivalent to congruence distributivity.

PROOF In view of Lemma 1.2.19, all that is required is a proof that if K satisfies (4) then it satisfies (1). Let $\mathbf{A} \in K$ and assume that all quotient algebras of $\mathbf{A}$ satisfy (4). We show that $\mathbf{L} = \mathbf{Con\,A}$ satisfies condition (3) of Lemma 1.2.17, thus establishing that $\mathbf{Con\,A}$ is distributive. Thus, assume that $\rho, \beta, \gamma \in L$, ρ is completely meet irreducible, and $\beta \wedge \gamma \le \rho$.

Now in the algebra $\mathbf{B} = \mathbf{A}/\beta \wedge \gamma$, the congruence $\rho' = \rho/\beta \wedge \gamma$ is completely meet irreducible—i.e., $\mathbf{B}/\rho'$ is SI—and $\beta' \wedge \gamma' = 0_B$ where $\beta' = \beta/\beta \wedge \gamma$ and $\gamma' = \gamma/\beta \wedge \gamma$. By condition (4) for $\mathbf{B}$, we have, say, $\beta' \le \rho'$. This is equivalent to $\beta \le \rho$, which shows that $\mathbf{L}$ satisfies condition (3) of Lemma 1.2.17 and hence is a distributive lattice.   •

Corollary 1.2.21 *If V is a locally finite variety and the finite members of V have no skew congruences then V is CD.*

PROOF Since the class K of homomorphic images of the free algebra on 3 generators in V has property (2) the result follows from Corollary 1.2.4.   •

Jónsson's proof that (1) $\Rightarrow$ (J) results in the following more tractable version of Birkhoff's theorem (Theorem 1.1.7).

Theorem 1.2.22 (a) *If K is a class of algebras of the same type and $V(K)$ is congruence distributive, then*

$$V(K) = IP_S H S P_U(K).$$

(b) *If $\mathbf{A}$ is a finite algebra and $V(\mathbf{A})$ is congruence distributive, then*

$$V(\mathbf{A}) = IP_S H S(\mathbf{A}).$$

PROOF Statement (b) follows from (a) since ultrapowers of a finite algebra are isomorphic with the given algebra. Statement (a) first appeared in [Jónsson, 1967] while statement (b) is implicit in [Foster, Pixley, 1964]. •

Statement (b) has several important consequences. First, it follows from (b) that a finitely generated CD variety is residually finite, and if the variety is of finite type then it contains only finitely many nonisomorphic SIs. Also it follows from (b) that if **A** and **B** are finite SI algebras, each generating the same CD variety, then **A** and **B** are isomorphic. Put another way, up to isomorphism, a finite SI algebra in a CD variety is completely determined by its equational theory. Finally it follows from (b) that if a residually finite CD variety has only finitely many subvarieties then it is generated by a finite algebra.

We close this section with a result (Corollary 1.2.24) showing how the existence of skew congruences can in certain situations be ruled out by assumptions weaker than congruence distributivity. Its importance will become clear when it is applied in Chapter 2. First we need the following technical lemma.

Lemma 1.2.23 *Let $\rho, \beta, \gamma \in \mathrm{Con}\,\mathbf{A}$ where $\mathbf{A}$ is any algebra.*

(i) *If $\mathbf{A}/\rho$ generates a CD variety then*
$$(\rho \vee \beta) \wedge (\rho \vee \gamma) = \rho \vee (\beta \wedge (\rho \vee \gamma)).$$

(ii) *If $\mathbf{A}/(\rho \wedge \beta)$ generates a CD variety then*
$$(\rho \vee \beta) \wedge (\rho \vee \gamma) = \rho \vee (\beta \wedge \gamma).$$

PROOF (i) Obviously $(\rho \vee \beta) \wedge (\rho \vee \gamma) \geq \rho \vee (\beta \wedge (\rho \vee \gamma))$. Assume that $(a, b) \in (\rho \vee \beta) \wedge (\rho \vee \gamma)$; in particular there are $c_0, c_1, \ldots, c_m \in A$ such that $a = c_0$, $c_m = b$ and $(c_{i-1}, c_i) \in \rho \cup \beta$ for every $i = 1, \ldots, m$.

Let $d_0, d_1, \ldots, d_n$ be Jónsson terms for $\mathbf{A}/\rho$ and define the elements $e_{ij} = d_i(a, c_j, b)$ where $i = 0, 1, \ldots, n$ and $j = 0, 1, \ldots, m$. Now consider the sequence

$$a, e_{00}, e_{01}, \ldots, e_{0m}, e_{1m}, e_{1,m-1}, \ldots, e_{10}, e_{20}, \ldots, e_{nm}, b. \qquad (1.10)$$

Obviously $e_{00} = d_0(a, a, b) \equiv_\rho a$, $e_{nm} = d_n(a, b, b) \equiv_\rho b$ and all pairs of the form $(e_{i,j-1}, e_{ij})$ are in $\rho \cup \beta$. As for the pairs of form

$(e_{i-1,m}, e_{im})$ and $(e_{i-1,0}, e_{i0})$, they all are in ρ by the definition of Jónsson terms. On the other hand, if $\theta = \mathrm{Cg}(a,b)$ then $\theta \leq \rho \vee \gamma$ and

$$e_{ij} = d_i(a, c_j, b) \equiv_\theta d_i(a, c_j, a) \equiv_\rho a$$

which implies $(e_{ij}, a) \in \rho \vee \gamma$ for every i and j. This means that all members of the sequence (1.10) are contained in a single $(\rho \vee \gamma)$-block. Now the adjacent members of this sequence are congruent either modulo $\rho \wedge (\rho \vee \gamma) = \rho$ or modulo $\beta \wedge (\rho \vee \gamma)$. Hence we have $(a,b) \in \rho \vee (\beta \wedge (\rho \vee \gamma))$.

(ii) In view of the first statement of the lemma it suffices to prove the inequality $\beta \wedge (\rho \vee \gamma) \leq (\beta \wedge \rho) \vee (\beta \wedge \gamma)$. This can be done in a fairly similar way as the proof of the preceding statement. We take an arbitrary pair $(a, b) \in \beta \wedge (\rho \vee \gamma)$ and pick the elements $c_0, c_1, \ldots, c_m \in A$ such that $a = c_0$, $c_m = b$ and $(c_{i-1}, c_i) \in \rho \cup \gamma$ for every $i = 1, \ldots, m$. Then we take Jónsson terms $d_0, d_1, \ldots, d_n$ for $A/(\rho \wedge \beta)$ and form again the sequence (1.10). Now

$$e_{00} = d_0(a, a, b) \equiv_{\rho \wedge \beta} a, \quad e_{nm} = d_n(a, b, b) \equiv_{\rho \wedge \beta} b$$

and all pairs of the form $(e_{i,j-1}, e_{ij})$ are in $\rho \cup \gamma$. As for the pairs of form $(e_{i-1,m}, e_{im})$ and $(e_{i-1,0}, e_{i0})$, they all are in $\rho \wedge \beta$ by the definition of Jónsson terms. On the other hand, if $\theta = \mathrm{Cg}(a,b)$ then $\theta \leq \beta$ and $e_{ij} = d_i(a, c_j, b) \equiv_\theta d_i(a, c_j, a) \equiv_{\rho \wedge \beta} a$ which implies $e_{ij} \equiv_\beta a$ for every i and j. This means that all members of the sequence (1.10) are contained in a single β-block. Now the adjacent members of this sequence are congruent either modulo $\beta \wedge \rho$ or modulo $\beta \wedge \gamma$. Hence $(a,b) \in (\beta \wedge \rho) \vee (\beta \wedge \gamma)$.   •

Corollary 1.2.24 *Let* A *be a subdirect product in* $B \times C$. *If the variety* $V(A/\rho, B)$ *is congruence distributive for every* $\rho \in \mathrm{Con}\,A$ *with subdirectly irreducible* A/ρ *then* A *has no skew congruences with respect to the given subdirect decomposition.*

PROOF Let β and γ be the kernels of the projections $A \to B$ and $A \to C$, respectively. Then whenever A/ρ is SI we have that $A/(\rho \wedge \beta)$ generates a CD variety and by Lemma 1.2.23

$$(\rho \vee \beta) \wedge (\rho \vee \gamma) = \rho \vee (\beta \wedge \gamma) = \rho,$$

since $\beta \wedge \gamma = 0_A$. Now if δ is any congruence of $\mathbf{A}$ then we have $\delta = \bigwedge_{i \in I} \rho_i$ where all $\mathbf{A}/\rho_i$ are SI. Thus

$$(\delta \vee \beta) \wedge (\delta \vee \gamma) \leq (\rho_i \vee \beta) \wedge (\rho_i \vee \gamma) = \rho_i \quad \text{for all } i$$

and consequently $(\delta \vee \beta) \wedge (\delta \vee \gamma) = \delta$. •

Exercises

1. Without appealing to Theorem 1.2.22 prove directly that if $\mathbf{A}$ is a finite algebra and $V(\mathbf{A})$ is CD, then

$$V(\mathbf{A})_{fg} \subseteq IP_S H S(\mathbf{A}).$$

2. Let $\mathbf{A}$ be any finite algebra which satisfies the inclusion of Exercise 1 above ($V(\mathbf{A})$ is *not* required to be CD); show that

$$V(\mathbf{A}) = IP_S H S(\mathbf{A}).$$

Combined with Exercise 1 this proves part (b) of Theorem 1.2.22.

This result can be proved in several different ways, all requiring an axiom of set theory at least as strong as the prime ideal theorem. One approach is to use the theory of inverse limits. A discussion of this topic and a proof of a special case (which generalizes to the present result) can be found in [Grätzer, 1979]. Another simple proof uses the full Axiom of Choice in the form of the Tychonoff theorem and starts from the observation that the universes of the members of $HS(\mathbf{A})$ are compact with the discrete topology. The proof we suggest below is inspired by this but is perhaps the most direct since it uses no machinery beyond the prime ideal theorem. First, since $\mathbf{A}$ is finite, the collection $\{\mathbf{A}_1, \ldots, \mathbf{A}_k\}$ of members of $HS(\mathbf{A})$ (one of each isomorphism type) is finite. To prove that $V(\mathbf{A}) = IP_S H S(\mathbf{A})$ it suffices to show, for each $\mathbf{B} \in V(\mathbf{A})$ and pair of distinct elements $a, b \in B$, that there is a homomorphism f from $\mathbf{B}$ onto some $\mathbf{A}_i$ such that $f(a) \neq f(b)$, (i.e., altogether there are "enough" homomorphisms to separate each pair of distinct elements). To do this we consider the Boolean algebra $\mathbf{P}$ of all subsets of the set

$$A_1^B \cup \cdots \cup A_k^B.$$

For each finite subset $F \subseteq B$ containing a and b let C_F be the set of all $f \in A_1^B \cup \cdots \cup A_k^B$ such that the restriction of f to $\mathrm{Sg}^{\mathbf{B}}(F)$ is a homomorphism into some $\mathbf{A}_i$.

By our assumption that $\mathbf{A}$ satisfies the inclusion of Exercise 1, each C_F is nonempty. If F_1 and F_2 are two finite subsets of B, each containing a and b, then clearly

$$C_{F_1} \cap C_{F_2} \supseteq C_{F_1 \cup F_2} \neq \emptyset,$$

so that the set of all C_F has the finite intersection property and hence is contained in an ultrafilter of $\mathbf{P}$. The proof is completed by applying ultrafilter properties to show that the intersection of all of the members of the ultrafilter is nonempty and hence the family of all C_F has a nonempty intersection.

3. Show that if V is a minimal CD variety then either $V = IP_S(\mathbf{A})$ for some finite $\mathbf{A}$ in V, or V has no finite members.

4. Give an example of a minimal CD variety which has no finite members.

5. Use Exercise 2 to prove: if V is a finitely generated variety and contains, up to isomorphism, only finitely many finite SIs, then V contains no infinite SI. This shows, contrapositively, that if a finitely generated variety contains an infinite SI then it contains an infinite number of finite SIs. (R. Quackenbush, Algebra Universalis **1**, 1971.)

6. Prove that if V is a CD variety, then V is generated by a finite algebra iff V is locally finite and has only finitely many subvarieties. (A. F. Pixley, Algebra Universalis, **4**, 1974.)

1.2.6   The commutator

An extremely important feature of group theory is the existence of binary commutator function on its congruence lattice, that is, on the lattice of its normal subgroups. Recall that if $\mathbf{A}$ and $\mathbf{B}$ are normal subgroups of a group $\mathbf{G}$ then their **commutator subgroup** is the subgroup of G generated by all commutators $[a, b] = a^{-1}b^{-1}ab$ where $a \in A$ and $b \in B$. By means of this operation various classes of groups (abelian, nilpotent, soluble) and special subgroups (center, centralizers) can be defined. In ring theory the similar role is played by the ideal multiplication operation.

One of the major achievements during the nineteen eighties in universal algebra was the creation of a satisfactory commutator theory for congruence modular varieties. We survey here a minimal number of notions and facts of this theory which will be needed later.

Definition Let ρ, σ and η be congruences of an algebra $\mathbf{A}$. We say that ρ **centralizes** σ modulo η if for every positive integer k, k-ary term t, elements $a, b \in A$ and $(k-1)$-tuples $\mathbf{c}, \mathbf{d} \in A^{k-1}$, the following implication is true:

$$\left. \begin{array}{rl} a & \equiv_\rho b \\ \mathbf{c} & \equiv_\sigma \mathbf{d} \\ t(a, \mathbf{c}) & \equiv_\eta t(a, \mathbf{d}) \end{array} \right\} \implies t(b, \mathbf{c}) \equiv_\eta t(b, \mathbf{d}).$$

If ρ and σ are two congruences of an algebra $\mathbf{A}$ then their **commutator** $[\rho, \sigma]$ is the smallest congruence η of $\mathbf{A}$ such that ρ centralizes σ modulo η.

The commutator $[\rho, \sigma]$ always exists; that is, for any algebra $\mathbf{A}$ there is a function

$$[\,,\,] : \mathrm{Con}\,\mathbf{A} \times \mathrm{Con}\,\mathbf{A} \to \mathrm{Con}\,\mathbf{A}$$

satisfying the definition. For the proof of this fact we refer to the book [McKenzie, McNulty, Taylor, 1987].

The following lemma is an easy exercise for a reader.

Lemma 1.2.25 *Let $\mathbf{A}$ be an arbitrary algebra. Then the following are true:*

(i) *for every $\rho, \sigma \in \mathrm{Con}\,\mathbf{A}$, $[\rho, \sigma] \leq \rho \wedge \sigma$;*

(ii) *the commutator function is order preserving with respect to both of its arguments.*

An algebra $\mathbf{A}$ is called **abelian** if $[1_\mathbf{A}, 1_\mathbf{A}] = 0_\mathbf{A}$ and **neutral**, if $[\rho, \sigma] = \rho \wedge \sigma$ holds for every two congruence relations ρ and σ of $\mathbf{A}$. These two classes of algebras are extremal cases in the sense of commutator. If we introduce the natural order relation on the set of all functions from $\mathrm{Con}\,\mathbf{A} \times \mathrm{Con}\,\mathbf{A}$ to $\mathrm{Con}\,\mathbf{A}$ then the commutator always lies in the interval $[0_A, \wedge]$. Thus abelian algebras are ones with the smallest possible commutator function and neutral algebras are ones with the largest possible commutator function.

The following lemma simplifies checking the neutrality of a given algebra.

Lemma 1.2.26 *An algebra* **A** *is neutral iff* $[\rho, \rho] = \rho$ *for every congruence* ρ *of* **A**.

PROOF The necessity of the condition is obvious. Assume that $[\rho, \rho] = \rho$ identically holds in Con **A** and take $\rho, \sigma \in$ Con **A**. Then in view of Lemma 1.2.25 we have

$$\rho \wedge \sigma = [\rho \wedge \sigma, \rho \wedge \sigma] \leq [\rho, \sigma] \leq \rho \wedge \sigma$$

so **A** is neutral.   •

The following examples describe the commutator operation in groups, rings, and modules and demonstrate that the concept has the expected meaning.

Example 1.2.27 *The commutator in groups.*

Let **A** be a group. The commutator of two elements $a, b \in A$ is $[a, b] = a^{-1}b^{-1}ab \in A$. The commutator of two subgroups **R** and **S** is the subgroup $[\mathbf{R}, \mathbf{S}] < \mathbf{A}$ generated by all elements $[r, s]$, $r \in R$, $s \in S$. It is easy to show that the commutator of normal subgroups is normal. We denote by $\rho_{\mathbf{R}}$ the congruence of **A** determined by a normal subgroup **R**. Hence $(a, b) \in \rho_{\mathbf{R}}$ iff $a^{-1}b \in R$. We want to show that the general commutator defined above for arbitrary universal algebras corresponds to the commutator we now defined for groups. In other words, we shall prove that for every two normal subgroups **R** and **S** of the group **A** the following formula holds:

$$\rho_{[\mathbf{R},\mathbf{S}]} = [\rho_{\mathbf{R}}, \rho_{\mathbf{S}}].$$

First observe that $\rho_{[\mathbf{R},\mathbf{S}]} \leq [\rho_{\mathbf{R}}, \rho_{\mathbf{S}}]$. Let $[\rho_{\mathbf{R}}, \rho_{\mathbf{S}}] = \eta = \rho_{\mathbf{T}}$ where **T** is a suitable normal subgroup of **A**. We must show that $[\mathbf{R}, \mathbf{S}] \subseteq T$, and it suffices to show that $[r, s] \in T$ for every $r \in R$, $s \in S$. The latter easily follows from the definition of the commutator if we take for term t the commutator, that is, $t(x, y) = x^{-1}y^{-1}xy$. Obviously

$t(1,1) = t(1,s)$, hence also $t(1,1) \equiv_\eta t(1,s)$. By the definition of the commutator this yields $1 = t(r,1) \equiv_\eta t(r,s)$, but this is equivalent to $[r,s] \in T$.

It remains to prove that $\rho_\mathbf{R}$ centralizes $\rho_\mathbf{S}$ modulo $\rho_{[\mathbf{R},\mathbf{S}]}$. It is sufficient to show, that given arbitrary $a,b \in A$, $\mathbf{c},\mathbf{d} \in A^{k-1}$ such that $a^{-1}b \in R$, $c_i^{-1}d_i \in S$, $i = 1,\ldots,k-1$, we have

$$t(b,\mathbf{d})^{-1}t(b,\mathbf{c})t(a,\mathbf{c})^{-1}t(a,\mathbf{d}) \in [\mathbf{R},\mathbf{S}] \tag{1.11}$$

for every k-ary term t. It is a good exercise to prove (1.11) by induction on the length of term t.

Now it is clear that a group is abelian iff it is an abelian algebra in the sense of the above definition. Lemma 1.2.26 implies that a group $\mathbf{A}$ is neutral iff $\mathbf{R} = [\mathbf{R},\mathbf{R}]$ for every normal subgroup $\mathbf{R}$ of $\mathbf{A}$. The latter is equivalent to the condition that none of the normal subgroups of $\mathbf{A}$ has nontrivial abelian homomorphic images. In particular all simple nonabelian groups are neutral. •

Example 1.2.28 *The commutator in rings.*

Let $\mathbf{A}$ be an arbitrary ring. Here $\mathbf{A}$ may have no identity element and may even be nonassociative. We denote by $\rho_\mathbf{R}$ the congruence of $\mathbf{A}$ determined by an ideal $\mathbf{R}$. Hence $(a,b) \in \rho_\mathbf{R}$ iff $a - b \in R$. It turns out that the general commutator defined above for universal algebras corresponds to the operation $\mathbf{RS} + \mathbf{SR}$ on the set of ideals of $\mathbf{A}$. In other words, we shall prove that for every two ideals $\mathbf{R}$ and $\mathbf{S}$ of the ring $\mathbf{A}$ the following formula holds:

$$\rho_{\mathbf{RS}+\mathbf{SR}} = [\rho_\mathbf{R},\rho_\mathbf{S}].$$

Note that the product of ideals $\mathbf{RS}$ is defined as the subgroup of the additive group of $\mathbf{R}$ generated by all products rs where $r \in R$, $s \in S$.

Let $I = \mathbf{RS} + \mathbf{SR}$ and $[\rho_\mathbf{R},\rho_\mathbf{S}] = \eta = \rho_\mathbf{J}$ where $\mathbf{J}$ is a suitable ideal of $\mathbf{A}$. We first prove that $I \subseteq J$, which is equivalent to $\rho_\mathbf{I} \leq \eta$ and it suffices to show that $rs \in J$ for every $r \in R$, $s \in S$. The latter easily follows from the definition of the commutator if we take for term t the multiplication operation, that is, $t(x,y) = xy$. Obviously $t(0,0) = t(0,s)$, hence also $t(0,0) \equiv_\eta t(0,s)$. By the definition of the commutator this yields $0 = t(r,0) \equiv_\eta t(r,s) = rs$, but this is equivalent to $rs \in J$.

It remains to prove that $\rho_{\mathbf{R}}$ centralizes $\rho_{\mathbf{S}}$ modulo $\rho_{\mathbf{I}}$. It is sufficient to show that for given arbitrary $a, b \in A$, $\mathbf{c}, \mathbf{d} \in A^{k-1}$ such that $a - b \in R$, $c_i - d_i \in S$, $i = 1, \ldots, k - 1$, we have

$$t(b, \mathbf{d}) - t(b, \mathbf{c}) + t(a, \mathbf{c}) - t(a, \mathbf{d}) \in I \tag{1.12}$$

for every m-ary term t. We prove (1.12) by induction on the length of t. This formula obviously holds if t is just a variable. Suppose it holds for two terms u and v. Then it is easy to see that it also holds for $u - v$. The following calculations, where the congruence indicated is modulo I, show that (1.12) also holds for uv:

$$
\begin{aligned}
&u(b, \mathbf{d})v(b, \mathbf{d}) - u(b, \mathbf{c})v(b, \mathbf{c}) + u(a, \mathbf{c})v(a, \mathbf{c}) - u(a, \mathbf{d})v(a, \mathbf{d}) \\
={} & u(b, \mathbf{d})(v(b, \mathbf{d}) - v(b, \mathbf{c})) + (u(b, \mathbf{d}) - u(b, \mathbf{c}))v(b, \mathbf{c}) \\
& + u(a, \mathbf{c})(v(a, \mathbf{c}) - v(a, \mathbf{d})) + (u(a, \mathbf{c}) - u(a, \mathbf{d}))v(a, \mathbf{d}) \\
\equiv{} & u(a, \mathbf{d})(v(b, \mathbf{d}) - v(b, \mathbf{c})) + (u(b, \mathbf{d}) - u(b, \mathbf{c}))v(a, \mathbf{c}) \\
& + u(a, \mathbf{c})(v(b, \mathbf{c}) - v(b, \mathbf{d})) + (u(b, \mathbf{c}) - u(b, \mathbf{d}))v(a, \mathbf{d}) \\
={} & (u(a, \mathbf{d}) - u(a, \mathbf{c}))(v(b, \mathbf{d}) - v(b, \mathbf{c})) \\
& + (u(b, \mathbf{d}) - u(b, \mathbf{c}))(v(a, \mathbf{c}) - v(a, \mathbf{d})).
\end{aligned}
$$

The congruence above is proved as follows (for the first summand; the others are handled similarly): Let $u(b, d) - u(a, d) = r \in R$ and $v(b, d) - v(b, c) = s \in S$. Then

$$
\begin{aligned}
u(b, d)(v(b, d) - v(b, c)) &= (u(a, d) + r)s = u(a, d)s + rs \\
&\equiv u(a, d)(v(b, d) - v(b, c)).
\end{aligned}
$$

Now it is clear that a ring is an abelian algebra in the sense of the above definition iff it has the zero multiplication. By Lemma 1.2.26 a ring $\mathbf{A}$ is neutral iff $\mathbf{R} = \mathbf{RR}$ for every ideal $\mathbf{R}$ of $\mathbf{A}$. The latter is equivalent to the condition that none of the ideals of $\mathbf{A}$ has a nontrivial homomorphic image with the zero multiplication. Hence, all simple rings with nonzero multiplication (in particular, all division rings) are neutral. ●

Example 1.2.29 *The commutator in modules.*

Let $\mathbf{A}$ be a left unital $\mathbf{R}$-module where $\mathbf{R}$ is an associative ring with an identity element. It is easy to observe that $\mathbf{A}$ is an abelian algebra in the sense of the above definition. One simply has to take in account that the congruences of **A** are determined by its submodules, just as the ring congruences are determined by ideals, and terms have the form $r_1 x_1 + \cdots + r_m x_m$ where $r_1, \ldots, r_m$ are fixed elements of the ring **R**. Hence, there are no nonzero neutral modules. •

The algebras in a congruence distributive variety are easily seen to be neutral. Since we will mostly be concerned with congruence distributivity until Chapter 5 the commutator will not appear again until then.

Chapter 2

Characterizations of Equivalence Lattices

2.1 Introduction

In Section 1.1.2 we introduced, for a set A, the complete lattice **Eqv** A of all equivalence relations on A and, in Section 1.1.3, equivalence lattices and 0-1 equivalence lattices as sublattices of **Eqv** A. The purpose of this chapter is to collect results which give either necessary or sufficient conditions for arithmeticity or distributivity of equivalence lattices. If it is clear from the context that an equivalence lattice contains 0_A and 1_A we shall sometimes neglect calling it a 0-1 equivalence lattice. Also note that while distributivity is a property of abstract lattices, arithmeticity for an equivalence lattice means that the lattice is distributive and consists of permuting equivalence relations. (If we apply the adjective *arithmetical* to an abstract lattice, this would reasonably be interpreted to imply that the lattice is arithmetical as an algebra, i.e., is congruence distributive and congruence permutable. In the present context of equivalence lattices this possible ambiguity should not arise.)

Mostly the results of this chapter will be applied in Chapters 3 and 4. Typical equivalence lattices are the congruence lattices of algebras; so the algebraic structure always gives rise to an equivalence lattice. On the other hand, given an equivalence lattice **L** on a set A, we have the clone $F = clo(L)$ and the algebra $\langle A; F \rangle$. As in Section 1.1.3, we shall refer to the members of F as the **L**-compatible operations (or functions). Recall that this means that F is the set of all $f \in O(A)$ which preserve every $\theta \in L$. Our later study of affine complete algebras will fully justify the importance we attach to the study of the relationship between the equivalence lattice **L** and the clone of **L**-compatible operations. Finally, reviewing the terminology of Section 1.1.3, notice that for an algebra **A**, Comp $\mathbf{A} = clo(\text{Con } \mathbf{A})$.

Together with the clone $clo(L)$ of total operations we also need *partial* **L**-compatible operations (or functions). These are again defined as in Section 1.1.3.

If **L** is an equivalence lattice on A it is convenient to extend the equivalences $\phi \in L$ to m-tuples in A^m componentwise: $\mathbf{a} \equiv_\phi \mathbf{b}$ iff $a_i \equiv_\phi b_i$ for $i = 1, \ldots, m$. Also the equivalence classes $\mathbf{a}/\phi$, with $\mathbf{a} \in A^m$, are defined in a natural way. Now the condition that the function $f : A^m \to A$ is compatible with $\phi \in L$ can be expressed as follows:

$$(\forall \mathbf{a}, \mathbf{b} \in A^m)[\mathbf{a} \equiv_\phi \mathbf{b} \implies f(\mathbf{a}) \equiv_\phi f(\mathbf{b})]. \qquad (2.1)$$

We require some notation for equivalence lattices which is the same as is used in a similar context for algebras. For an equivalence lattice **L** on A and $\phi \in L$, let $\mathbf{L}_\phi$ denote the equivalence lattice on A/ϕ naturally isomorphic with the interval of elements θ of **L** with $\theta \geq \phi$. For $\theta, \phi \in L$, and $\theta \geq \phi$, θ/ϕ denotes the element of L_ϕ corresponding to θ in **L**. It is easy to observe that if the equivalence lattice **L** on A is permutable (distributive, complete) then so is the equivalence lattice $\mathbf{L}_\phi$ on A/ϕ for every $\phi \in L$.

If f is an m-ary operation on A which is compatible with some relation $\phi \in \text{Eqv } A$, let f_ϕ denote the m-ary operation on A/ϕ which is induced by f, i.e., defined by

$$f_\phi(\mathbf{x}/\phi) = f(\mathbf{x})/\phi.$$

Occasionally, when no ambiguity can arise, we may drop the subscript on f_ϕ, and simply write, say $f(\mathbf{x}/\phi)$, since it should be clear that the induced function is intended.

It is easy to see that if $f \in clo(L)$ then $f_\phi \in clo(L_\phi)$ for every $\phi \in L$.

We shall often require **L** to be a complete sublattice of **Eqv** A, most often when it is a congruence lattice. As usual, we assume that a complete equivalence lattice contains 0_A and 1_A. When this is so, then, just as for congruence lattices, principal equivalence relations exist. In particular, because $1_A \in L$, for $a, b \in A$ the **principal equivalence relation** $\text{Eg}^{\mathbf{L}}(a, b)$ determined by a and b always exists and is the intersection of all members of L containing the pair (a, b). The condition $0_A \in L$ guarantees that $\text{Eg}^{\mathbf{L}}(a, a) = 0_A$ for every $a \in A$. The notion of principal equivalence relations extends to members of A^m as follows. If $\mathbf{a} = (a_1, \ldots, a_m)$ and $\mathbf{b} = (b_1, \ldots, b_m)$ then

$$\text{Eg}^{\mathbf{L}}(\mathbf{a}, \mathbf{b}) = \text{Eg}^{\mathbf{L}}(a_1, b_1) \vee \cdots \vee \text{Eg}^{\mathbf{L}}(a_m, b_m) .$$

(As with principal congruences we omit the superscript when the context is unambiguous.) When principal equivalence relations exist and $f : A^m \to A$ we have

$$f \text{ is } \mathbf{L}\text{-compatible} \iff (\forall \mathbf{a}, \mathbf{b} \in A^m)[(f(\mathbf{a}), f(\mathbf{b})) \in \text{Eg}^{\mathbf{L}}(\mathbf{a}, \mathbf{b})].$$

The following formula is essentially from [Köhler, Pigozzi 1980] where it was established for congruence lattices of algebras.

$$\begin{aligned}
\text{Eg}(c, d) &\leq \text{Eg}(a, b) \vee \phi \text{ in } \mathbf{L} \\
&\iff \text{Eg}(c/\phi, d/\phi) \leq \text{Eg}(a/\phi, b/\phi) \text{ in } \mathbf{L}_\phi
\end{aligned} \tag{2.2}$$

holds for every $\phi \in L$ and $a, b, c, d \in A$.

Exercise

Verify formula (2.2) for any complete equivalence lattice $\mathbf{L}$ on a set A.

2.2 Arithmeticity

In the present section we discuss several important special properties which may be satisfied by an equivalence lattice. All of these properties are equivalent to arithmeticity in case $\mathbf{L}$ is a finite 0-1 equivalence lattice; several are equivalent in less restrictive circumstances. The most well known of these properties specializes to the Chinese remainder theorem for the ring of integers. Since this is the strongest form of this property we call it the strong Chinese remainder condition abbreviated as the (SCRC).

2.2.1   The strong Chinese remainder condition for equivalence lattices

Definition An equivalence lattice **L** on a set A satisfies the **strong Chinese remainder condition** (briefly, the (SCRC)), if for each finite set $\theta_1,\ldots,\theta_n$ of equivalence relations in L, and each element $\mathbf{a} = (a_1,\ldots,a_n) \in A^n$, the system

$$x \equiv a_i\,(\theta_i), \quad i = 1,\ldots,n, \qquad (2.3)$$

is solvable for $x \in A$ iff for all $1 \leq i < j \leq n$ the conditions

$$a_i \equiv a_j\,(\theta_i \vee \theta_j) \qquad (2.4)$$

are satisfied.

For any system such as (2.3) to be solvable it is certainly necessary that the conditions (2.4) are satisfied. Later, in Chapter 3 (Section 3.2) we shall be concerned with conditions similar to the (SCRC) and naturally called **weak Chinese remainder conditions**, because they are defined the same as the above except that the conditions of solvability (2.4) are replaced by one of the conditions: solvability of every two (three, four, etc.) members of (2.3). These are successively stronger solvability conditions and hence result in, formally, successively weaker weak Chinese remainder conditions.

The following theorem is the basic characterization of the strong Chinese remainder condition. In Section 1.2 we have already proved the equivalence of statements (2), (3) and (4) of the theorem for congruence lattices and the same argument applies here. To complete the proof we must show that the (SCRC) is equivalent to arithmeticity. To prove that the (SCRC) implies arithmeticity, first take $n = 2$ in the (SCRC) and infer permutability directly. For distributivity, it is easy to apply the (SCRC) to directly verify the inequality

$$(\theta \circ \phi) \wedge (\theta \circ \psi) \leq \theta \circ (\phi \wedge \psi).$$

Conversely, assuming both permutability and distributivity obtain the (SCRC) for the case $n = 2$ from permutability alone. Obtain the general case $n > 2$ by induction using distributivity. We leave the straightforward details for the reader; as noted in Section 1.2.1 the proof is, except for notation, the same as for Dedekind domains as given in [Zariski, Samuel, 1958].

Theorem 2.2.1 *If* **L** *is an equivalence lattice, then the following conditions are equivalent:*

(1) **L** *satisfies the strong Chinese remainder condition;*

(2) **L** *is arithmetical;*

(3) *For all* $\theta, \phi, \psi \in L$,

$$\theta \wedge (\phi \circ \psi) = (\theta \wedge \psi) \circ (\theta \wedge \phi);$$

(4) *For all* $\theta, \phi, \psi \in L$,

$$\theta \wedge (\phi \circ \psi) \le (\theta \wedge \psi) \circ (\theta \wedge \phi).$$

The equation in statement (3) of the theorem is an example of an **equivalence equation**, also often called a **congruence equation**. Formally, an equivalence equation is an equation $e \approx f$ where e and f are terms in the algebraic language having operation symbols $\wedge$, $\vee$, and $\circ$. If no $\circ$ occurs the equivalence equation is a **lattice equation**. (Notice that statement (4) is equivalent to an equivalence equation.) Thus if **L** is an equivalence lattice, $\mathbf{L} \models e \approx f$ means that $e \approx f$ is satisfied for every assignment of members of L to the variables and where $\wedge$, $\vee$, and $\circ$ are interpreted respectively as meet, join, and relation product. The strong Chinese remainder condition is significant among equivalence equations in that it logically implies every nontrivial equivalence equation. This is so since if $e \approx f$ is any nontrivial equivalence equation and **L** satisfies the (SCRC), then **L** is distributive, and hence every nontrivial lattice equation holds in **L** and, in particular, the lattice equation obtained from $e \approx f$ by replacing $\circ$ by $\vee$. But since $\circ = \vee$ in **L**, it follows that $e \approx f$ holds in **L**.

Because of the strength of the strong Chinese remainder condition we may anticipate that it has some remarkable consequences. We now consider three of these which, under additional hypotheses, in fact characterize the strong Chinese remainder condition. Each of these is stated in terms of **L**-compatible operations.

2.2.2 The compatible function extension property

Definition An equivalence lattice on a set A satisfies the **compatible function extension property** (briefly, the (CFE) property) if for any positive integer m and finite subsets $X, Y \subseteq A^m$, with $X \subseteq Y$, any **L**-compatible function $f : X \to A$ has an **L**-compatible extension from Y to A.

Theorem 2.2.2 *If* **L** *is a complete equivalence lattice on the set* A, *then* **L** *satisfies the strong Chinese remainder condition iff it satisfies the compatible function extension property.*

PROOF First suppose that **L** satisfies the (SCRC) and that f is **L**-compatible with domain $X = \{\mathbf{a}^1, \ldots, \mathbf{a}^n\}$. Obviously we need only consider the case where $Y = X \cup \{\mathbf{b}\}$, i.e., the addition of a single element to the domain of f. If $n = 1$ then defining $f(\mathbf{b}) = f(\mathbf{a}^1)$ is clearly a compatible extension. For $n \geq 2$ we can extend f compatibly by defining $f(\mathbf{b}) = w$ where w is a solution of the system

$$w \equiv f(\mathbf{a}^i) \ (\mathrm{Eg}(\mathbf{a}^i, \mathbf{b})), \ i = 1, \ldots, n.$$

By the (SCRC) the system is solvable provided the conditions

$$f(\mathbf{a}^i) \equiv f(\mathbf{a}^j) \ (\mathrm{Eg}(\mathbf{a}^i, \mathbf{b}) \vee \mathrm{Eg}(\mathbf{a}^j, \mathbf{b}))$$

are satisfied. But **L**-compatibility implies $f(\mathbf{a}^i) \equiv f(\mathbf{a}^j) \ (\mathrm{Eg}(\mathbf{a}^i, \mathbf{a}^j))$ and obviously

$$\mathrm{Eg}(\mathbf{a}^i, \mathbf{a}^j) \leq \mathrm{Eg}(\mathbf{a}^i, \mathbf{b}) \vee \mathrm{Eg}(\mathbf{a}^j, \mathbf{b}),$$

so the extension is possible.

Conversely, suppose that **L** satisfies the compatible function extension property. We suppose that $(x, z) \in \theta \wedge (\phi \circ \psi)$ so that we have $(x, z) \in \theta, (x, y) \in \phi, (y, z) \in \psi$ for some $y \in A$. Let

$$X = \{(x, y, x), (x, y, y), (z, y, z), (y, y, z)\} \subseteq A^3$$

and define $f : X \to A$ by assigning value x for the first two triples in X and z for the last two. To check the **L**-compatibility of f we have to show that for each of the six possible choices of $\{\mathbf{a}, \mathbf{b}\} \subseteq X$, and for every $\phi \in L$ the implication (2.1) holds. Let for example $\mathbf{a} = (x, y, y)$, $\mathbf{b} = (y, y, z)$ and $\mathbf{a} \equiv_\phi \mathbf{b}$ so that $x \equiv_\phi y$ and $y \equiv_\phi z$. Then by transitivity, $f(\mathbf{a}) = x \equiv_\phi z = f(\mathbf{b})$.

The other five cases are immediate since in each of these the value of

$$(f(\mathbf{a}), f(\mathbf{b})) = f((a_1, b_1), (a_2, b_2), (a_3, b_3))$$

is one of (a_i, b_i). Therefore f is **L**-compatible and can be extended to an **L**-compatible function on $X \cup \{(x, y, z)\}$. Then we have

$$x = f(x, y, x) \equiv_\theta f(x, y, z) \equiv_\theta f(z, y, z) = z \text{ and}$$
$$x = f(x, y, y) \equiv_\psi f(x, y, z) \equiv_\phi f(y, y, z) = z,$$

so that $(x, z) \in (\theta \wedge \psi) \circ (\theta \wedge \phi)$. Therefore, by Theorem 2.2.1, **L** satisfies the (SCRC). $\bullet$

Note that in the proof that the compatible function extension property implies the strong Chinese remainder theorem, the completeness assumption was not used. For subsequent use it is worth noticing that in the argument above we have actually established the following:

Lemma 2.2.3 *If* **L** *is any equivalence lattice on* A *and*

$$X = \{(x, y, x), (x, y, y), (z, y, z), (y, y, z)\} \subseteq A^3,$$

then the function $f : X \to A$, *defined by*

$$f(x, y, x) = f(x, y, y) = x \quad and \quad f(z, y, z) = f(y, y, z) = z,$$

is **L***-compatible. If such* f *can always be* **L***-compatibly extended to* $X \cup \{(x, y, z)\}$, *then* **L** *satisfies the (SCRC).*

2.2.3 Existence of compatible Pixley functions

The following is a simple but important observation.

Lemma 2.2.4 *If* **L** *is any equivalence lattice on the set* A *and* **L** *supports a Mal'cev function, Jónsson functions, or Pixley function on* A, *then* **L** *satisfies the corresponding condition: permutability, distributivity, arithmeticity.*

PROOF The proof of Theorem 1.2.2 applies here to show that if **L** supports a Pixley function then **L** satisfies the (SCRC). The corresponding proofs in Theorems 1.2.1 and 1.2.3 likewise apply directly to equivalence lattices. $\bullet$

It is remarkable that in the case of arithmeticity, the converse of Lemma 2.2.4 holds with some cardinality restrictions. The following theorem establishes this for the most important case for us, namely when A is finite. It is remarkable that it exactly parallels Theorem 1.2.2 which was for varieties and is another example of the strength of the (SCRC) as an equivalence equation.

Theorem 2.2.5 *If A is a finite set and* **L** *is an equivalence lattice on A then* **L** *satisfies the strong Chinese remainder condition iff* **L** *supports a Pixley function on A.*

PROOF Assuming the (SCRC) and hence (by the first part of the proof of Theorem 2.2.2) the compatible function extension property, we take X to be the set of all triples of the forms $(x, y, y), (x, y, x)$ or (y, y, x) in A^3 and define f to have value x in all of these cases. Then just as in the proof of Lemma 2.2.3 we can easily see that f is **L**-compatible. Finally, extend f to the remaining triples (x, y, z), where x, y, z are all different, by the compatible function extension property. •

A consequence of these results is that the Mal'cev conditions for arithmeticity determined for varieties have a counterpart for single algebras. (So far we have proven it here only for finite algebras, but the next theorem (Theorem 2.2.6) shows that we need only the finiteness of the congruence lattice.) H. P. Gumm [Gumm, 1978] has shown that arithmeticity is the only Mal'cev condition expressible as a congruence equation with this property. In Example 2.3.13 at the end of this chapter we describe the beginning part of Gumm's proof.

I. Korec ([Korec, 1978]) showed that the finiteness of A in Theorem 2.2.5 can be dropped provided $|A| \leq \omega$ and **L** is complete. He further showed ([Korec, 1981]) that, in general, the Theorem is false if $|A| > \omega$. The exercises in the end of the subsection present further extensions of Theorem 2.2.5 due to K. Kaarli.

We now present an extended version of Theorem 2.2.5. In what follows we shall call a Pixley function $f(x, y, z)$ on a set A a **principal** Pixley function for an equivalence lattice **L** on A if it is **L**-compatible and also has the following property:

$$x \equiv y \ (\mathrm{Eg}^{\mathbf{L}}(u, v)) \quad \Longleftrightarrow \quad f(u, v, x) = f(u, v, y), \qquad (2.5)$$

i.e., if f defines principal equivalence relations in the manner indicated; notice that this is in the same form as in principal arithmetical varieties as discussed in Section 1.2.2.

Notice that if f is any L-compatible Pixley function on A, then for any $\phi \in L$ the right to left implication of (2.5) holds for $\mathbf{L}_\phi$ on A/ϕ with f_ϕ replacing f. Also, if $f(x,y,z)$ is a principal Pixley function then all induced functions f_ϕ, $\phi \in L$, are also principal Pixley functions, i.e., the left to right implication of (2.5) also holds in L_ϕ. To see this we need only observe that, by (2.2), $x/\phi \equiv y/\phi$ ($\mathrm{Eg}^{\mathbf{L}\phi}(u/\phi, v/\phi)$) implies $x \equiv y$ ($\mathrm{Eg}^{\mathbf{L}}(u,v) \vee \phi$) so that, by permutability, there is a $z \in A$ with $x \equiv z$ ($\mathrm{Eg}^{\mathbf{L}}(u,v)$) and $z \equiv_\phi y$. Hence we have both $f(u,v,x) = f(u,v,z)$ and $z \equiv_\phi y$ and from this we obviously have

$$f_\phi(u/\phi, v/\phi, x/\phi) = f_\phi(u/\phi, v/\phi, y/\phi).$$

Since $\mathbf{L}$ is completely determined by its principal equivalence relations, it follows that any principal Pixley function for $\mathbf{L}$ (there may be many) uniquely determines $\mathbf{L}$. Together with the obvious analogy with principal arithmetical varieties this is the significance of the following theorem. An earlier incorrect version appeared in [Pixley, 1984].

Theorem 2.2.6 *Let A be a set and $\mathbf{L}$ be a finite 0-1 equivalence lattice on A. Then $\mathbf{L}$ is arithmetical iff it supports at least one principal Pixley function $f(x,y,z)$.*

PROOF In view of Lemma 2.2.4 we need to only prove that the arithmeticity implies the existence of a principal Pixley function.

First observe that if the height of $\mathbf{L}$ is 1 then we may define f to be the discriminator function on A. Then obviously f is a principal Pixley function. Also, in this case (height 1) if f is any principal Pixley function, then $u \neq v$ implies $\mathrm{Eg}(u,v) = 1_A$ so by (2.5), $f(u,v,x) = f(u,v,u) = u$ for all x, i.e., f is the discriminator. Hence for height 1 the discriminator is the unique principal Pixley function.

In general, if the height of $\mathbf{L}$ is greater than 1, we construct f inductively, starting from the top. As a matter of fact, we construct, step by step, the functions f_ρ, $\rho \in L$, each of which is a principal Pixley function on A/ρ for $\mathbf{L}_\rho$, and eventually come to f_0 which is a principal Pixley function for $\mathbf{L}$.

As a first step we define f_θ for all coatoms θ of **L**. Since the lattices $\mathbf{L}_\theta$ are of height 1, all f_θ must be the discriminator functions on A/θ. To proceed, we assume that we already have the functions f_ρ for all $\rho \in L$ such that the height of the interval $[\rho, 1]$ is less than n ($n \geq 2$). Moreover we assume that the system of these functions is compatible in the sense that each of them induces on every higher level the functions belonging to our function system. More precisely, if f_ρ and f_σ are defined already, and $\rho < \sigma$, then f_ρ induces the function f_σ on A/σ. We show how to define the functions f_ρ for the next level (i.e., for $\rho \in L$ such that the height of $[\rho, 1]$ is n), so that the resulting function system will still satisfy the same compatibility conditions.

Let $\rho \in L$ be any such equivalence and let $\phi(1), \ldots, \phi(k)$ be all the members of L which cover ρ in **L**. Now, by factoring over ρ the construction of f_ρ reduces to the case $\rho = 0$. Correspondingly, the $\phi(1), \ldots, \phi(k)$ will be atoms of **L**. We distinguish two cases.

<u>Case 1.</u> $k > 1$. (Notice that this is the only case which will ever occur if **L** is complemented.) For each $u, v, x \in A$ pick

$$w_i \in f_{\phi(i)}(u/\phi(i), v/\phi(i), x/\phi(i)), \quad i = 1, \ldots, k.$$

Since every $f_{\phi(i)}$ and $f_{\phi(j)}$ induce the same function $f_{\phi(i) \vee \phi(j)}$ on $A/(\phi(i) \vee \phi(j))$, it follows that

$$w_i \equiv w_j \ (\phi(i) \vee \phi(j)) \quad \text{for all} \quad 1 \leq i < j \leq k.$$

Therefore, by the (SCRC), there is a $w \in A$ such that $w \equiv w_i \ (\phi(i))$ for $i = 1, \ldots, k$, and thus

$$w \in w/\phi(i) = f_{\phi(i)}(u/\phi(i), v/\phi(i), x/\phi(i))$$

for $i = 1, \ldots, k$. Since the $\phi(i)$ are atoms and $k > 1$, we see that w is unique and is given by

$$\{w\} = \tag{2.6}$$
$$f_{\phi(i)}(u/\phi(i), v/\phi(i), x/\phi(i)) \cap f_{\phi(j)}(u/\phi(j), v/\phi(j), x/\phi(j))$$

for any $i \neq j$. Hence we define $f(u, v, x) = w$ and, since the $f_{\phi(i)}$ are Pixley functions, conclude that f is also. To establish (2.5) use distributivity, formula (2.2), and the fact that the $f_{\phi(i)}$ are principal to see that

$$x \equiv y \ (\mathrm{Eg}(u,v)) \quad \text{in } \mathbf{L}$$
$$\implies \quad x \equiv y \ (\mathrm{Eg}(u,v) \vee \phi(i)) \quad \text{in } \mathbf{L} \quad \text{for } i = 1, \dots, k,$$
$$\iff x/\phi(i) \equiv y/\phi(i) \ (\mathrm{Eg}(u/\phi(i), v/\phi(i))) \quad \text{in } \mathbf{L}_{\phi(i)}$$
$$\iff f_{\phi(i)}(u/\phi(i), v/\phi(i), x/\phi(i)) = f_{\phi(i)}(u/\phi(i), v/\phi(i), y/\phi(i)).$$

Since the latter holds for every $i = 1, \dots, k$, the formula (2.6) implies $f(u, v, x) = f(u, v, y)$.

It remains to notice that by construction the function f induces the functions f_σ for all $\sigma \in L$, $\sigma \geq \rho$, because it induces all functions $f_{\phi(i)}$, $i = 1, \dots, k$, and the latter induce all functions f_σ for higher levels.

<u>Case 2.</u> $k = 1$. Here $\phi = \phi(1)$ is the sole atom of $\mathbf{L}$. (This case will occur at least once iff $\mathbf{L}$ is not complemented.) On the set of all ordered triples (u, v, x) of elements of A define the equivalence relation $\sim$ by

$$(u, v, x) \sim (u', v', x') \text{ iff } u = u', v = v', \quad \text{and} \quad x \equiv x' \ (\mathrm{Eg}^{\mathbf{L}}(u, v)).$$

Concerning the relation $\sim$ we have the following remarks:

1. For all (u, v, x) in a given $\sim$-block $f_\phi(u/\phi, v/\phi, x/\phi)$ has the same value:

$$(u, v, x) \sim (u', v', x') \implies$$
$$f_\phi(u/\phi, v/\phi, x/\phi) = f_\phi(u'/\phi, v'/\phi, x'/\phi).$$

This is true since

$$x \equiv x' \ (\mathrm{Eg}^{\mathbf{L}}(u, v)) \implies x/\phi \equiv x'/\phi \ (\mathrm{Eg}^{\mathbf{L}}(u/\phi, v/\phi))$$

and the fact that f_ϕ is a principal Pixley function.

2. If $u = v$ then for each x, $\{(u, v, x)\}$ obviously constitutes a one element $\sim$-block.

Now we define f on A to be constant on each $\sim$-block by the following three clauses:

$$f(u, v, x) = \begin{cases} u & \text{if } u \neq v \text{ and } u \equiv x \ (\mathrm{Eg}^{\mathbf{L}}(u, v)) \ \text{(first)}, \\ x & \text{if } u = v \ \text{(second)}, \\ \text{any fixed element in } f_\phi(u/\phi, v/\phi, x/\phi) \\ \quad \text{if } u \neq v \text{ and } u \not\equiv x \ (\mathrm{Eg}^{\mathbf{L}}(u, v)) \ \text{(third)}. \end{cases}$$

The three clauses of the definition are clearly disjoint and the above remarks show that each is independent of the particular representative of the $\sim$-block chosen.

Obviously, by the first two clauses, f is a Pixley function on A. Next we verify the inclusion

$$f(u, v, x) \in f_\phi(u/\phi, v/\phi, x/\phi)$$

which shows that f is compatible with ϕ and induces the function f_ϕ on A/ϕ. Then, since ϕ is the sole atom of $\mathbf{L}$ and f_ϕ is compatible on A/ϕ for $\mathbf{L}_\phi$, it follows that f is an $\mathbf{L}$-compatible function on A. For the verification we consider the three clauses:

For the first clause, if $u \equiv x$ ($\mathrm{Eg}^{\mathbf{L}}(u, v)$) then $(u, v, x) \sim (u, v, u)$ so

$$
\begin{aligned}
f(u, v, x) = u \in u/\phi &= f_\phi(u/\phi, v/\phi, u/\phi) \\
&= f_\phi(u/\phi, v/\phi, x/\phi)
\end{aligned}
$$

by remark 1 and since f_ϕ is a Pixley function. For the second clause, if $u = v$ then $u/\phi = v/\phi$ so

$$f(u, v, x) = x \in x/\phi = f_\phi(u/\phi, v/\phi, x/\phi)$$

since f_ϕ is a Pixley function. For the third clause the inclusion is immediate by the definition of $f(u, v, x)$.

Finally, to verify (2.5) suppose $x \equiv y$ ($\mathrm{Eg}^{\mathbf{L}}(u, v)$). Then we have $(u, v, x) \sim (u, v, y)$ and $f(u, v, x) = f(u, v, y)$ since f is defined to be constant on each $\sim$-block.   ●

In Case 2 of the proof above, the third clause of the definition of f may allow for some choice. The following examples illustrate this.

Example 2.2.7 *An arithmetical equivalence lattice on a four element set with many compatible principal Pixley functions.*

Let $A = \{a, b, c, d\}$, take the partition $\theta = \{\{a, b\}, \{c, d\}\}$ and let $\mathbf{L}$ be the three element chain with elements $\{0_A, \theta, 1_A\}$. The function f_θ is defined to be the discriminator on A/θ. According to Case 2 of the proof $f(u, v, x)$ is uniquely defined on A in the following cases:

1. $\mathrm{Eg}^{\mathbf{L}}(u, v) = 1_A$ yielding $f(u, v, x) = u$ for all x (first clause).

2. $\mathrm{Eg}^{\mathbf{L}}(u, v) = \theta$ with $u \equiv x \ (\theta)$ yielding $f(u, v, x) = u$ for all x (first clause).

3. $\mathrm{Eg}^{\mathbf{L}}(u, v) = 0_A$ yielding $u = v$ so that $f(u, v, x) = x$ for all x (second clause).

On the other hand if $\mathrm{Eg}^{\mathbf{L}}(u, v) = \theta$ with $u \not\equiv x \ (\theta)$ then the third clause applies and this leads to the following four two-element $\sim$-blocks:

$$\{(a, b, c), (a, b, d)\}, \quad \{(b, a, c), (b, a, d)\},$$
$$\{(c, d, a), (c, d, b)\}, \quad \{(d, c, a), (d, c, b)\}.$$

On these blocks

$f(a, b, c) = f(a, b, d)$ can be either c or d,
$f(b, a, c) = f(b, a, d)$ can be either c or d, and likewise
$f(c, d, a) = f(c, d, b)$ can be either a or b, and
$f(d, c, a) = f(d, c, b)$ can be either a or b.

Hence there are 2^4 principal Pixley functions. There are many more nonprincipal ones; for example we could take $f(a, c, b) = b$ instead of a and still have an **L**-compatible Pixley function which is not principal.

Example 2.2.8 *An arithmetical equivalence lattice on a three element set with unique compatible principal Pixley function though the third clause occurs.*

Let $A = \{a, b, c\}$, take the partition $\theta = \{\{a, b\}, \{c\}\}$, and let **L** be the chain with elements $\{0_A, \theta, 1_A\}$. Again the definition requires that f_θ be defined as the discriminator on A/θ. If $\mathrm{Eg}^{\mathbf{L}}(u, v) = 1_A$ or 0_A, or θ with $u \equiv x \ (\theta)$, then $f(u, v, x)$ is again uniquely defined as in Example 2.2.7. On the other hand if $\mathrm{Eg}^{\mathbf{L}}(u, v) = \theta$ and $u \not\equiv x \ (\theta)$ we must have $f(a, b, c) = c$ and $f(b, a, c) = c$, i.e., there is no choice.

Now we examine the question of when the principal Pixley function constructed in the proof of Theorem 2.2.6 is unique. First it is obvious that nonuniqueness occurs only if the third clause of the definition of f in Case 2 occurs. In particular if Case 2 never occurs, so that **L** is Boolean, then we have uniqueness. Also if the third clause never applies even if Case 2 does occur we will still have uniqueness. But even if the third clause does occur, choice will still be possible if and only if there is more than one element of A contained in $f_\phi(u/\phi, v/\phi, x/\phi)$; this means that uniqueness will occur iff in every occurrence of the third clause $f_\phi(u/\phi, v/\phi, x/\phi)$ contains just one element of A, i.e., we must choose f (constant on each $\sim$-block) by taking $f(u, v, x)$ to be the sole element contained in $f_\phi(u/\phi, v/\phi, x/\phi)$. This is exactly what Example 2.2.8 above displays: in that example if $\mathrm{Eg}^{\mathbf{L}}(u, v) = \theta$ and $u \not\equiv x \; (\theta)$ (i.e., the third clause applies), then we must have $\{u, v\} = \{a, b\}$ and $x = c$ and

$$f_\theta(a/\theta, b/\theta, c/\theta) = c/\theta = \{c\}$$

which forces $f(a, b, c) = f(b, a, c) = c$.

Summarizing, we have:

Corollary 2.2.9 *The principal Pixley function constructed by Theorem 2.2.6 is unique if and only if whenever an element $\rho \in L$ has a unique cover ϕ, then for all triples (u, v, x) such that*

$$u/\rho \neq v/\rho \quad and \quad u/\rho \not\equiv x/\rho \, (\mathrm{Eg}(u/\phi, v/\phi)),$$

$f_\phi(u/\phi, v/\phi, x/\phi)$ *equals some ρ-block.*

Exercises

1. Suppose an equivalence lattice **L** on A has the (CFE) property and we are given subsets $X \subseteq Y \subseteq A^m$ with X finite and Y countable. Show that any **L**-compatible function $f : X \to A$ can be extended compatibly to Y. ([Kaarli, 1983])

2. In Exercise 1 suppose that **L** is finite. Show that the result still holds if the subsets $X \subseteq Y \subseteq A^m$ are allowed to have arbitrary cardinality. ([Kaarli, 1983])

2.2.4 Applications of principal Pixley functions

1. Principal congruences In Chapter 3 we shall show (Theorem 3.3.3) that arithmetical varieties V can be characterized by the following interpolation condition:

for each algebra $\mathbf{A}$ of V and finite partial operation f on A, if f is congruence compatible, where defined, then f can be interpolated by a polynomial of $\mathbf{A}$.

We shall use this characterization now to show how Theorem 2.2.6 yields the following description of principal congruence formulas in arithmetical varieties. Recall that principal congruence formulas satisfy (1.3) of Section 1.2.2.

Note how the following two theorems compare with Theorem 1.2.6.

Theorem 2.2.10 *Let V be an arithmetical variety. For each $\mathbf{A} \in V$ and finite subset $F \subseteq A$ there is an integer $m \geq 0$ and an $(m+3)$-ary term $t_F(x, y, z, \mathbf{w})$ such that the following are true:*

(i) *there exists $\mathbf{a} \in A^m$ such that the restriction of the polynomial function $t_F(x, y, z, \mathbf{a})$ to F^3 is a partial Pixley function on A;*

(ii) *the formula*

$$\phi_F(x, y, u, v) = (\exists \mathbf{w})\, (t_F(u, v, x, \mathbf{w}) \approx t_F(u, v, y, \mathbf{w}))$$

is a principal congruence formula for F in the sense that (1.3) holds for ϕ_F and all $a, b, c, d \in F$.

PROOF Let $\mathbf{L}$ be the sublattice of $\mathbf{Con}\,\mathbf{A}$ generated by the set of congruences

$$\{0_{\mathbf{A}}, 1_{\mathbf{A}}\} \cup \{\mathbf{Cg}^{\mathbf{A}}(a, b) : a, b \in F, a \neq b\}.$$

Since $\mathbf{L}$ is a finitely generated distributive lattice it is finite, and as a sublattice of $\mathbf{Con}\,\mathbf{A}$ it is arithmetical. Consequently, by Theorem 2.2.6, $\mathbf{L}$ supports a principal Pixley function. Let g be its restriction to F^3 so that g is a finite partial function on A. The function g is $\mathbf{Con}\,\mathbf{A}$-compatible (where defined) for if (x, y, z) and (x', y', z') are in F^3, then

$$(g(x, y, z), g(x', y', z')) \in \mathbf{Cg}(x, x') \vee \mathbf{Cg}(y, y') \vee \mathbf{Cg}(z, z')$$

since this congruence is in $\mathbf{L}$. Therefore, by Theorem 3.3.3, g has an interpolating polynomial function. This means that there is a term $t_F(x, y, z, \mathbf{w})$ satisfying the requirements (i) and (ii). •

For locally finite arithmetical varieties, for each integer m, we can find a single congruence formula which defines principal congruences for all m generated algebras of the variety:

Theorem 2.2.11 *If V is a locally finite arithmetical variety and m a nonnegative integer then there is an $(m+3)$-ary term $t_m(x, y, x, \mathbf{w})$ of V such that if $\mathbf{A} \in V$ and the components of $\mathbf{a} \in A^m$ generate $\mathbf{A}$, then for all $u, v, x, y \in A$,*

$$x \equiv y \; (\mathrm{Cg}^{\mathbf{A}}(u, v)) \iff t_m(u, v, x, \mathbf{a}) = t_m(u, v, y, \mathbf{a}).$$

Thus the polynomial

$$t_m(u, v, x, \mathbf{a})$$

is a principal Pixley function for **Con A**.

PROOF Let $\mathbf{F}$ be the V-free algebra with m free generators. Then **Con F** is finite so it supports a principal Pixley polynomial f on F (by Theorems 2.2.6 and 3.3.3). This means that there is an $(m+3)$-ary term $t_m(x, y, z, \mathbf{w})$ which is easily seen to satisfy the requirements of the theorem.   •

2. Algebras with prescribed congruence lattices Recall that the theorem of Quackenbush and Wolk (Chapter 1, Section 1.1.3) asserts that each finite distributive 0-1 equivalence lattice is the congruence lattice of some algebra. The following result sharpens this for the case of an arithmetical equivalence lattice.

Theorem 2.2.12 *If $\mathbf{L}$ is a finite arithmetical 0-1 equivalence lattice on the set A and t is a principal Pixley function for $\mathbf{L}$, then for the algebra $\mathbf{A} = \langle A; t \rangle$, **Con A**$=\mathbf{L}$. Also the variety $V(\mathbf{A})$ is a principal arithmetical variety.*

In the language of clones the first statement of the theorem asserts that if A is a finite set and $\mathbf{L}$ is an arithmetical 0-1 equivalence lattice on A with principal Pixley function t, then $inv(clo(L)) \cap \mathrm{Eqv}\, A = L$. Later (Section 3.4.2) we shall also see that if F is the clone generated by t and the elements of A (constant functions), then $clo(L) = F$.

PROOF Just as in the proof of Theorem 1.2.6 we see (using only that t is a Pixley function) that for any $u, v \in A$,

$$R(u, v) = \{(x, y) : t(u, v, x) = t(u, v, y)\}$$

is an equivalence relation which contains (u, v) and

$$R(u, v) \leq \mathrm{Cg}^{\mathbf{A}}(u, v). \tag{2.7}$$

Since t is a principal Pixley function for $\mathbf{L}$, by Theorem 2.2.6,

$$R(u, v) = \mathrm{Eg}^{\mathbf{L}}(u, v),$$

so that from (2.7) we have

$$\mathrm{Eg}^{\mathbf{L}}(u, v) \leq \mathrm{Cg}^{\mathbf{A}}(u, v).$$

But since t is $\mathbf{L}$-compatible, $L \subseteq \mathrm{Con}\,\mathbf{A}$, and from this it follows that

$$\mathrm{Cg}^{\mathbf{A}}(u, v) \leq \mathrm{Eg}^{\mathbf{L}}(u, v).$$

Therefore $\mathrm{Eg}^{\mathbf{L}}(u, v) = \mathrm{Cg}^{\mathbf{A}}(u, v)$. Since $\mathbf{L}$ and $\mathbf{Con}\,\mathbf{A}$ are generated by their principal members, it follows that $\mathbf{L}=\mathbf{Con}\,\mathbf{A}$.

To show that $V(\mathbf{A})$ is a principal arithmetical variety, according to Theorem 1.2.6, we need to show that the following is an identity of $\mathbf{A}$, and hence of $V(\mathbf{A})$:

$$t(u, v, t(x, y, z)) = t(u, v, t(t(u, v, x), t(u, v, y), t(u, v, z))). \tag{2.8}$$

Since $R(u, v) = \mathrm{Cg}^{\mathbf{A}}(u, v)$, this is equivalent to the requirement that

$$t(x, y, z) \equiv t(t(u, v, x), t(u, v, y), t(u, v, z)) \ (\mathrm{Cg}^{\mathbf{A}}(u, v)) \tag{2.9}$$

holds for all $u, v, x, y, z \in A$. But in $\mathbf{A}$, for any u, v, w,

$$w = t(u, u, w) \equiv t(u, v, w) \ (\mathrm{Cg}^{\mathbf{A}}(u, v)).$$

Setting $w = x, y, z$ successively we obtain (2.9) by the substitution property, and hence (2.8). ●

Finally, the preceding theorem provides an interesting extension of Theorem 2.2.6:

Theorem 2.2.13 *Let $\mathbf{L}$ be a finite arithmetical 0-1 equivalence lattice on a set A. A Pixley function f for $\mathbf{L}$ is principal iff it satisfies the identity*

$$f(u, v, f(x, y, z)) = f(u, v, f(f(u, v, x), f(u, v, y), f(u, v, z))) \tag{2.10}$$

(for all $u, v, x, y, z \in A$). In this case f completely determines $\mathbf{L}$ via

$$(x, y) \in \mathrm{Eg}^{\mathbf{L}}(u, v) \quad \Longleftrightarrow \quad f(u, v, x) = f(u, v, y).$$

The lattice $\mathbf{L}$ is Boolean iff f satisfies the additional identity

$$f(x, f(x, y, z), y) = y \tag{2.11}$$

(for all $x, y, z \in A$). In this case the Pixley function f and the Boolean equivalence lattice $\mathbf{L}$ completely determine each other.

The theorem is another remarkable example of the strength of the strong Chinese remainder condition: it asserts that the same identities which characterize principal arithmetical varieties and discriminator varieties (for a language with a single ternary operation) as given by Theorems 1.2.6 and 1.2.10 respectively, also characterize principal Pixley functions and Boolean arithmetical equivalence lattices.

PROOF We consider the algebra $\mathbf{A} = \langle A; f \rangle$. First suppose f is a principal Pixley function for $\mathbf{L}$. Then Theorem 2.2.12 asserts that $\mathbf{L} = \mathrm{Con}\,\mathbf{A}$ and $V(\mathbf{A})$ is a principal arithmetical variety relative to f. Therefore (2.10) is an identity by Theorem 1.2.6.

Now suppose f satisfies (2.10). Then $\mathbf{A}$ is such that $V(\mathbf{A})$ is a principal arithmetical variety by Theorem 1.2.6; so principal congruences are defined on $\mathbf{A}$ by

$$(x, y) \in \mathrm{Cg}^{\mathbf{A}}(u, v) \quad \Longleftrightarrow \quad f(u, v, x) = f(u, v, y).$$

But $\mathrm{Con}\,\mathbf{A} = \mathbf{L}$ so $\mathrm{Cg}^{\mathbf{A}}(u, v) = \mathrm{Eg}^{\mathbf{L}}(u, v)$; therefore

$$(x, y) \in \mathrm{Eg}^{\mathbf{L}}(u, v) \quad \Longleftrightarrow \quad f(u, v, x) = f(u, v, y).$$

Therefore f is a principal Pixley function for $\mathbf{L}$, and f completely determines $\mathbf{L}$.

If $\mathbf{L}$ is Boolean then $\mathbf{A}$ is isomorphic to $\mathbf{A}/\theta_1 \times \cdots \times \mathbf{A}/\theta_n$ where the θ_i are the maximal elements of $\mathbf{L}$. Then the construction in the proof of Theorem 2.2.6 shows that f is unique (and is determined by inducing the discriminator on each $\mathbf{A}/\theta_i$). Theorem 1.2.22 shows that the $\mathbf{A}/\theta_i$ are the only SIs of $V(\mathbf{A})$ and hence this is a discriminator variety so that (by Theorem 1.2.10) (2.11) is an identity.

Conversely, if (2.11) is satisfied by f, then it is an identity of $V(\mathbf{A})$; so $V(\mathbf{A})$ is a discriminator variety. Hence all SIs of $V(\mathbf{A})$ are simple and hence all meet irreducible elements of $\mathbf{L}$ are maximal. But this implies that $\mathbf{L}$ is Boolean. $\qquad\bullet$

3. **Prescribed automorphisms** Theorem 2.2.12 gives a prescription for constructing an algebra with a given finite arithmetical congruence lattice and having a principal Pixley function as the only basic operation. Now we briefly discuss the problem of prescribing certain permutations as automorphisms as well. It is natural to attempt this by extending the construction of Theorem 2.2.6.

Let $\mathbf{L}$ be a 0-1 equivalence lattice on A and $x \mapsto x^\sigma$ be a permutation of A. Then for each $\theta \in L$,

$$\theta^\sigma = \{(a^\sigma, b^\sigma) : (a,b) \in \theta\}$$

is again an equivalence relation on A and the set of all such θ^σ yields an isomorphic equivalence lattice $\mathbf{L}^\sigma$ under the mapping $\theta \mapsto \theta^\sigma$. This isomorphism entails

$$\mathrm{Eg}^{\mathbf{L}}(u,v)^\sigma = \mathrm{Eg}^{\mathbf{L}^\sigma}(u^\sigma, v^\sigma).$$

The restriction of σ to the θ-block x/θ is a bijection onto the θ^σ-block x^σ/θ^σ. We denote this θ^σ-block by $(x/\theta)^\sigma$. Thus

$$(x/\theta)^\sigma = \{y^\sigma : y \in x/\theta\} = x^\sigma/\theta^\sigma.$$

It is also clear that for $\theta, \phi \in L$,

$$(x/\theta \cap y/\phi)^\sigma = (x/\theta)^\sigma \cap (y/\phi)^\sigma. \tag{2.12}$$

Now if $\mathbf{A}$ is an algebra with universe A and congruence lattice $\mathbf{L}$, and σ is an automorphism of $\mathbf{A}$, then $\mathbf{L}^\sigma = \mathbf{L}$, $\theta \mapsto \theta^\sigma$ is an automorphism of $\mathbf{L}$ and $x/\theta \mapsto (x/\theta)^\sigma$ is an isomorphism between the quotient algebras $\mathbf{A}/\theta$ and $\mathbf{A}/\theta^\sigma$. (The graph of this isomorphism is a "square" graph subuniverse in $A \times A$ which contains the graph $\{(x, x^\sigma) : x \in A\}$ of the automorphism.)

For this reason we shall say that a permutation $x \mapsto x^\sigma$ of A is **L-admissible** if $\mathbf{L} = \mathbf{L}^\sigma$. Clearly L-admissibility is a necessary but not usually sufficient condition for a permutation to be an automorphism of an algebra with $\mathbf{Con\,A} = \mathbf{L}$.

Let us go through the proof of Theorem 2.2.6 to see if we can modify the construction so that a given **L**-admissible permutation σ will be an automorphism of the algebra $\langle A; f \rangle$; to do this we add to the statement of the theorem that for every $\rho \in L$

$$f((u/\rho)^\sigma, (v/\rho)^\sigma, (x/\rho)^\sigma) = f_\rho(u/\rho, v/\rho, x/\rho)^\sigma, \qquad (2.13)$$

for then if $\rho^\sigma = \rho$ and in particular at the bottom of the lattice where $0_A = \rho$, this statement asserts that $x \mapsto x^\sigma$ is an automorphism with respect to f_ρ, which is what we wish. For the maximal ϕ the requirement (2.13) is met trivially since the discriminator is a pattern function as defined in Section 1.2. At the induction step, if Case 1 $(k > 1)$ applies we proceed just as before and observe that since

$$f_\rho(u/\rho, v/\rho, x/\rho) =$$
$$f_{\phi(i)}(u/\phi(i), v/\phi(i), x/\phi(i)) \cap f_{\phi(j)}(u/\phi(j), v/\phi(j), x/\phi(j))$$

for any $i \neq j$, (2.13) follows directly from (2.12).

A problem then arises in the induction step if Case 2 $(k = 1)$ occurs and in particular only when we encounter the third clause of the definition of f_ρ where choice may be allowed. The equivalence relation $\sim$ on A/ρ^σ behaves the same as on A/ρ so ϕ^σ is the unique cover of ρ^σ and we want to choose $f_\rho(u/\rho, v/\rho, x/\rho)$ (constant over a given $\sim$-block) to be a ρ block in $f_\phi(u/\phi, v/\phi, x/\phi)$ and $f_{\rho^\sigma}((u/\rho)^\sigma, (v/\rho)^\sigma, (x/\rho)^\sigma)$ to be the ρ^σ-block in

$$f_{\phi^\sigma}((u/\phi)^\sigma, (v/\phi)^\sigma, (x/\phi)^\sigma)$$

equal to $f_\rho(u/\rho, v/\rho, x/\rho)^\sigma$. In general this may not be possible, for the bijection $x/\rho \mapsto x^\sigma/\rho^\sigma$ is already uniquely determined by the given σ. We return to Example 2.2.7 to illustrate.

In this example the nontrivial **L**-admissible permutations of A are the following (written as the products of disjoint cycles):

$$(ab), \quad (cd), \quad (ab)(cd), \quad (ac)(bd), \quad (ad)(bc), \quad (acbd), \quad (adbc).$$

(This is just the eight element dihedral group.) For $\sigma = (ab)$ (and $\rho = 0_A, \phi = \theta$), we have $f(c^\sigma, d^\sigma, a^\sigma) = f(c, d, b)$ which must equal $f(c, d, a)$, since they are in the same $\sim$-block, and must be either a or $b = a^\sigma$. Therefore we cannot have $f(c, d, a)^\sigma = f(c^\sigma, d^\sigma, a^\sigma)$. Likewise $\sigma = (cd)$ is not an automorphism for any choice of f. For the remaining permutations several successful choices are possible. For example we see that for the choices

$$
\begin{aligned}
f(a,b,c) = f(a,b,d) &= c, \\
f(b,a,c) = f(b,a,d) &= d, \\
f(c,d,a) = f(c,d,b) &= a, \\
f(d,c,a) = f(d,c,b) &= b,
\end{aligned}
$$

the permutations $(ab)(cd)$, $(ac)(bd)$, $(ad)(bc)$ together with the identity function are the only automorphisms. Hence these constitute the automorphism group of the algebra $\langle A; f \rangle$ for this choice of f.

In Example 2.2.8, the permutation (ab) is the only L-admissible permutation and it evidently is the unique nontrivial automorphism of $\langle A; f \rangle$. In fact this illustrates the following theorem which indicates the significance of the existence of a unique principal Pixley function.

Theorem 2.2.14 *If a finite equivalence lattice* **L** *on A has a unique principal Pixley function f then every* **L**-*admissible permutation is an automorphism of* $\langle A; f \rangle$.

PROOF We apply Corollary 2.2.9. Suppose the function f is unique and σ is an arbitrary permutation of A. We give the proof by induction as indicated above (including the requirement (2.13)). As noted earlier the construction proceeds as before unless we encounter a situation where ρ has the unique cover ϕ. But by Corollary 2.2.9, we have for all (u,v,x), both

$$
f_\rho(u/\rho, v/\rho, x/\rho) = f_\phi(u/\phi, v/\phi, x/\phi)
$$

and

$$
\begin{aligned}
f_{\rho\sigma}((u/\rho)^\sigma, (v/\rho)^\sigma, (x/\rho)^\sigma) = \\
f_{\phi\sigma}((u/\phi)^\sigma, (v/\phi)^\sigma, (x/\phi)^\sigma).
\end{aligned}
$$

Then since, by the induction hypothesis, we have

$$
\begin{aligned}
f_\phi(u/\phi, v/\phi, x/\phi)^\sigma = \\
f_{\phi\sigma}((u/\phi)^\sigma, (v/\phi)^\sigma, (x/\phi)^\sigma),
\end{aligned}
$$

so that $f_\rho(u/\rho, v/\rho, x/\rho)^\sigma = f_{\rho^\sigma}((u/\rho)^\sigma, (v/\rho)^\sigma, (x/\rho)^\sigma)$.   •

The approach illustrated above can be applied in other simple cases to prescribe certain automorphisms for principal Pixley functions even when Theorem 2.2.14 does not apply.

2.2.5   Compatible choice functions

We conclude this section with an interesting result like Theorem 2.2.6 and which provides yet another characterization of finite arithmetical equivalence lattices. Given an equivalence relation θ on A, a function $f : A \to A$ is called a **choice function modulo** θ if for each θ-block B, there is a $b \in B$ such that $f(x) = b$ for all $x \in B$, i.e., if f chooses a fixed representative in every θ-block of A. Thus for every $a, b \in A$, we have $a \equiv_\theta f(a)$ and $f(a) = f(b)$ iff $a \equiv_\theta b$. Clearly all choice functions are idempotent.

If **L** is an equivalence lattice supporting a principal Pixley function $f(x, y, z)$ then obviously the function $g(x) = f(a, b, x)$ is a compatible choice function modulo $Eg(a, b)$ and $g(a) = a$; that is, a is a fixed point of g. We shall show that actually for finite arithmetical equivalence lattices such functions exist modulo every member of L, and with arbitrarily prescribed fixed points. Moreover, this property characterizes arithmeticity.

Definition An equivalence lattice **L** on A satisfies the **compatible choice function condition** if for each $a \in A$ and $\theta \in L$, there exists a compatible choice function modulo θ having a as a fixed point.

Theorem 2.2.15 *A finite equivalence lattice* **L** *on a set* A *is arithmetical iff it satisfies the compatible choice function condition.*

PROOF We first prove that the (compatible choice function) condition implies arithmeticity. Suppose $\theta, \phi \in L$ and $(a, c) \in \theta \circ \phi$. Then there is a $b \in A$ such that $a \equiv_\theta b$ and $b \equiv_\phi c$. Let f be a compatible choice function modulo θ such that $f(a) = a$. Since $a \equiv_\theta b$, we have $f(b) = a$. Hence $b \equiv_\phi c$ implies $(a, f(c)) = (f(b), f(c)) \in \phi$. On the other hand $f(c) \equiv_\theta c$. This proves $(a, c) \in \phi \circ \theta$; that is, **L** is permutable.

Let now $(a, c) \in \theta \wedge (\phi \circ \psi)$ and let $b \in A$ be such that $a \equiv_\phi b$ and $b \equiv_\psi c$. We have to prove that $(a, c) \in (\theta \wedge \phi) \circ (\theta \wedge \psi)$. Let g be a compatible choice function modulo θ with fixed point b. Then $b \equiv_\phi a$ implies $(b, g(a)) = (g(b), g(a)) \in \phi$. Hence by transitivity, $a \equiv_\phi g(a)$. Since g is a choice function modulo θ, we also have $a \equiv_\theta g(a)$, thus $(a, g(a)) \in \theta \wedge \phi$. Similar arguments show that $(g(a), c) \in \theta \wedge \psi$. Hence, $(a, c) \in (\theta \wedge \phi) \circ (\theta \wedge \psi)$. This proves that **L** is distributive.

Now assume that **L** is a finite arithmetical equivalence lattice on A. Then clearly $L' = L \cup \{0_A\}$ is a subuniverse of **Eqv** A and the equivalence lattice **L'** is finite and arithmetical also. Obviously, if we succeed in proving that **L'** satisfies the condition then so does **L**. Thus, we may assume, without loss of generality, that $0_A \in L$.

We prove by induction on the height of **L** that **L** satisfies the condition. Obviously the statement holds if **L** is of height 0, that is, $L = \{0_A\}$. Suppose it holds in case of all equivalence lattices on A which contain 0_A and whose height is less than the height of **L**. Let $a \in A$ and $0 \ne \theta \in L$. Then choose an atom σ of **L** such that $\sigma \le \theta$. Clearly the lattice $\mathbf{L}_\sigma$ is finite and arithmetical, it contains $0_{A/\sigma}$, and its height is less than the height of **L**. Thus, by the induction hypothesis there exists an $\mathbf{L}_\sigma$-compatible choice function g modulo θ/σ on A/σ such that $g(a/\sigma) = a/\sigma$.

Since **L** is finite and distributive, the atom σ has a unique pseudocomplement τ in **L**. Let $B = \{b_i : i \in I\}$ be a complete set of representatives of $(\sigma \vee \tau)$-blocks such that $a \in B$ and, for every $i \in I$, put $T_i = b_i/\tau$. Then, by permutability of **L**, every σ-block of A intersects exactly one T_i and, because of $\sigma \wedge \tau = 0$, each of these intersections contains just one element.

Now define $f : A \to A$ by the rule: $f(x)$ is the unique common element of $g(x/\sigma)$ and a suitable T_i. Obviously f is a choice function modulo θ and it preserves a. It remains to prove that f is **L**-compatible. Assume that $x, y \in A$ and $(x, y) \in \rho \in L$. Then by the $\mathbf{L}_\sigma$-compatibility of g it follows that $(f(x), f(y)) \in \rho \vee \sigma$. Hence $f(x) \equiv_\rho f(y)$ if $\sigma \le \rho$. Otherwise we have $\rho \le \tau$ and therefore $(f(x), f(y)) \in \tau \vee \sigma$. Hence $f(x)$ and $f(y)$ must be the members of the same set T_i implying $f(x) \equiv_\tau f(y)$. Since

$$(\rho \vee \sigma) \wedge \tau = (\rho \wedge \tau) \vee (\sigma \wedge \tau) = \rho \vee 0 = \rho,$$

we have $f(x) \equiv_\rho f(y)$. $\bullet$

Remark An equivalence lattice **L** need not be arithmetical if there only exist compatible choice functions modulo every member of L, but which fail to have prescribed fixed points. The simplest counter-example is $A = \{a, b, c\}$, $L = \{0_A, \rho, \sigma, 1_A\}$ where ρ and σ are given by the partitions of A: $\{\{a\}, \{b, c\}\}$ and $\{\{a, b\}, \{c\}\}$, respectively. It is easy to check that the functions $f(a) = a$, $f(b) = f(c) = b$ and $g(a) = g(b) = b$, $g(c) = c$ are **L**-compatible choice functions modulo ρ and σ, respectively. Obviously the identity function is a choice function modulo 0_A and any constant function is a choice function modulo 1_A. Also, the identity function and all constant functions preserve all equivalence relations. However, there is no **L**-compatible choice function modulo ρ having a fixed point c; in fact, if f is any such function then $f(c) = b$. Indeed, $f(c) = c$ would imply $f(b) = c$ contradicting $f(a) \equiv_\sigma f(b)$.

2.3   Compatible function lifting

In this section we shall introduce important concepts leading us, most importantly, to Theorems 2.3.9 and 2.3.11, which will play a critical role in the theory of affine complete varieties developed in Chapter 4. Most of the results of this section are from [Kaarli, 1997] and [Kaarli, McKenzie, 1997].

Let **L** be an equivalence lattice on A and $\rho \in L$. As we mentioned in the introduction to the present chapter, every member f of the clone $clo(L)$ determines a member f_ρ of the clone $F_\rho = clo(L_\rho)$. Hence, there is a canonical mapping $F \to F_\rho$ that takes the function f to the function it induces modulo ρ. Throughout this section F and F_ρ have the meaning introduced here. For future use the following is the most important new concept.

Definition The equivalence lattice **L** on the set A is said to have the **compatible function lifting property** (briefly, the (CFL) property) if all of the canonical mappings $F \to F_\rho$, $\rho \in L$, are surjective.

From results to be established in Chapters 3 and 4 it will become apparent that every arithmetical equivalence lattice on a finite set A has the (CFL) property. Since we cannot verify this as a general proposition at this early point in our development we shall instead give an example now (from [Kaarli, Pixley, 1987] and [Kaarli, 1990]) for which we can immediately establish the (CFL) property, and which will also be useful later as a source of further important examples.

Example 2.3.1 *A class of Boolean equivalence lattices which have the compatible function lifting property and which need not be arithmetical.*

Let $A_1, \ldots, A_n$ be finite sets, each having at least two elements. Let $\mathbf{c}$ be a fixed element of $A_1 \times \cdots \times A_n$ and let $A \subseteq A_1 \times \cdots \times A_n$ be a subset which is arbitrary except for satisfying the following two properties:

(i) A is subdirect in the product, i.e., it projects onto each of the factors A_i.

(ii) For all $\mathbf{a} = (a_1, \ldots, a_n) \in A$, A also contains all elements $(x_1, \ldots, x_n)$ where $x_i \in \{a_i, c_i\}$.

Finally, let $\mathbf{L}$ be the lattice whose universe $L \subseteq \mathrm{Eqv}\, A$ consists of the kernels of all of the projections of A into $A_{i_1} \times \cdots \times A_{i_k}$ for some $1 \leq i_1 < \cdots < i_k \leq n$. This means that if the kernel of the projection of A onto A_i is θ_i, then L consists of all intersections $\theta_{i_1} \cap \cdots \cap \theta_{i_k}$ and $(\mathbf{a}, \mathbf{b}) \in \theta_{i_1} \cap \cdots \cap \theta_{i_k}$ simply means that $\mathbf{a}$ and $\mathbf{b}$ agree on components $i_1, \ldots, i_k$. Note that the condition (ii) implies that L is closed under joins.

Hence $\mathbf{L}$ is Boolean with coatoms θ_i. It is important, for future use, to see that many choices of A are possible and that in general the elements of the L need not permute under relation product (and hence $\mathbf{L}$ need not be arithmetical). As an extreme example, the choice

$$A = \bigcup_{i=1}^{n} \{c_1\} \times \cdots \times \{c_{i-1}\} \times A_i \times \{c_{i+1}\} \times \cdots \times \{c_n\}$$

certainly yields that A satisfies the requirements and has nonpermutable θ_i.

It is easy to see that $\mathbf{L}$ has the (CFL) property: let $\phi \in L$ and let $g_\phi \in F_\phi = clo(L_\phi)$. For simplicity let us suppose that g_ϕ is unary and that $\phi = \theta_1 \cap \cdots \cap \theta_k$ for some $k \leq n$. Then for some functions $g_i : A_i \to A_i$, $i = 1, \ldots, k$, $g_\phi = (g_1, \ldots, g_k)$. Then we define $g : A \to A$ by

$$
\begin{aligned}
g(x_1, \ldots, x_n) &= (g_\phi(x_1, \ldots, x_k), c_{k+1}, \ldots, c_n) \\
&= (g_1(x_1), \ldots, g_k(x_k), c_{k+1}, \ldots, c_n).
\end{aligned}
$$

From the definition it is clear that g takes values in A and is **L**-compatible, for let $x \equiv_\psi y$ where $\psi = \theta_{i_1} \cap \cdots \cap \theta_{i_m}$; so we have $x_{i_1} = y_{i_1}, \ldots, x_{i_m} = y_{i_m}$ and obviously $g(x)$ and $g(y)$ also agree on these same components, i.e., $g(x) \equiv_\psi g(y)$. Hence g is **L**-compatible and induces g_ϕ on A_ϕ. Therefore **L** has the (CFL) property.

Shortly we are going to define a technical auxiliary property which formally is very much weaker than the (CFL) property, but which, for finite sets, will be proven equivalent to it. To do this we first need the technical notion of a compatible function system.

Definition Let **L** be an equivalence lattice on A and for any $\rho \in L$ let f_ρ be an m-ary partial function on A/ρ. Then $f = (f_\rho)_{\rho \in L}$ is said to be a **compatible function system** (briefly, a CFS) if it satisfies the following two conditions.

$(\mathcal{A})$ For every $\rho_1, \rho_2 \in L$ and $\mathbf{a}^i/\rho_i \in dom\ f_{\rho_i}$,

$$f_{\rho_i}(\mathbf{a}^i/\rho_i) = b^i/\rho_i\ (i = 1, 2) \implies (b^1, b^2) \in \rho_1 \vee \rho_2 \vee Eg(\mathbf{a}^1, \mathbf{a}^2).$$

$(\mathcal{B})$ If $f_{\rho_i}(\mathbf{a}^i/\rho_i) = b^i/\rho_i$, $i = 1, \ldots, n$, and there exists $\mathbf{a} \in A^m$ and $\sigma_i \in L$, such that $\rho_i \leq \sigma_i$ and $\mathbf{a} \equiv \mathbf{a}^i\ (\sigma_i)$, $i = 1, \ldots, n$, then $(b^1/\sigma_1, \ldots, b^n/\sigma_n) \in A/\sigma_1 \wedge \cdots \wedge \sigma_n$.

Remarks

1. We shall identify $a/\sigma_1 \wedge \cdots \wedge \sigma_n$ and $(a/\sigma_1, \ldots, a/\sigma_n)$ via the canonical embedding. Hence, the last inclusion in condition $(\mathcal{B})$ means that there exists $b \in A$ such that $b \equiv b^i\ (\sigma_i)$, $i = 1, \ldots, n$.

2. We do not require in the condition $(\mathcal{B})$ that $\sigma_i \neq \sigma_j$ if $i \neq j$. However, it is easy to see that if $(\mathcal{B})$ is satisfied in this special case then it is also satisfied in general. Indeed, if for example $\sigma_1 = \sigma = \sigma_2$ then $\mathbf{a}^1 \equiv_\sigma \mathbf{a}^2$ and by the compatibility of f also $b^1 \equiv_\sigma b^2$. Obviously then

$$(b^2/\sigma_2, \ldots, b^n/\sigma_n) \in A/\sigma_2 \wedge \cdots \wedge \sigma_n$$

which implies

$$(b^1/\sigma_1, \ldots, b^n/\sigma_n) \in A/\sigma_1 \wedge \cdots \wedge \sigma_n.$$

3. A natural way to obtain a CFS is the following: take any **L**-compatible function g on A and select in every quotient set A/ρ,

$\rho \in L$, a subset X_ρ. Then define f_ρ to be the restriction of g_ρ to X_ρ. Recall that g_ρ is the function induced by g on A/ρ. Then the system $f = (f_\rho)_{\rho \in L}$ is a CFS. We say that this CFS is induced by g.

Now we can define the technical auxiliary property mentioned above.

Definition An equivalence lattice **L** on the set A has the **compatible function system extension property** (briefly, the (CFSE) property) if every compatible function system on A is induced by some (total) **L**-compatible function.

If f_ρ is any $\mathbf{L}_\rho$-compatible function on A/ρ then obviously we obtain a compatible function system if for each $\sigma \neq \rho$ we define f_σ to be the empty function. Hence the (CFSE) property implies the (CFL) property. We shall now set about proving that under an appropriate finiteness condition these properties are equivalent and they imply the distributivity of **L**. Moreover, if the set A is finite then we shall show that the (CFL) property implies the existence of an **L**-compatible near unanimity function.

2.3.1 Compatible function lifting implies distributivity (finite case)

In this subsection we are going to show that if an equivalence lattice **L** on a finite set A has the (CFL) property, then **L** is distributive. If **L** has the (CFL) property, then so does **L**$'$ where $L' = \{0,1\} \cup L$ and **L** is distributive iff so is **L**$'$. Thus we may assume that $0,1 \in L$. The proof uses an induction argument based on the next lemma. Note that the first claim in this lemma is a variation of the well-known lemma of P. P. Pálfy and P. Pudlák (see [Hobby, McKenzie, 1988]).

Lemma 2.3.2 *Let **L** be an equivalence lattice on the set A, e be a unary idempotent **L**-compatible function on A and $U = e(A)$. Then*

(i) *the restriction map is a lattice homomorphism from **L** into* **Eqv** U *and the range* **L**$|_U$ *of this map is an equivalence lattice of the set U;*

(ii) *if **L** has the compatible function lifting property, then so does* **L**$|_U$.

PROOF Obviously the restriction map preserves meets so we have to show that $(\phi \vee \theta)|_U = \phi|_U \vee \theta|_U$ for every $\phi, \theta \in L$. Obviously we need only prove the inequality $(\phi \vee \theta)|_U \leq \phi|_U \vee \theta|_U$. Let $u, v \in U$ and $u \equiv v\,(\phi \vee \theta)$. Then there exist $a_0, a_1, \ldots, a_n \in A$ such that $a_0 = u$, $a_n = v$ and $(a_{i-1}, a_i) \in \phi \cup \theta$ for every i. Now consider the sequence

$$u = e(a_0), e(a_1), \ldots, e(a_n) = v.$$

Obviously all members of this sequence are in U and because of the L-compatibility of e, its adjacent members are in $\phi \cup \theta$. Hence we have $u \equiv v\,(\phi|_U \vee \theta|_U)$. Since the restrictions of 0_A and 1_A are obviously 0_U and 1_U, respectively, we have proven (i).

Now suppose that **L** has the (CFL) property. Let $\rho \in L$ and let g be an m-ary $(\mathbf{L}|_U)_{\rho|_U}$-compatible function on $U/\rho|_U$. Consider the function $f : (A/\rho)^m \rightarrow A/\rho$ defined via

$$f(a_1/\rho, \ldots, a_m/\rho) = g(e(a_1)/\rho, \ldots, e(a_m)/\rho).$$

Clearly $f \in L_\rho$.

Hence, by the (CFL) property, there exists a function $h \in F$ such that

$$h(a_1, \ldots, a_m)/\rho = f(a_1/\rho, \ldots, a_m/\rho)$$

for all $a_1, \ldots, a_m \in A$. In particular, for $u_1, \ldots, u_m \in U$ we have

$$h(u_1, \ldots, u_m)/\rho = g(u_1/\rho, \ldots, u_m/\rho) = u/\rho$$

for some $u \in U$. Thus,

$$eh(u_1, \ldots, u_m)/\rho = e(u)/\rho = u/\rho$$

and the function $eh : U^m \rightarrow U$ induces the function g on $U/\rho|_U$, as required. ●

Theorem 2.3.3 *Let A be a finite set and* **L** *be an equivalence lattice on A. If* **L** *has the compatible function lifting property, then it is distributive.*

PROOF Our proof is by induction on the size of A. The claim is obvious if $|A| = 1$. Assume that our claim is valid for all sets of size less than $|A|$. Denote by Φ the smallest congruence of **L** with $\mathbf{L}/\Phi$ distributive. This exists since if ϕ and ψ are congruences such that each of the quotients $\mathbf{L}/\phi$, $\mathbf{L}/\psi$ is distributive, then $\mathbf{L}/\phi \wedge \psi$ is isomorphic to a subdirect product of the quotients and hence is distributive. We prove that $\rho \not\equiv \mu$ (Φ) for all ρ and μ in L such that ρ is meet irreducible and covered by μ. Then, since there is a homomorphism of $\mathbf{L}/\Phi$ onto $\mathbf{D}_2$ separating ρ/Φ and μ/Φ, Lemma 1.2.17 applies to show that $\mathbf{L}$ is distributive.

First consider the case with $|A/\rho| \geq 3$. Let $a, b \in A$ be such that $(a, b) \in \mu \backslash \rho$. Now choose an arbitrary idempotent surjective function $g : A/\rho \rightarrow \{a/\rho, b/\rho\}$. Obviously $g \in F_\rho$; hence by the (CFL) property it is induced by some function $f \in F$. Since A is finite, an appropriate power of f is idempotent. Denote this idempotent function by e and let $U = e(A)$. Obviously U is contained in the union of just the two ρ-blocks: a/ρ and b/ρ, so $|A/\rho| \geq 3$ implies $U \neq A$. Denote the restriction mapping $L \rightarrow L|_U$ by π. By statement (i) of Lemma 2.3.2, π is a lattice homomorphism. Obviously π maps L onto $L|_U$ and separates ρ and μ. Applying statement (ii) of Lemma 2.3.2 and the induction hypothesis (which applies since $U \neq A$), we have that the lattice $\mathbf{L}|_U$ is distributive. Since Φ is the least congruence for which the quotient is distributive, $\Phi \leq \ker \pi$, and we therefore have $\rho \not\equiv \mu$ (Φ).

The assumption $|A/\rho| \geq 3$ was needed in order to construct a unary idempotent nonidentical function $f \in F$ separating ρ and μ. Hence we must handle separately the case with $|A/\rho| = 2$, $\mu = 1$ and where the only unary idempotent function $f \in F$ separating ρ and μ is the identity function. We shall show that this is possible only if A is a two element set, which obviously yields that $\mathbf{L}$ is distributive. The argument to show this comes from tame congruence theory ([Hobby, McKenzie, 1988]).

Let $f(x, y) \in F$ be a binary function on A which induces a semilattice operation on A/ρ. Any function on a two element set is compatible; so by the (CFL) property there exists such a function $f(x, y)$.

Let X and Y be the ρ-blocks and assume that $f(X, Y) \subseteq Y$. Putting $g(x) = f(x, x)$, we have that $g \in F$ and an appropriate power g^n is idempotent. Note that $g(X) \subseteq X$ and $g(Y) \subseteq Y$, hence g^n separates ρ from μ and by our assumption is the identity function. Put $h(x, y) = g^{n-1}(f(x, y))$. Now h also induces the semilattice operation on A/ρ and $h(x, x) = x$ for every $x \in A$.

Now put $k(x, y) = h(x, h(x, \ldots, h(x, y)) \ldots)$ where h occurs $m!$ times with $m = |A|$. Then we have that $k(x, k(x, y)) = k(x, y)$ for all $x, y \in A$. Also, k induces the semilattice operation on A/ρ and $k(x, x) = x$ for every $x \in A$. Moreover, for all b in X, z in A, we have $k(b, z) = z$; this is because the map $z \mapsto k(b, z)$ is idempotent and preserves the subsets X and Y.

Now put

$$p(x, y) = k(k(\ldots k((x, y), y)\ldots), y)$$

so that $p(p(x, y), y) = p(x, y)$. Then p induces the semilattice operation on A/ρ and $p(b, x) = x = p(x, b)$ for all $b \in X$ and $x \in A$. From these equations it follows that $b_1 = p(b_1, b_2) = b_2$ for all $b_1, b_2 \in X$, hence $|X| = 1$. The equality $|Y| = 1$ can be proved by similar arguments. Hence A is a two element set.   •

2.3.2 Compatible function lifting implies compatible function system extension

Our aim in this section is to show that if an equivalence lattice **L** on a finite set A has the (CFL) property then it has the (CFSE) property as well. By Theorem 2.3.3 we know that such a lattice is distributive. In fact the first technical results depend only on the distributivity of **L**. The (CFL) property, in its full strength, will be needed only at the final step of the proof.

Let each of f and g be a CFS on A. We say that g is an **extension** of f if every g_ρ extends $f_\rho, \rho \in L$. In the sequel we usually omit the subscript ρ if this does not cause ambiguities. Thus, we write $f(\mathbf{a}/\rho)$ instead of $f_\rho(\mathbf{a}/\rho)$ and $\mathbf{a}/\rho \in dom\ f$ instead of $\mathbf{a}/\rho \in dom\ f_\rho$. Also, whenever we write $f(\mathbf{a}/\rho) = b/\rho$ then it will always be assumed that $\mathbf{a}/\rho \in dom\ f$. The next lemma lists the basic properties of compatible function systems which we shall require.

Lemma 2.3.4 *Let f be a CFS on A. Then*

(i) *every f_ρ, $\rho \in L$, is $\mathbf{L}_\rho$-compatible;*

(ii) *if $\rho \le \sigma$, $\mathbf{a}/\rho$ and $\mathbf{a}/\sigma$ are in dom f, and $f(\mathbf{a}/\rho) = b/\rho$ then $f(\mathbf{a}/\sigma) = b/\sigma$;*

(iii) *if $\mathbf{a}/\rho_1, \ldots, \mathbf{a}/\rho_n, \mathbf{a}/\rho_1 \wedge \cdots \wedge \rho_n \in dom\ f$ then*

$$(f(\mathbf{a}/\rho_1), \ldots, f(\mathbf{a}/\rho_n)) = f(\mathbf{a}/\rho_1 \wedge \cdots \wedge \rho_n).$$

PROOF The first two properties are direct consequences from $(\mathcal{A})$. The third property follows from the second one:

if $f(\mathbf{a}/\rho_1 \wedge \cdots \wedge \rho_n) = b/\rho_1 \wedge \cdots \wedge \rho_n$ then by (ii),

$$(f(\mathbf{a}/\rho_1), \ldots, f(\mathbf{a}/\rho_n)) = (b/\rho_1, \ldots, b/\rho_n) = b/\rho_1 \wedge \cdots \wedge \rho_n$$

which is what was needed. ●

A CFS f is said to be **total** if all f_ρ are total functions, i.e., $dom\ f_\rho = (A/\rho)^m$ for every $\rho \in L$. It follows from Lemma 2.3.4 that a total CFS is completely determined by f_0. The statement (ii) of Lemma 2.3.4 asserts that if $\mathbf{a}/\rho \in dom\ f$ and we wish to extend f so that $\mathbf{a}/\sigma$ is in domain of the extension, then there is at most one way to do so. Similarly, (iii) asserts that if $\mathbf{a}/\rho_1, \ldots, \mathbf{a}/\rho_n \in dom\ f$ and we wish to extend f so that $\mathbf{a}/\rho_1 \wedge \ldots \wedge \rho_n$ is in domain of the extension, then again there is at most one way to do this. Hence, when trying to prove that a given CFS has a global extension, it is essential first to prove that it has an extension which is closed in the sense of the next definition.

Definition A compatible function system f is said to be **closed** if the following two conditions are satisfied:

$(\mathcal{C})$ $\mathbf{a}/\rho \in dom\ f$, $\rho \leq \sigma$ $\implies$ $\mathbf{a}/\sigma \in dom\ f$;

$(\mathcal{D})$ $\mathbf{a}/\rho, \mathbf{a}/\sigma \in dom\ f$ $\implies$ $\mathbf{a}/\rho \wedge \sigma \in dom\ f$.

It is useful to notice that actually the conditions $(\mathcal{A})$, $(\mathcal{C})$ and $(\mathcal{D})$ imply $(\mathcal{B})$. Indeed, let f be a function system on A which satisfies the aforementioned three conditions and let $\mathbf{a}, \mathbf{a}^i, b^i, \rho_i$ and σ_i be as in condition $(\mathcal{B})$. Then, because of $(\mathcal{C})$ and $(\mathcal{D})$, $\mathbf{a}^1/\sigma_1, \ldots, \mathbf{a}^n/\sigma_n$ and $\mathbf{a}/\sigma_1 \wedge \cdots \wedge \sigma_n$ are in $dom\ f$. Hence, by Lemma 2.3.4,

$$(b^1/\sigma_1, \ldots, b^n/\sigma_n) = (f(\mathbf{a}^1/\sigma_1), \ldots, f(\mathbf{a}^n/\sigma_n)) =$$
$$(f(\mathbf{a}/\sigma_1), \ldots, f(\mathbf{a}/\sigma_n)) =$$
$$f(\mathbf{a}/\sigma_1 \wedge \cdots \wedge \sigma_n) \in A/\sigma_1 \wedge \cdots \wedge \sigma_n.$$

Now we prove (Lemma 2.3.7) that a CFS has a closed extension provided the lattice **L** is distributive. This will require the next two lemmas.

Lemma 2.3.5 *For every* CFS f *there exists a* CFS *which extends f and satisfies* (C).

PROOF Let f be a CFS on a set A and define $g = (g_\rho)_{\rho \in L}$ as follows: $g(\mathbf{a}/\rho) = b/\rho$ if and only if

> there exist $\tau \in L$ and $\mathbf{c}/\tau \in dom\ f$ such that
> $\tau \le \rho$, $\mathbf{a} \equiv_\rho \mathbf{c}$ and $f(\mathbf{c}/\tau) = b/\tau$.

It follows directly from the condition (A) for f that all g_ρ are well defined. Obviously g extends f and satisfies (C).

Let $g(\mathbf{a}^i/\rho_i) = b^i/\rho_i$, $i = 1, 2$. Then there exist $\tau_i \in L$ and $\mathbf{c}^i/\tau_i \in dom\ f$ such that

$$\tau_i \le \rho_i,\ \mathbf{a}^i \equiv \mathbf{c}^i\ (\rho_i) \quad \text{and} \quad f(\mathbf{c}^i/\tau_i) = b^i/\tau_i, \quad i = 1, 2.$$

Obviously

$$\mathrm{Eg}(\mathbf{c}^1, \mathbf{c}^2) \le \rho_1 \vee \rho_2 \vee \mathrm{Eg}(\mathbf{a}^1, \mathbf{a}^2)$$

and the compatibility of f implies

$$(b^1, b^2) \in \tau_1 \vee \tau_2 \vee \mathrm{Eg}(\mathbf{c}^1, \mathbf{c}^2) \le \rho_1 \vee \rho_2 \vee \mathrm{Eg}(\mathbf{a}^1, \mathbf{a}^2).$$

Hence g satisfies (A).

It remains to prove that g satisfies (B). Suppose that

$$g(\mathbf{a}^i/\rho_i) = b^i/\rho_i,\ \rho_i \le \sigma_i,\ \mathbf{a} \equiv \mathbf{a}^i\ (\sigma_i), \quad i = 1, \ldots, n.$$

Then there exist $\tau_i \in L$ and $\mathbf{c}^i \in dom\ f$ such that

$$\tau_i \le \rho_i,\ \mathbf{a}^i \equiv \mathbf{c}^i\ (\rho_i)\ \text{and}\ f(\mathbf{c}^i/\tau_i) = b^i/\tau_i, \quad i = 1, \ldots, n.$$

Now obviously $\mathbf{a} \equiv \mathbf{c}^i\ (\sigma_i), i = 1, \ldots, n$, and since f satisfies (B),

$$(b^1/\sigma_1, \ldots, b^n/\sigma_n) \in A/\sigma_1 \wedge \cdots \wedge \sigma_n.$$

Hence, g satisfies (B) also.   ●

Lemma 2.3.6 *If the lattice* **L** *is distributive then for every* CFS f *on A there exists a* CFS *which extends f and satisfies* (D).

PROOF Let f be a CFS on A and define $g = (g_\rho)_{\rho \in L}$ as follows: $g(\mathbf{a}/\rho) = b/\rho$ if and only if there exist $\rho_1, \ldots, \rho_n \in L$ with $\rho_1 \wedge \cdots \wedge \rho_n = \rho$ and $f(\mathbf{a}/\rho_i) = b/\rho_i$, $i = 1, \ldots, n$.

First check that all g_ρ are well defined. Suppose that there exist

$$\rho_1, \ldots, \rho_k, \sigma_1, \ldots, \sigma_l \in L$$

and $b, c \in A$ such that

$$\rho_1 \wedge \cdots \wedge \rho_k = \rho = \sigma_1 \wedge \cdots \wedge \sigma_l$$

and

$$f(\mathbf{a}/\rho_i) = b/\rho_i, \quad f(\mathbf{a}/\sigma_j) = c/\sigma_j, \quad i = 1, \ldots, k, \ j = 1, \ldots, l.$$

Since $\rho_1 \wedge \cdots \wedge \rho_k \wedge \sigma_1 \wedge \cdots \wedge \sigma_l = \rho$ and f satisfies $(\mathcal{B})$,

$$(b/\rho_1, \ldots, b/\rho_k, c/\sigma_1, \ldots, c/\sigma_l) \in A/\rho,$$

let it be equal to d/ρ. Then

$$b \equiv d \ (\rho_i) \quad \text{and} \quad d \equiv c \ (\sigma_j), \quad i = 1, \ldots, k, \ j = 1, \ldots, l.$$

Hence both (b, d) and (d, c) are in ρ implying $b/\rho = c/\rho$.

The next step is to show that g is a CFS. Let $g(\mathbf{a}^1/\rho) = b^1/\rho$ and $g(\mathbf{a}^2/\sigma) = b^2/\sigma$. Then there exist $\rho_1, \ldots, \rho_k, \sigma_1, \ldots, \sigma_l \in L$ such that

$$\rho = \rho_1 \wedge \cdots \wedge \rho_k, \quad \sigma = \sigma_1 \wedge \cdots \wedge \sigma_l$$

and

$$f(\mathbf{a}^1/\rho_i) = b^1/\rho_i, \quad f(\mathbf{a}^2/\sigma_j) = b^2/\sigma_j, \quad i = 1, \ldots, k, \ j = 1, \ldots, l.$$

Since f satisfies $(\mathcal{A})$,

$$(b^1, b^2) \in \rho_i \vee \sigma_j \vee \mathrm{Eg}(\mathbf{a}^1, \mathbf{a}^2)$$

for all i and j. Applying the distributivity of $\mathbf{L}$, we get

$$(b^1, b^2) \in \rho \vee \sigma \vee \mathrm{Eg}(\mathbf{a}^1, \mathbf{a}^2),$$

which means that g satisfies $(\mathcal{A})$ as well.

Now we prove that g satisfies $(\mathcal{B})$. Suppose that

$$g(\mathbf{a}^i/\rho_i) = b^i/\rho_i, \quad \rho_i \leq \sigma_i, \quad \mathbf{a} \equiv \mathbf{a}^i \ (\sigma_i), \quad i = 1, \ldots, n.$$

Then by definition of g, there exist $\rho_{ij} \in L$ such that

$$\rho_i = \bigwedge_j \rho_{ij} \text{ and } f(\mathbf{a}^i/\rho_{ij}) = b^i/\rho_{ij}, \quad i = 1,\ldots,n, j = 1,\ldots,k.$$

(Since repeating ρ_{ij} are not excluded, we may assume that every ρ_i is a meet of the same number k equivalences ρ_{ij}.) Obviously $\mathbf{a} \equiv \mathbf{a}^i\ (\sigma_i \vee \rho_{ij})$ for every i and j. Since f satisfies $(\mathcal{B})$, we have

$$(b^1/\sigma_1 \vee \rho_{11}, \ldots, b^1/\sigma_1 \vee \rho_{1k}, \ldots,$$
$$b^n/\sigma_n \vee \rho_{n1}, \ldots, b^n/\sigma_n \vee \rho_{nk}) \in A/\bigwedge_{i,j}(\sigma_i \vee \rho_{ij}).$$

This means that there is $d \in A$ such that

$$d \equiv b^i\ (\sigma_i \vee \rho_{ij}) \quad \text{for all}\ \ i = 1,\ldots,n, j = 1,\ldots,k.$$

Due to the distributivity of $\mathbf{L}$ then also $(d, b^i) \in \sigma_i,\ i = 1,\ldots,n$, implying

$$(b^1/\sigma_1, \ldots, b^n/\sigma_n) \in A/\sigma_1 \wedge \cdots \wedge \sigma_n.$$

Finally, it remains to prove that g satisfies $(\mathcal{D})$. Thus suppose that $g(\mathbf{a}/\rho) = b/\rho$ and $g(\mathbf{a}/\sigma) = c/\sigma$. Then, by definition of g, there exist $\rho_1, \ldots, \rho_k, \sigma_1, \ldots, \sigma_l \in L$ such that

$$\rho = \rho_1 \wedge \cdots \wedge \rho_k, \quad \sigma = \sigma_1 \wedge \cdots \wedge \sigma_l$$

and

$$f(\mathbf{a}/\rho_i) = b/\rho_i, \ f(\mathbf{a}/\sigma_j) = c/\sigma_j, \quad i = 1,\ldots,k, j = 1,\ldots,l.$$

Since f satisfies $(\mathcal{B})$,

$$(b/\rho_1, \ldots, b/\rho_k, c/\sigma_1, \ldots, c/\sigma_l)$$
$$= (f(\mathbf{a}/\rho_1), \ldots, f(\mathbf{a}/\rho_k), f(\mathbf{a}/\sigma_1), \ldots, f(\mathbf{a}/\sigma_l)) \in A/\rho \wedge \sigma.$$

Hence there exists $d \in A$ such that

$$f(\mathbf{a}/\rho_i) = d/\rho_i, \ f(\mathbf{a}/\sigma_j) = d/\sigma_j, \quad i = 1,\ldots,k, j = 1,\ldots,l.$$

Since $\rho \wedge \sigma = \rho_1 \wedge \cdots \wedge \rho_k \wedge \sigma_1 \wedge \cdots \wedge \sigma_l$, in consequence we have that $\mathbf{a}/\rho \wedge \sigma \in dom\ g$.   ●

Lemma 2.3.7 *If the lattice* **L** *is distributive then every* CFS *on* A *has a closed extension.*

PROOF Since A is finite, every CFS on A has a maximal extension. Due to Lemmas 2.3.5 and 2.3.6, the latter must be closed. •

Now we are ready to prove the main result of this subsection. We already remarked earlier that the (CFSE) property always implies the (CFL) property. From the next theorem we see that for A finite the two properties are equivalent.

Theorem 2.3.8 *Let* **L** *be an equivalence lattice on a finite set* A *and assume that* **L** *has the* (CFL) *property. Then* **L** *has the* (CFSE) *property as well.*

PROOF Let f be a CFS on A for **L**. Our proof uses induction on the height of **L**. If **L** is trivial then the assertion is trivial. By Theorem 2.3.3, **L** is distributive. Hence in view of Lemma 2.3.7 there exists a closed CFS g on A which extends f. We may assume that g is a maximal closed CFS which extends f. If *dom* $g_0 = A$ then we are done. So we assume, to the contrary, that *dom* $g_0 \neq A$ and show that this leads to a contradiction. Under this assumption there must exist an atom α of **L** such that *dom* $g_\beta \neq (A/\beta)^m$ where β is the pseudo-complement of α. Indeed, the meet of all of the pseudocomplements of all of the atoms of **L** is zero. Note that obviously the system $(f_{\rho/\alpha})_{\alpha \leq \rho \in L}$ is a CFS on A/α for $\mathbf{L}_\alpha$. Hence by the induction hypothesis and the (CFL) property there is a function $q \in F = clo(L)$ which induces all g_ρ with $\alpha \leq \rho$.

Now define $h = (h_\rho)_{\rho \in L}$ as follows. First pick an element $\mathbf{a}^0 \in A^m$ such that $\mathbf{a}^0/\beta \notin dom\ g$ and put then $h_\rho = g_\rho$ if $\rho \neq \beta$, and

$$h(\mathbf{a}/\beta) = \begin{cases} g(\mathbf{a}/\beta) & \text{if} \quad \mathbf{a}/\beta \in dom\ g, \\ q(\mathbf{a}/\beta) & \text{if} \quad \mathbf{a}/\beta = \mathbf{a}^0/\beta. \end{cases}$$

Thus, *dom* $h_\rho = dom\ g_\rho$ if $\rho \neq \beta$ and *dom* $h_\beta = dom\ g_\beta \cup \{\mathbf{a}^0/\beta\}$. We are going to prove that h is a CFS. Then by Lemma 2.3.7 it admits a closed extension which contradicts the maximality of g and hence shows that *dom* $g_0 = A$.

As a first step, we show that h satisfies $(\mathcal{A})$.

Let $h(\mathbf{a}^i/\rho_i) = b^i/\rho_i$, $i = 1, 2$. Obviously the only nontrivial case is $\mathbf{a}^1/\rho_1 \notin dom\ g$ and $\mathbf{a}^2/\rho_2 \in dom\ g$. Then $\rho_1 = \beta$, $\mathbf{a}^1/\beta = \mathbf{a}^0/\beta$

and $b^1/\beta = q(\mathbf{a}^0)/\beta$. If $\alpha \leq \rho_2$ then $b^2/\rho_2 = q(\mathbf{a}^2)/\rho_2$ and, since $q \in F$,

$$(b^1, b^2) \in \rho_1 \vee \rho_2 \vee \mathrm{Eg}(\mathbf{a}^1, \mathbf{a}^2);$$ (2.14)

so $(\mathcal{A})$ is satisfied. If $\alpha \not\leq \rho_2$ then $\rho_2 \leq \beta$ combined with the fact that g is closed yields $\mathbf{a}^2/\beta \in dom\ g$. Since $\mathbf{a}^1/\beta \notin dom\ g$, we have $\mathbf{a}^1 \not\equiv \mathbf{a}^2$ (β). The other consequence of g being closed is that $\mathbf{a}^2/\rho_2 \vee \alpha \in dom\ g$. Hence the condition $(\mathcal{A})$ for g implies

$$b^2/\rho_2 \vee \alpha = g(\mathbf{a}^2/\rho_2 \vee \alpha) = q(\mathbf{a}^2)/\rho_2 \vee \alpha,$$

and then, by the L-compatibility of q,

$$(b^1, b^2) \in \rho_1 \vee \rho_2 \vee \alpha \vee \mathrm{Eg}(\mathbf{a}^1, \mathbf{a}^2).$$ (2.15)

However, $\mathrm{Eg}(\mathbf{a}^1, \mathbf{a}^2) \not\leq \beta$ and therefore $\alpha \leq \mathrm{Eg}(\mathbf{a}^1, \mathbf{a}^2)$. Hence (2.15) implies (2.14); so in this case $(\mathcal{A})$ is also satisfied by h, and we are done.

Finally, we have to prove that h satisfies $(\mathcal{B})$. Suppose that $h(\mathbf{a}^i/\rho_i) = b^i/\rho_i$, $\rho_i \leq \sigma_i$, and $\mathbf{a} \equiv \mathbf{a}^i$ $(\sigma_i), i = 1,\ldots,n$. Since g satisfies $(\mathcal{B})$, we may assume without loss of generality that there is at least one i such that $\rho_i = \beta$ and $\mathbf{a}^i = \mathbf{a}^0$. The case with all $\mathbf{a}^i/\rho_i$ equal to $\mathbf{a}^0/\beta$ is trivial: then

$$(b^1/\sigma_1,\ldots,b^n/\sigma_n) = q(\mathbf{a}^0)/\sigma_1 \wedge \cdots \wedge \sigma_n \in A/\sigma_1 \wedge \cdots \wedge \sigma_n.$$

Let us first handle the case with $n = 2$, $\mathbf{a}^1/\rho_1 = \mathbf{a}^0/\beta$ and $\mathbf{a}^2/\rho_2 \in dom\ g$, and then show that the general case reduces to this one.

If $\alpha \leq \sigma_2$ then $b^1/\rho_1 = q(\mathbf{a}^1/\rho_1)$ implies

$$b^1/\sigma_1 = q(\mathbf{a}^1/\sigma_1) = q(\mathbf{a}/\sigma_1)$$

and, because g is closed,

$$b^2/\sigma_2 = g(\mathbf{a}^2/\sigma_2) = q(\mathbf{a}^2/\sigma_2) = q(\mathbf{a}^2)/\sigma_2 = q(\mathbf{a})/\sigma_2.$$

Hence

$$(b^1/\sigma_1, b^2/\sigma_2) = q(\mathbf{a})/\sigma_1 \wedge \sigma_2 \in A/\sigma_1 \wedge \sigma_2.$$

If $\alpha \not\leq \sigma_2$ then $\sigma_2 \leq \beta \leq \sigma_1$. Hence $(b^1/\sigma_1, b^2/\sigma_2) \in A/\sigma_1 \wedge \sigma_2$ is equivalent to $b^1 \equiv b^2$ (σ_1). Since h satisfies $(\mathcal{A})$, we have

$$(b^1, b^2) \in \rho_1 \vee \rho_2 \vee \mathrm{Eg}(\mathbf{a}^1, \mathbf{a}^2) \leq \sigma_1 \vee \mathrm{Eg}(\mathbf{a}^1, \mathbf{a}^2).$$

However, $\mathbf{a} \equiv \mathbf{a}^1 \; (\sigma_1)$ and $\mathbf{a} \equiv \mathbf{a}^2 \; (\sigma_2)$ imply $\mathbf{a}^1 \equiv \mathbf{a}^2 \; (\sigma_1)$ and we are done.

Now consider the general case. Assume that

$$\mathbf{a}^i/\rho_i = \begin{cases} \mathbf{a}^0/\beta & \text{if} \quad i = 1,\ldots,k, \\ \mathbf{a}^i/\rho_i \in dom\, g & \text{if} \quad i = k+1,\ldots,n \end{cases}$$

and denote $\tau_1 = \sigma_1 \wedge \cdots \wedge \sigma_k$, $\tau_2 = \sigma_{k+1} \wedge \cdots \wedge \sigma_n$. Since g satisfies $(\mathcal{B})$,

$$(b^{k+1}/\sigma_{k+1},\ldots,b^n/\sigma_n) = g(\mathbf{a}/\tau_2) = b/\tau_2 \in A/\tau_2.$$

We apply what was proved above for $n = 2$ to the situation with

$\mathbf{a}^0, \mathbf{a}, \mathbf{a}, b^1, b, \beta, \tau_2, \tau_1, \tau_2$ in the roles of $\mathbf{a}^1, \mathbf{a}^2, \mathbf{a}, b^1, b^2, \rho_1, \rho_2, \sigma_1, \sigma_2$,

respectively. This is possible, since $\mathbf{a}^i \equiv \mathbf{a} \; (\sigma_i)$ for every $i = 1,\ldots,k$ implies $\mathbf{a}^0 \equiv \mathbf{a} \; (\tau_1)$.

Hence there exists $d \in A$ such that $(b^1/\tau_1, b/\tau_2) = d/\tau_1 \wedge \tau_2$. Then obviously

$$(d, b^1) \in \tau_1 \le \sigma_i, \quad i = 1,\ldots,k,$$

and in view of $b \equiv b^j \; (\sigma_j), d \equiv b \; (\tau_2)$ implies $d \equiv b^j \; (\sigma_j)$ for the integers $j = k+1,\ldots,n$. Thus $(b^1/\sigma_1,\ldots,b^n/\sigma_n) = d/\sigma_1 \wedge \cdots \wedge \sigma_n$ and we are done.   •

2.3.3 Compatible function lifting implies distributivity (residually finite case)

In this section we shall show that distributivity of the equivalence lattice **L** follows from the (CFL) property under a much weaker assumption than the finiteness of the ground set A. To do this, we first have to prove that in the case of a finite A the (CFL) property implies not just the distributivity of **L** but the congruence distributivity of the variety generated by $\langle A; clo(L) \rangle$. In fact we prove, using the results of the last section, that $clo(L)$ contains a near unanimity function.

Theorem 2.3.9 *Let* **L** *be an equivalence lattice on a finite set A. If* **L** *has the compatible function lifting property then the clone of* **L**-*compatible functions contains a near unanimity function.*

PROOF If L has fewer than 3 elements then there is nothing to prove. So suppose $|L| = m \geq 3$, and define $f = (f_\rho)_{\rho \in L}$ as follows: $f_\rho = \emptyset$ if $\rho \neq 0$ and f_0 is the partial m-ary near unanimity function on $A = A/0_A$ with the domain consisting of all vectors of the form $(a, \ldots, a, b, a, \ldots, a)$ where $a, b \in A$ and b may have an arbitrary position. Obviously f satisfies $(\mathcal{A})$.

To check the condition $(\mathcal{B})$ suppose that $f(\mathbf{a}^i) = b^i$ and that $\sigma_i \in L$ are distinct equivalences such that for some $\mathbf{a} \in A^m$, we have $\mathbf{a} \equiv \mathbf{a}^i \ (\sigma_i)$, $i = 1, \ldots, n$. If some σ_i is equal to 0_A then it follows from $(\mathcal{A})$ that $b^i \equiv b^j \ (\sigma_j)$ for every $j = 1, \ldots, n$ implying $(b^1/\sigma_1, \ldots, b^n/\sigma_n) = b^i \in A$. Thus we may assume that 0_A does not occur among σ_i, implying $n < m$. Hence by the definition of a near unanimity function, there must exist $j \in \{1, \ldots, m\}$ such that $b^i = a^i_j$ for all $i = 1, \ldots, n$. Then

$$(b^1/\sigma_1, \ldots, b^n/\sigma_n) = (a^1_j/\sigma_1, \ldots, a^n_j/\sigma_n) = (a_j/\sigma_1, \ldots, a_j/\sigma_n)$$
$$= a_j/\sigma_1 \wedge \cdots \wedge \sigma_n \in A/\sigma_1 \wedge \cdots \wedge \sigma_n.$$

Thus f is a CFS and by Theorem 2.3.8 it must be induced by some L-compatible total function on A.   •

Now we prove the extended version of Theorem 2.3.3. We say that an equivalence relation $\rho \in \mathrm{Eqv}\,A$ is of **finite index** if the set A/ρ is finite. An equivalence lattice L on A is said to be **residually finite** if every $\rho \in L$ is the meet of equivalence relations $\sigma \in L$ of finite index. We require the following technical lemma.

Lemma 2.3.10 *Let L be a 0-1 equivalence lattice on the set A and assume that L is residually finite and has the (CFL) property. Let $\beta, \rho, \mu \in L$ be such that $\beta \not\leq \rho$, $\beta \wedge \rho = 0$, ρ is completely meet irreducible in L, and μ is its cover in L. Then, given any two element set U contained in some of the μ/ρ-blocks of A/ρ, there exists a unary L-compatible function h on A which induces a bijective function on U and a constant function on A/β.*

PROOF Let $a_1, a_2 \in A$ be such that $(a_1, a_2) \in \beta \setminus \rho$. Because $\beta \wedge \rho = 0$ we have the canonical subdirect embedding $A \subseteq A/\rho \times A/\beta$. Hence we may write $a_i = (b_i, c)$, for some $b_i \in A/\rho$, $c \in A/\beta$, $i = 1, 2$.

Since ρ is completely meet irreducible and L is residually finite, ρ is of finite index. Clearly, the induced equivalence lattice L_ρ on

A/ρ has the (CFL) property and μ/ρ is its only atom. Therefore every function from A/ρ to U is $\mathbf{L}_\rho$-compatible and can be lifted to an L-compatible function on A. In particular there exists a unary function $f \in F = clo(L)$ that induces an idempotent function e on A/ρ such that $e(A/\rho) = e(\{b_1, b_2\}) = U$. Denote $e(b_i) = u_i$, $i = 1, 2$. Then $f(a_i) = (u_i, d)$ for some $d \in A/\beta$, $i = 1, 2$.

Now let $h = (e, g)$ be the function on $A/\rho \times A/\beta$ where g is a constant function: $g(A/\beta) = \{d\}$. Clearly $h(A) = \{(u_1, d), (u_2, d)\} \subseteq A$. We prove that the restriction of h to A is L-compatible by showing that this function (which we also denote by h) preserves all $\sigma \in L$ of finite index. For any such σ, the equivalence relation $\rho \wedge \sigma \in L$ is of finite index too, and the equivalence lattice $\mathbf{L}_{\rho \wedge \sigma}$ on $A/\rho \wedge \sigma$ has the (CFL) property. Hence by Theorem 2.3.9 there exists a $\mathbf{L}_{\rho \wedge \sigma}$-compatible near unanimity function for $\mathbf{A}/\rho \wedge \sigma$. Now if $\mathbf{A} = \langle A; F \rangle$, then, by Theorem 1.2.11, the variety $V(\mathbf{A}/\rho \wedge \sigma) = V(\mathbf{A}/\sigma, \mathbf{A}/\rho)$ is CD and we may apply Corollary 1.2.24 to see that the equivalence σ cannot be skew with respect to the subdirect decomposition $A \subseteq A/\rho \times A/\beta$. However, the function h obviously preserves all nonskew equivalences $\tau \in L$. Hence the restriction of h to A is L-compatible. $\bullet$

Theorem 2.3.11 *Let $\mathbf{L}$ be an equivalence lattice of the set A. If $\mathbf{L}$ is residually finite and has the compatible function lifting property, then $\mathbf{L}$ is distributive.*

PROOF We are going to prove that the lattice $\mathbf{L}$ satisfies condition (3) from Lemma 1.2.17. (It is true that every member of L is the meet of completely meet irreducible elements, due to the fact that every member is the meet of elements of finite index.) Thus let $\rho, \beta, \gamma \in L$ be such that ρ is completely meet irreducible and $\beta \wedge \gamma \leq \rho$.

Let μ be the cover of ρ in $\mathbf{L}$ and $U = \{u_1, u_2\}$ be a two element set contained in some μ/ρ-blocks of A/ρ. Assume that $\beta \nleq \rho$, i.e., there exists a pair $(a_1, a_2) \in \beta \backslash \rho$. Then by Lemma 2.3.10 there exists a function $f \in F_{\rho \wedge \beta}$, where $F = clo(L)$, which induces a bijection on U and a constant map on A/β. Similarly, if $\gamma \nleq \rho$, then there exists a function $g \in F_{\rho \wedge \gamma}$ which induces a bijection on U and a constant map on A/γ. By the (CFL) property the functions f and g are induced by some L-compatible functions f_1 and g_1, respectively.

Now the composed function $f_1 g_1$ still induces a bijection on U but a constant map on $A/\beta \wedge \gamma$. Hence $f_1 g_1$ induces a constant on A/ρ as well, since $\beta \wedge \gamma \leq \rho$. Since $U \subseteq A/\rho$, this is a contradiction.   •

We conclude the chapter by picturing for finite sets the hierarchy of the properties of equivalence lattices we considered. Three of them (distributivity, existence of a compatible near unanimity function, and arithmeticity) have been known for a long time. Our new contribution is the (CFL) property, which in the case of finite sets, we proved to be equivalent to the (CFSE) property and, because of this, to imply the existence of a compatible near unanimity function if the ground set is finite. The importance of this property and of Theorem 2.3.11 will become apparent in Chapter 4.

We mentioned in the beginning of the present section that arithmetical equivalence lattices on finite sets have the (CFL) property. This will follow from Theorems 2.2.6 and 3.3.3. Indeed, if **L** is an arithmetical equivalence lattice on a finite set A then by Theorem 2.2.6 there exists an **L**-compatible principal Pixley function p on A. Hence the algebra $\mathbf{A} = \langle A; p \rangle$ generates an arithmetical variety V and $L = \mathrm{Con}\, \mathbf{A}$ by Theorem 2.2.12. Now, if $\rho \in L$ and $f \in F_\rho$ then f is a compatible function on finite algebra $\mathbf{A}/\rho \in V$. From Theorem 3.3.3 we shall see that f is a polynomial function of $\mathbf{A}/\rho$. Clearly then f is induced by a polynomial function of $\mathbf{A}$ so that **L** has the (CFL) property.

Since the (CFL) property implies distributivity we will conclude (after Theorem 3.3.3 is established) yet another characterization of arithmetical equivalence lattices:

Theorem 2.3.12 *An equivalence lattice on a finite set is arithmetical iff it has the compatible function lifting property and consists of permuting equivalence relations.*

The following picture describes the hierarchy for the case of equivalence lattices on finite sets: moving downwards means going to strictly smaller classes.

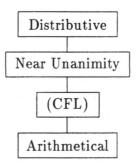

Figure 2.1

The following example, showing that the distributive class is a strictly larger class than the near unanimity class, can be found in [Gumm, 1978] (which was previously cited in Section 2.2.3). Recall that an operation f is *idempotent* if $f(x,\ldots,x) = x$ for all x. Also recall from Section 2.2.1 that congruence equations may involve relation product $\circ$ as well as meet and join, while lattice equations involve only meet and join.

Example 2.3.13 *An equivalence lattice on a finite set which is distributive but which supports no near unanimity operation.*

Let the algebra $\mathbf{A_6} = \langle\{0,\ldots,5\}; f,g\rangle$ where f and g are unary defined by $f(x) = x + 2 \pmod 6$ and

$$g(0) = g(5) = 1,\; g(1) = g(4) = 0,\; g(2) = g(3) = 5\,.$$

Then $\mathbf{Con\,A_6}$ is the four element distributive lattice with atoms

$$\theta = \{\{0,1\},\{2,3\},\{4,5\}\} \text{ and } \phi = \{\{1,2\},\{3,4\},\{0,5\}\}.$$

These congruences do not permute and in fact

$$\theta \circ \phi \circ \theta = \phi \circ \theta \circ \phi = \theta \vee \phi = 1\,.$$

Then the following can be shown:

1. For every n there is no n-ary $\mathbf{Con\,A_6}$-compatible operation f which is idempotent other than the projections.

2. $\mathbf{Con\,A_6}$ satisfies every congruence equation which does not imply permutability, but (obviously) does satisfy every nontrivial lattice equation.

Since any near unanimity function is idempotent **Con A_6** can support none.

We conclude by showing that the (CFL) property (finite sets) lies strictly between arithmeticity and the existence of a compatible near unanimity function. First note that the (CFL) property does not imply arithmeticity. For example, take a special case of Example 2.3.1 with $n = 2$, $A_1 = A_2 = \{0,1\}$ and $A = \{(0,0),(0,1),(1,0)\}$. Obviously in this case θ_1 and θ_2 do not permute. The next example shows that the existence of a compatible near unanimity function does not imply the (CFL) property.

Example 2.3.14 *An equivalence lattice on a finite set having a compatible majority function and not having the* (CFL) *property.*

Let $A = \{0,1,2\}$ and consider the subset

$$B = \{(0,1),(0,2),(1,1),(1,0),(2,0)\} \subseteq A^2 .$$

Let θ_1 and θ_2 be the kernels of the two projections $B \rightarrow A$ and $L = \{0_B, \theta_1, \theta_2, 1_B\}$. It is easy to check that L is a subuniverse in **Eqv** B and that $F = clo(L)$ contains a majority function. However, **L** does not satisfy the (CFL) property. Indeed, let $f_1 : A \rightarrow A$ be a function such that $f_1(2) = 2$ and $f_1(1) = 0$. We consider f_1 as an element of F_{θ_1} and want to find its lift $f = (f_1, f_2) \in F$ where $f_2 : A \rightarrow A$ is a suitable function. Since $(2,x) \in B$ implies $x = 0$, we must have $f_2(0) = 0$. On the other hand, because $(0,x) \in B$ implies $x \neq 0$, we must have $f_2(0) \neq 0$. This contradiction shows that such an extension does not exist.

Exercise

Verify the details of Example 2.3.13. ([Gumm, 1978])

Chapter 3

Primality and Generalizations

3.1 Primality and functional completeness

Historically, primal algebras were first introduced as particular algebraic versions (called Post algebras) of multiple valued logics, analogous to Boolean algebras as an algebraic version of classical two valued logic. We will briefly discuss Post algebras and their development in Section 3.1.2 below.

3.1.1 Primality

Recall from Section 1.1.3 that a finite algebra $\mathbf{A}$ is primal if all operations on A are representable by term functions and that this can be expressed formally by the requirement that for all $n > 0$, $\mathrm{Clo}_n \mathbf{A} = O_n(A)$. Primal algebras were named and first studied systematically by A. L. Foster, who developed their study as an extension of the theory of Boolean algebras. Boolean algebras can be approached in several different ways; Foster started with the two element Boolean algebra

$$\mathbf{B_2} = \langle \{0,1\}; \vee, \wedge,', 0, 1 \rangle.$$

The variety of Boolean algebras is then just the variety generated by $\mathbf{B_2}$ and is in fact described as

$$V(\mathbf{B_2}) = IP_S(\mathbf{B_2}),$$

i.e., every Boolean algebra is isomorphic to a subdirect power of $\mathbf{B_2}$. Furthermore finite Boolean algebras are just copies of direct powers of $\mathbf{B_2}$:

$$V_f(\mathbf{B_2}) = IP_f(\mathbf{B_2}).$$

From Foster's and our point of view the key property of Boolean algebras is the primality of $\mathbf{B_2}$: all operations on $\{0,1\}$ are term functions. The primality of $\mathbf{B_2}$ is often called the "fundamental theorem of switching circuit theory" and is of great importance in that subject since it asserts that every switching function (truth function) can be constructed from the Boolean operations $\vee, \wedge, ', 0, 1$. A more general concept is also of interest in the construction of switching circuits, that of functionally completeness. Recall that $\mathbf{A}$ is functionally complete if A is finite and each operation on A is a polynomial, i.e., $\mathrm{Pol}\,\mathbf{A} = O(A)$. Clearly any primal algebra is functionally complete. The most important examples of functionally complete algebras which are not primal are the Galois fields. That they are functionally complete is a direct consequence of the multivariable version of the Lagrange interpolation theorem together with their finiteness.

A significant event in the development of universal algebra occurred when Foster ([Foster, 1953]) recognized that essentially all of the general algebraic theory of Boolean algebras is shared by the variety generated by any primal algebra. Most important for our interests, Foster established the following result.

Theorem 3.1.1 *If* $\mathbf{P}$ *is any primal algebra then*

$$V(\mathbf{P}) = IP_S(\mathbf{P}) \ and \ V_f(\mathbf{P}) = IP_f(\mathbf{P}).$$

Another related result in the theory of Boolean algebras is the Stone duality theorem which states, in the language of category theory, that as a category, the variety of Boolean algebras is equivalent to the dual of the category of Boolean spaces: compact Hausdorff spaces with a base of topology consisting of clopen sets. (Any variety is a category where the morphisms are just the homomorphisms between algebras in the variety.) Thus algebraic statements about Boolean algebras can be translated into topological statements about continuous functions, and *vice versa*. T. K. Hu ([Hu, 1969]) directly generalized Stone duality and in fact showed that for any primal algebra $\mathbf{P}$ the variety $V(\mathbf{P})$ is equivalent as a category to the dual of the variety of Boolean spaces. Therefore any two primal algebras generate categorically equivalent varieties. Even more, Hu showed that if **P** is any primal algebra, then a variety W is categorically equivalent to $V(\mathbf{P})$ iff W is the variety generated by some primal algebra. Put another way, the class of all algebras generating varieties categorically equivalent to any primal generated variety is just the class of all primal algebras. Hu's important results thus constitute a significant precise validation of Foster's original insight.

3.1.2 History: Post algebras

In 1920 E. L. Post defined n-valued propositional logics for integers $n \geq 2$ as direct generalizations of classical 2-valued logic. During the years 1934-36 D. L. Webb showed both that all n-valued truth functions were expressible by the basic connectives of n-valued logics and also defined n element algebras playing the same role for n-valued logics as that played by the two element Boolean algebra for classical logic; thus he showed essentially that these n element algebras were primal. P. C. Rosenbloom ([Rosenbloom, 1942]) introduced the name **Post algebras** for what turns out to be the algebras in the various varieties $V(n)$ generated by each of these n element algebras and gave axioms for these varieties. $V(n)$ is usually called the variety of n-valued Post algebras. Just as in the case of Boolean algebras there are many possible sets of operations for Post algebras and many different ways of describing them.

Following Rosenbloom (except for a notation change), for each integer $n \geq 2$, we define the algebra

$$\mathbf{P_n} = \langle P_n; \vee, \wedge, ' \rangle,$$

where $P_n = \{0, \ldots, n-1\}$, $\vee$ and $\wedge$ are the lattice operations which are just "max" and "min" with P_n taken as a chain with 0 the least element. Further define $x' = x + 1(\bmod\ n)$. The algebra $\mathbf{P_n}$ is often called the basic Post algebra of order n and for $n = 2$ is term equivalent to the two element Boolean algebra $\mathbf{B_2}$. As in the case of $\mathbf{B_2}$ it is also common to take the basic operations of $\mathbf{P_n}$ to be just $\wedge$ and $'$, for example. In [Rosenbloom, 1942] it is proven that all finite members of $V(n)$ are isomorphic to direct powers of $\mathbf{P_n}$. L. I. Wade ([Wade, 1945]) showed that the only SI member of $V(n)$ is $\mathbf{P_n}$ and then using Birkhoff's Theorem 1.1.3 inferred that $V(n) = IP_S(\mathbf{P_n})$.

Since at the time of Wade's result it was already known from Webb's work that $\mathbf{P_n}$ is primal, and since clearly any two primal algebras of the same size are term equivalent, Wade could have concluded Theorem 3.1.1 directly, but perhaps did not fully appreciate that primality was the critical property of $\mathbf{P_n}$ and the obvious fact that by studying one particular primal algebra of order n, one, in effect, studies all primal algebras of order n.

The study of varieties generated by primal algebras via their Post representation continues to some degree today. Rosenbloom's axioms for the variety $V(n)$, taken with the few operations $\vee, \wedge, '$, were, however, rather complicated, resulting in a rather complicated theory. Today $V(n)$ is usually taken with more operations; an approach which has the advantages that it allows Post algebras to be construed as certain pseudocomplemented distributive lattices (called Stone lattices) and which also allows easy access to the topological representation theory via that for Boolean algebras. This approach is well represented in the book [Balbes, Dwinger, 1974].

In the present book we wish to stress far reaching generalizations of primal and functionally complete algebras and it is difficult to see how this might be effectively done by initially restricting the algebraic language used, i.e., by not simply starting with primality as the primitive concept (rather than with models of some set of axioms).

As noted above, A. L. Foster ([Foster, 1953]) started with the notion of primality and obtained an entirely different proof of Theorem 3.1.1 by introducing and applying the concept of Boolean powers for arbitrary algebras. (Boolean products and powers are thoroughly discussed in [Burris, Sankappanavar, 1981].) Moreover, in consequence of his proof by this method, he obtained canonical forms for primal polynomials, regardless of the language chosen, thus demonstrating the advantage of the universal algebra approach.

Remarks on terminology As noted above, primal algebra theory has grown up in two different mathematical cultures, the culture of traditional algebra on the one hand, as well as the culture of multiple valued logic and clone theory on the other. In addition the two different approaches often took place in two different political cultures as well, the western European and American on the one hand and the Soviet dominated culture eastward from middle Europe on the other. For these reasons as well as others less clear, the terminology of the subject has been confusing, unnecessarily profuse, and, at worst, conflicting. For example what we call "primal" algebras have been, and sometimes still are, called "complete" algebras, thus often leading to primality and functional completeness being confused. Thus readers are warned to take special care in reading journal articles, particularly older ones. In [Pixley, 1982] a table of terminology is given which may help.

On the brighter side, in the last two decades we have seen not just the opening of the eastern political culture but also the recognition within the universal algebra community of the fundamental role played by the theory of clones. Perhaps these events will help standardize the notation of the field; at any rate we no longer hold the view, expressed in [Pixley, 1982], that attempts to rationalize the terminology would likely be "both futile and thankless". In the present chapter we will adopt and use terminology (suggested in [Bergman, Berman, 1996]) which attempts this rationalization within practical bounds.

3.1.3 Primality and functional completeness: characterizations

An apparent shortcoming of defining primal algebras by the requirement that all operations be representable by term operations, despite its concrete nontechnical character, is that it bears no relation to the ordinary concepts of interest to algebraists, and for this reason, primality was for a time, despite Foster's striking results, considered algebraically anomalous. Our first task is to eliminate this anomaly. The following theorem ([Pixley, 1961]) characterizes primality in terms of the central concepts of universal algebra, and thus places primality at the center of primary algebraic considerations. Also, as we shall see, it naturally suggests generalizations of primality.

Theorem 3.1.2 *A finite algebra* **A** *is primal iff* **A** *is simple, rigid, has no proper subalgebras, and generates an arithmetical variety, i.e.,* **A** *has a Pixley term.*

(An algebra is **rigid** if it has only the trivial (identity) automorphism.)

PROOF Since we wish to use the theorem in the remainder of this and in the next section, we give here a direct proof from first principles. Later, after developing important general properties of arithmetical varieties, a short proof will be obvious.

First, suppose **A** is primal. Then the discriminator t on A, as defined in Section 1.2.1, is certainly a term function of **A** so, as observed in Section 1.2.1, $V(\mathbf{A})$ is arithmetical and also, as observed there, **A** is simple. Next, for each element $a \in A$ there must be unary term functions with constant value a, and from this it follows that **A** has no proper subalgebras and only the trivial automorphism. Thus a finite primal algebra satisfies all four conditions.

For the converse suppose the four conditions of the theorem hold for **A** and let $f : A^m \to A$ be any operation on A. Since **A** has no proper subalgebras the $V(\mathbf{A})$-free algebra $\mathbf{F} = \mathbf{F_A}(m)$ is isomorphic to a subdirect power in $\mathbf{A}^{|A|^m}$, one copy of **A** for each of the $|A|^m$ mappings of the set $\{x_1, \ldots, x_m\}$ of free generators into A. Each of these mappings, e_i, extends to a uniquely determined projection homomorphism p_i from **F** onto **A**. Representing F as equivalence classes of terms in the x_i, p_i is defined on F by

$$p_i(s(x_1, \ldots, x_m)) = s(e_i(x_1), \ldots, e_i(x_m))$$

where s is any representative of its equivalence class. Let π_i be the kernel of p_i. Since **A** is simple, each π_i is maximal in **Con A**. Hence $i \neq j$ implies either that $\pi_i \vee \pi_j = 1_{\mathbf{A}}$ or $\pi_i = \pi_j$. But this latter case is impossible for otherwise the distinct natural isomorphisms σ_i, σ_j of $\mathbf{F}/\pi_i = \mathbf{F}/\pi_j$ onto **A** induced by p_i and p_j would yield the nontrivial automorphism $\sigma_i^{-1}\sigma_j$ of **A**.

Finally, for each of the e_i, let s_i be a term for which

$$p_i(s_i(x_1, \ldots, x_m)) = f(e_i(x_1), \ldots, e_i(x_m)).$$

Since $\pi_i \vee \pi_j = 1_{\mathbf{A}}$ for $i \neq j$, by the strong Chinese remainder condition for **Con A** (Theorem 2.2.5) there is a term s in **F** such that $s \equiv s_i \ (\pi_i)$ for all i. This means that

$$s(e_i(x_1), \ldots, e_i(x_m)) = f(e_i(x_1), \ldots, e_i(x_m)) \text{ for all } i,$$

i.e., that f is a term function. Hence **A** is primal. (We have actually shown that **F** is isomorphic to $\mathbf{A}^{|A|^m}$; see Exercise 5 at the end of

Section 1.1.4.) •

In the proof of Theorem 3.1.2 we observed that if **A** is primal then the discriminator is a term function and this implies both that $V(\mathbf{A})$ is arithmetical and that **A** is simple. Hence we have the following alternative characterization of primality.

Corollary 3.1.3 *A finite algebra* **A** *is primal iff* **A** *is rigid, has no proper subalgebras, and the discriminator is a term function of* **A**.

Recall (Section 1.2.1) that a finite algebra is quasiprimal iff the discriminator is a term function. Thus quasiprimality directly generalizes primality.

As a consequence of the above characterizations of primality we also obtain the following characterizations of functional completeness due to H. Werner ([Werner, 1970]).

Theorem 3.1.4 *For a finite algebra* **A**, *the following are equivalent:*

(1) **A** *is functionally complete;*

(2) *the discriminator on* **A** *is a polynomial operation of* **A***;*

(3) **A** *is simple and has a Pixley polynomial.*

PROOF If **A** is functionally complete then certainly the discriminator is a polynomial of **A**. If the discriminator $t(x, y, z)$ is a polynomial of **A** then t itself is a Pixley polynomial for **A** and, by the same reasoning as above, **A** is simple. This proves the implications $(1) \Rightarrow (2) \Rightarrow (3)$.

For any algebra **A** recall that $\mathbf{A}^{+}$ is the algebra obtained from **A** by adding new nullary operations, one for each element of A and having constant value equal to that element. Notice that $\mathbf{A}^{+}$ has no proper subalgebras and is rigid. Also, we have observed that Con $\mathbf{A}^{+}$ = Con **A**. If **A** is simple and has a Pixley polynomial then by Theorem 3.1.2 $\mathbf{A}^{+}$ is primal; hence **A** is functionally complete. •

As an example to Theorem 3.1.4 recall from Section 1.2.1 that the term

$$t(x, y, z) = z + (x + (p - 1)z)(x + (p - 1)y)^{p^k - 1}$$

induces the discriminator on $GF(p^k)$; this directly demonstrates that finite fields are quasiprimal and hence functionally complete.

Finally, since the discriminator is a Pixley function, the variety generated by a primal algebra is arithmetical so that by Theorem 1.2.22 and Lemma 1.2.15, we have a direct short proof of Foster's Theorem (Theorem 3.1.1).

Further characterizations The characterizations above are essentially algebraic and are the most important for our subsequent purposes. There is, however, an extensive literature devoted to further characterizations; as suggested in Section 3.1.2 above, this work originated in the study of multiple valued logics. (The same is true for the origins of the theory of clones.) We survey some of these characterizations; as we might expect, they are typically more combinatorial and less algebraic than those above. The book [Pöschel, Kalužnin, 1979] includes many of these results. A good short survey, which we shall follow below, is due to B. Csákány ([Csákány, 1994]).

First, a simple *ad hoc* criterion.

Theorem 3.1.5 *A finite algebra* $\mathbf{A} = \langle A; F \rangle$ *is functionally complete iff the following conditions are satisfied:*

  (i) *There are elements* $0, 1$ *of* A *and binary polynomial operations* $+$ *and* $\cdot$ *of* $\mathbf{A}$ *satisfying*

$$x + 0 = 0 + x = x, \quad x \cdot 1 = x, \quad x \cdot 0 = 0$$

  for all $x \in A$;

  (ii) *For each* $a \in A$ *the characteristic function* χ_a *is a polynomial of* $\mathbf{A}$.

The algebra is primal iff in addition to (i) *and* (ii),

  (iii) *Each element of* A *is a term function.*

PROOF If the conditions (i) and (ii) hold and f is any m-ary operation on A, then for all $x_1, \ldots, x_m \in A$,

$$f(x_1, \ldots, x_m) = \sum_{a_1, \ldots, a_m \in A} \left(f(a_1, \ldots, a_m) \prod_{i=1}^{m} \chi_{a_i}(x_i) \right)$$

represents f as a polynomial. (We agree that in both the addition and multiplication, parentheses are nested from the left: $(((\cdots)\cdots)\cdots)$ so that no associative laws are needed.) If (iii) also holds then $+$ and as well as the constants $f(a_1,\ldots,a_m)$ in the formula above will be term functions of any desired arity. Hence f will be a term function so that **A** is primal. $\bullet$

As an example, in a Galois field $GF(p^k)$, the characteristic function of an element a is the polynomial $1 - (x - a)^{p^k - 1}$. If the ternary discriminator is a polynomial $t(x, y, z)$ of a nontrivial finite algebra we can choose 0 and 1 to be arbitrary distinct elements and define $x \cdot y = t(y, 1, x)$, $x + y = t(x, 0, y)$ and the characteristic function of an element $a \neq 0$ is $t(0, t(a, x, 0), 1)$ while $\chi_0(x) = t(0, x, 1)$. This is essentially Werner's proof of Theorem 3.1.4.

Theorem 3.1.5 leads directly to the following well-known theorem due to Sierpiński ([Sierpiński, 1945]):

Theorem 3.1.6 *If A is any nonempty set then every operation on A can be obtained as a composition of binary operations.*

PROOF For A finite this is a corollary of Theorem 3.1.5. If A is an infinite set the proof is totally different from the finite case and is generally nonconstructive, as might be expected. Specifically, since A is infinite, we can fix a particular (binary) bijection $b : A \times A \to A$. Having done this there then exists a pair of uniquely determined unary operations b_1^{-1}, b_2^{-1} on A (inverses of b) with the properties: for all $y, z, w \in A$,

$$b(y, z) = w \quad \Longleftrightarrow \quad y = b_1^{-1}(w) \text{ and } z = b_2^{-1}(w).$$

Now, for example, let $f(x, y, z)$ be a ternary operation on A. Then

$$f(x, y, z) = f(x, b_1^{-1}(b(y, z)), b_2^{-1}(b(y, z))),$$

so that $f(x, y, z)$ is given by the composition of $f(x, b_1^{-1}(w), b_2^{-1}(w))$ and $w = b(y, z)$, both binary. Likewise an operation of any arity can be produced first by a composition of a function of arity one less with $w = b(y, z)$, and continuing, finally by a composition of binary operations only. $\bullet$

Now let $P_n = \{0, 1, \ldots, n-1\}$ and let $x \cdot y = min(x, y)$ relative to the natural order and let $x' = x + 1 \pmod n$. Let 0 and $n-1$ in P_n play the respective roles of the 0 and 1 of Theorem 3.1.5. Letting $0^{(1)} = 0'$, $0^{(2)} = 0''$ etc., and defining $x + y = (x^{(n-1)} \cdot y^{(n-1)})'$, we see that $+$ and $\cdot$ satisfy (i) of Theorem 3.1.5 ($n-1$ replacing 1). Also $P_n = \{0, 0', \ldots, 0^{(n-1)}\}$ and 0 is defined by the unary term function $z(x) = x \cdot x' \cdots x^{(n-1)}$, so that all elements of P_n are given by term functions. Finally

$$\chi_{0^{(r)}}(0^{(k)}) = (z(x)' \cdot (0^{(k)})^{(n-r)})^{(n-1)},$$

(i.e., $\chi_{0^{(r)}}(0^{(k)}) = n-1$ for $k = r$ and 0 otherwise). Hence conditions (ii) and (iii) are also satisfied, and we conclude that the algebra $\langle P_n; \cdot, ' \rangle$ is primal. Therefore the basic Post algebra $\mathbf{P_n}$ of the last section is primal (Webb's result) and, as in the Boolean case, the operation $\vee$ is redundant.

Since any finite set can be totally ordered, Webb's discovery anticipated the finite part of Sierpiński's theorem. Even more is true: with $\cdot$ and $'$ defined as above, let f^k be the binary operations on the set P_n defined recursively by

$$f^1(x, y) = x \cdot y + 1 \pmod n \text{ and } f^k(x, y) = f^1(f^{k-1}(x, y), f^{k-1}(x, y)).$$

Then $f^1(x, x) = x'$ and $f^n(x, y) = x \cdot y$. This proves that the algebra $\langle P_n; f^1 \rangle$ is primal. It follows that for every finite set A every operation on A is representable by an operation generated from some single binary operation. Such an operation is usually called a *Sheffer operation*. (Of course we can then obviously find single generating operations for $O(A)$ of any higher arity, and these are also usually called Sheffer operations.) Hence to test a finite algebra for primality (or functional completeness) we need only check that some single Sheffer operation is a term function (or polynomial). On the two element Boolean algebra (which is term equivalent to $\mathbf{P_2}$) the operation $(x \wedge y)'$, the familiar "not-and" (nand) operation used in switching circuits, and its dual $(x \vee y)'$ (the "not-or" (nor) operation) are the only Sheffer operations. Notice that the operation f^1 defined above is just the nand.

Another class of functionally complete algebras arises from observing that the ternary discriminator is a pattern function, as defined in Section 1.2. The proof of the following theorem involves a fairly lengthy case-by-case analysis and we omit it.

Theorem 3.1.7 *If A is a finite set of more than two elements and f is any pattern function on A which is not a projection, then the algebra $\langle A; f \rangle$ is functionally complete.*

As an example, the ternary **dual discriminator function** on any set is the pattern function defined by $d(x, y, z) = x$ if $x = y$ and $d(x, y, z) = z$ if $x \neq y$. Hence any finite algebra of more than two elements having the dual discriminator as a polynomial is functionally complete. The requirement that the algebra have at least 3 elements is essential: on the two element lattice the majority term $(x \wedge y) \vee (x \wedge z) \vee (y \wedge z)$ is the dual discriminator, but this algebra is obviously not functionally complete—a sharp contrast with the ternary discriminator (Theorem 3.1.4). The dual discriminator was introduced in [Fried, Pixley, 1979]. We shall say more about it in Section 3.4.1.

Pattern functions on a set A obviously admit every permutation of A as an automorphism. Such functions, a larger class than the pattern functions, are called **homogeneous** and an algebra is called homogeneous if every permutation of its universe is an automorphism, which is the case iff each of its basic operations is homogeneous. A homogeneous algebra is nontrivial if its terms include some homogeneous operation which is not a projection. The following is a list of homogeneous algebras each of which is easily seen to fail to be functionally complete:

(1) $\langle \{0, 1\}; ' \rangle$ where $'$ is complementation;

(2) $\langle \{0, 1\}; s \rangle$ where s is the ternary *minority* function given by

$$s(x, y, z) = x + y + z \pmod 2;$$

(3) $\langle \{0, 1\}; s' \rangle$ where s' is the complement of s;

(4) $\langle \{0, 1\}; d \rangle$ where d is the dual discriminator;

(5) $\langle \{0, 1, 2\}; \circ \rangle$, the (unique) three element idempotent, commutative, nonassociative groupoid; $\circ$ can be defined by

$$x \circ y = 2x + 2y \pmod 3;$$

(6) the four element direct square $\langle\{0,1\}; s\rangle^2$ of the algebra (2) above.

Somewhat remarkably, this list is essentially complete; the following result is due to B. Csákány ([Csákány, 1994]). The proof, again, involves an exhaustive analysis of cases.

Theorem 3.1.8 *A finite homogeneous algebra which has at least one term function that is not a projection is functionally complete iff it is not weakly isomorphic to one of the algebras (1)–(6) above.*

In [Pálfy, Szabó, Szendrei, 1982] the authors begin with the remark that "as a rule a finite algebra with 'large' automorphism group is functionally complete" and develop this observation in a further significant extension of Theorem 3.1.8.

A final, and the most important, nonalgebraic characterization of primality follows from the description of all maximal clones on a finite set which is due to I. Rosenberg and which was described in Section 1.1.3. According to that classification (Theorem 1.1.5), for a finite set A, a clone M on A is maximal iff $M = clo(\{r\})$ for some relation r which is one of six specified types. Now it is not difficult to show that every proper clone on a finite set is contained in a maximal clone (Exercise 5 below). From this it follows that a set F of operations on A generates $O(A)$ iff for each maximal clone M on A there is an $f \in F$ with $f \notin M$. From this we have the following theorem, usually known as *Rosenberg's primality criterion*:

Theorem 3.1.9 *A finite algebra* **A** *is primal iff for each relation r of the six types of Theorem 1.1.5, there is a term function of* **A** *which does not belong to $clo(\{r\})$.*

Notice that since **A** is functionally complete iff $\mathbf{A}^+$ is primal, Rosenberg's criterion asserts that **A** is functionally complete iff for each relation r of the six types there is a polynomial function which is not a member of $clo(\{r\})$.

If the set has only two elements then the Post classification of maximal clones yields a corresponding simpler version of Rosenberg's criterion.

An algebra **A** on a finite set is called **preprimal** if Clo A is maximal, i.e., is of the form $clo(\{r\})$ for one of Rosenberg's six types.

Rosenberg's primality criterion has several significant applications, some of which we shall discuss a little later; a first and particularly interesting one is due to Rousseau ([Rousseau, 1967]). If the finite algebra **A** has only a single operation and this operation is at least binary, then Rousseau's result asserts that **A** is primal iff it is simple, rigid, and has no proper subalgebras, i.e., in the case of a single (at least binary) operation the condition that $V(\mathbf{A})$ be arithmetical is redundant in Theorem 3.1.2. This result is rather unexpected since the definition of primality apparently has nothing to do with the type of **A**. In the next section we shall give another application of Rosenberg's criterion.

Finally, since primal algebras have such special properties one might think that they almost never occur. But this is not so; in fact if we restrict our attention to groupoids (algebras with a single binary operation), let $I(n)$ be the total number of isomorphism types of groupoids of order n, and let $P(n)$ be the number of these which are primal, then an unpublished result of R. O. Davies shows that

$$\lim_{n \to \infty} \frac{P(n)}{I(n)} = \frac{1}{e}.$$

Thus, asymptotically, $1/e$, more than a third of all groupoids, are primal. Without going into the details of Davies' proof, we can (following an observation due to R. W. Quackenbush) get some insight into the result as follows: if groupoid $\langle A; \cdot \rangle$ has n elements then it has no one element subalgebras iff $a \cdot a \neq a$ for all $a \in A$. Hence the probability that a groupoid of n elements has no one element subalgebras is $(1 - 1/n)^n$ and the limit of this as $n \to \infty$ is $1/e$. For a groupoid to be primal more conditions than just the absence of one element subalgebras are necessary; hence if $\lim_{n \to \infty} P(n)/I(n)$ exists, it must be $\leq 1/e$. Davies thus shows that the limit does exist and has this upper bound as its value.

Continuing in this vein, V. L. Murskiǐ ([Murskiǐ, 1975]) obtained related results which imply that if one repeats this computation for quasiprimal algebras the resulting limit also exists and is 1. This obviously implies that, asymptotically, almost all finite algebras having at least one operation, which is at least binary, are quasiprimal and therefore almost all such finite algebras are functionally complete! Far from being rare, functionally complete algebras asymptotically predominate. We examine some aspects of their algebraic theory in the next section.

Exercises

1. Construct two (finite) primal algebras of the same finite type such that the variety they generate is neither CD nor CP. Construct two with the property that they generate a CD but not arithmetical variety.

2. Show that except in the case of Rousseau's result, i.e., of a type consisting of a single at least binary operation, the conditions CP and CD are each essential in Theorem 3.1.2.

3. Show that for infinite algebras the conditions of Theorem 3.1.2 are necessary but not sufficient for primality.

4. Let **A** be a finite primal algebra of finite type. Show how to explicitly construct a finite set Σ of equations of the given type such that all of the equations holding in **A** are logical consequences of the equations in Σ. (Compare Exercises 4 and 5 of Section 1.2.1.)

5. Show that every clone on a finite set A which is *proper* (i.e., not equal to $O(A)$) is contained in a maximal clone on A.

6. Show that every preprimal algebra is either quasiprimal or each element of the algebra is the constant value of some term function. (K. Denecke, *Preprimal algebras*, Akademie-Verlag, Berlin, 1982)

3.1.4   Functionally complete algebras

Werner's Theorem 3.1.4 asserts that quasiprimal algebras are special functionally complete algebras: the discriminator is not just a polynomial but is in fact a term function. Hence it is natural to ask about the relationship between quasiprimality and functional completeness. Also, functionally complete algebras are just the (finite) simple affine complete algebras. Hence we devote this section to functional completeness.

An obvious first question is: what can one say about the variety generated by a functionally complete algebra? The following example from [Kaarli, 1992] shows that in general we can say very little.

Example 3.1.10 *A functionally complete algebra* **A** *such that the variety* $V(\mathbf{A})$ *satisfies no nontrivial congruence equation.*

Let $A = \{a, b, c\}$ and define operations f, g and h on A as follows:

$$f(x, y, z, u) = \begin{cases} t(x, y, z) & \text{(the discriminator) if } u = b, \\ a & \text{if } u \neq b, \end{cases}$$

$$g(x) = \begin{cases} c & \text{if } x = b, \\ a & \text{if } x \neq b, \end{cases} \qquad h(x) = \begin{cases} b & \text{if } x = c, \\ a & \text{if } x \neq c. \end{cases}$$

Since the discriminator function is a polynomial of the algebra $\mathbf{A} = \langle A; f, g, h \rangle$, **A** is functionally complete. Evidently $\{a\}$ is the only proper subuniverse of **A**. Consider the subsets

$$B = \{(b, c), (a, a), (a, b), (a, c), (b, a), (c, a)\} \text{ and } C = B \backslash \{(b, c)\}$$

in $A \times A$. From the definitions of f, g and h, the inclusions

$$f(B^4), \ g(B), \ h(B) \subseteq C$$

easily follow. From these inclusions we see that B is a subuniverse of $\mathbf{A} \times \mathbf{A}$ and also that there is a congruence relation $\rho \in \operatorname{Con} \mathbf{B}$ such that $|B/\rho| = 2$ (specifically $(b, c)/\rho = \{(b, c)\}$ and $(a, a)/\rho = C$). Obviously a two element subalgebra cannot be a subdirect power of the three element algebra **A**; so certainly the variety generated by **A** is not $IP_S S(\mathbf{A})$ as would be so if **A** were quasiprimal. But more important, it is also apparent that in the algebra $\mathbf{A}/\rho$ all operations collapse to a single constant operation with value $(b, c)/\rho$. Hence $\mathbf{B}/\rho$ generates a variety term equivalent to the variety of pointed sets. It is well known that the latter variety does not satisfy any nontrivial congruence equation.

On the other hand, if we disallow all proper subalgebras we have the following strongly contrasting result. The proof here is from [Kaarli, 1992]. Another proof, an immediate corollary of a sharpened version of Rosenberg's primality criterion (Theorem 3.1.9), appears in [Szendrei, 1992].

Theorem 3.1.11 *A functionally complete algebra having no proper subalgebras generates an arithmetical variety.*

Later (Theorem 3.4.4) we shall see that, in fact, a functionally complete algebra having no proper subalgebras is even quasiprimal.

Before proving Theorem 3.1.11 we first prove the following simple proposition which distinguishes between primal and functionally complete algebras.

Proposition 3.1.12 *A functionally complete algebra* **A** *is primal iff every subuniverse of* **A** $\times$ **A** *contains the diagonal* Δ.

PROOF Let **A** be a primal algebra and $S < $ **A** $\times$ **A**. If $(a, b) \in S$ then, by primality, there is a unary term u such that $u(a) = u(b)$ and $(u(a), u(b)) \in S$. Hence S contains a diagonal element of $A \times A$. Since **A** has no proper subalgebras, S contains the whole diagonal Δ.

Conversely suppose that **A** is functionally complete and every subuniverse of **A** $\times$ **A** contains Δ. In order to prove that **A** is primal, it is sufficient to find a constant term function of **A**. Let u be a unary term of **A** with $u(A)$ of minimal size. If $|u(A)| \neq 1$ we pick distinct elements $a, b \in u(A)$. Since the subalgebra generated by (a, b) in **A** $\times$ **A** must contain the diagonal, there exists a unary term v such that $v(a) = v(b)$. Then however, the size of $vu(A)$ is strictly less than that of $u(A)$, contradicting the choice of u. Hence $|u(A)| = 1$ so u is a constant term function.   •

Proposition 3.1.12 suggests the following definition which will turn out to be important later: An algebra **A** is **weakly diagonal** if every subuniverse of **A** $\times$ **A** contains the graph of an automorphism of **A**. It is clear that a weakly diagonal algebra can have no proper subalgebras.

Now the proof of Theorem 3.1.11 follows. Since it introduces some important ideas, we present it in detail via a sequence of four lemmas.

Lemma 3.1.13 *Let* **A** *be a functionally complete algebra and let* **C** *be any subdirect product in* **A** $\times$ **A** *which is nontrivial (in the sense that neither of the projections is an isomorphism). Then for some* $b \in A$, *C contains all elements* (a, b) *with* $a \in A$. *(Geometrically, this means that C contains a horizontal line which projects onto A.)*

PROOF Since the projection of **C** onto the second subdirect factor is not an isomorphism, there are distinct a_1' and a_2' in A and a $b' \in A$

such that both (a_1', b') and (a_2', b') are in C. Choose the integer n so large that $2^n \geq |A|$ and let $f : A^n \to A$ be any function such that $f(\{a_1', a_2'\}^n) = A$. Since $\mathbf{A}$ is functionally complete f is a polynomial, i.e., for some $m \geq 0$, $\mathbf{a} \in A^m$, and $(n+m)$-ary term t, $t(\mathbf{x}, \mathbf{a}) = f(\mathbf{x})$ for all $\mathbf{x} \in A^n$. Since $\mathbf{C}$ is subdirect in $\mathbf{A} \times \mathbf{A}$, for each component a_i of $\mathbf{a}$ there is a $b_i \in A$ such that all (a_i, b_i), $i = 1, \ldots, n$, are in C. Finally, for any preassigned $a \in A$, choose $\mathbf{x}$ with components x_i in $\{a_1', a_2'\}$ and so that $f(\mathbf{x}) = a$. Then, since each $(x_i, b') \in C$, we see that for $b = t(b', \ldots, b', \mathbf{b})$,

$$
\begin{aligned}
(a, b) &= (t(\mathbf{x}, \mathbf{a}), t(b', \ldots, b', \mathbf{b})) \\
&= (t(x_1, \ldots, x_n, a_1, \ldots, a_m), t(b', \ldots, b', \mathbf{b})) \\
&= t((x_1, b'), \ldots, (x_n, b'), (a_1, b_1), \ldots, (a_m, b_m))
\end{aligned}
$$

is an element of C. $\bullet$

Lemma 3.1.14 *If $\mathbf{A}$ is functionally complete with no proper subalgebras then any subuniverse C in $A \times A$ is either the graph of an automorphism of $\mathbf{A}$ or C properly contains Δ. In particular, $\mathbf{A}$ is weakly diagonal.*

PROOF Since $\mathbf{A}$ has no proper subalgebras C is subdirect. If C is not the graph of an automorphism of $\mathbf{A}$, then C contains either (a_1, b) and (a_2, b) with $a_1 \neq a_2$ or (a, b_1) and (a, b_2) with $b_1 \neq b_2$. By symmetry we need only consider the first case. Then from the preceding lemma we have that for suitable b', all (a, b') with $a \in A$ are in C. Hence $(b', b') \in C$ and since $\mathbf{A}$ has no proper subalgebras, Δ is contained in C. $\bullet$

Lemma 3.1.15 *If $\mathbf{A}$ is functionally complete and weakly diagonal then there exists a unary term $u(x)$ with the property that all elements of $u(A)$ are automorphic images of each other.*

PROOF Let $u(x)$ be a unary term with the property that $u(A)$ has minimal possible size and suppose there exist different elements a and b in $u(A)$ which are not automorphic images of each other. Then the subuniverse in $A \times A$ generated by (a, b) is not the graph of an automorphism. By Lemma 3.1.14 this subuniverse must therefore contain Δ. Hence there is a unary term v such that $v(a) = v(b)$. But then $|vu(A)| < |u(A)|$, contradicting the choice of u. •

Lemma 3.1.16 *Every weakly diagonal functionally complete algebra generates an arithmetical variety.*

PROOF Let **A** be a weakly diagonal functionally complete algebra and let f be any Pixley function on A (for example the discriminator). We choose, by Lemma 3.1.15, a unary term $u(x)$ with the property that every two elements of $u(A)$ are automorphic images of each other. Since **A** has no proper subalgebras, for an arbitrary element $a \in u(A)$ we can find a 4-ary term $t(x, y, z, w)$ such that $f(x, y, z) = t(x, y, z, a)$ for all $x, y, z \in A$.

Now let b, c be arbitrary elements of A and choose an automorphism σ such that $u(b) = \sigma a$. Then we have

$$
\begin{aligned}
t(b, b, c, u(b)) = t(b, b, c, \sigma a) &= t(\sigma\sigma^{-1}b, \sigma\sigma^{-1}b, \sigma\sigma^{-1}c, \sigma a) \\
&= \sigma t(\sigma^{-1}b, \sigma^{-1}b, \sigma^{-1}c, a) \\
&= \sigma\sigma^{-1}c = c.
\end{aligned}
$$

Likewise we can also prove the identities

$$
t(x, y, x, u(x)) \approx x \quad \text{and} \quad t(x, y, y, u(x)) \approx x.
$$

Hence the term $t(x, y, z, u(x))$ induces a Pixley function on A. Therefore by Theorem 1.2.2 $V(\mathbf{A})$ is arithmetical. This completes the proof of the lemma and Theorem 3.1.11. •

In the remainder of this section we shall describe the functionally complete members of the varieties of groups and rings, both of which are congruence permutable. To obtain functionally complete algebras in these varieties it is convenient here to make use of the following lemma which can be found in [Burris, Sankappanavar, 1981] (page 179, Lemma 11.9).

Lemma 3.1.17 *Let $S = \{\mathbf{A}_1, \ldots, \mathbf{A}_n\}$ be a set of algebras in a congruence modular variety such that any subdirect product of any two (not necessarily distinct) members is skew free. Then any subdirect product of any subset of the algebras in S is skew free.*

The next theorem is a criterion for functional completeness in CP varieties due to Werner ([Werner, 1971]).

Theorem 3.1.18 *If* **A** *is a finite simple algebra in a congruence permutable variety then* **A** *is functionally complete iff* **A** × **A** *has no skew congruences (i.e.,* Con*(***A** × **A***) contains only the two projection kernels and* $0_{A \times A}$ *and* $1_{A \times A}$*).*

PROOF For the "if" direction we need only show that $\mathbf{A}^+$ is primal. Since $\mathbf{A}^+$ is simple, has no proper subalgebras and is rigid, we need only show that $V(\mathbf{A}^+)$ is CD. Now the free algebra **F** in $V(\mathbf{A}^+)$ with three free generators is isomorphic to a subdirect product in $(\mathbf{A}^+)^{|A|^3}$. (Since $\mathbf{A}^+$ is simple in a congruence permutable variety, **F** is actually isomorphic to some direct power $(\mathbf{A}^+)^n$.) By Theorem 1.2.13 the only subuniverses of the product $\mathbf{A}^+ \times \mathbf{A}^+$ are Δ and $A \times A$. Hence from the preceding lemma it follows that **F** has no skew congruences so that **Con F** is distributive (isomorphic with $\mathbf{2}^n$). By Corollary 1.2.4 it follows that $V(\mathbf{A}^+)$ is CD.

The "only if" direction follows from the fact that **A** functionally complete implies that $V(\mathbf{A}^+)$ is CD and in a CD variety no finite subdirect product can have skew congruences. •

Finally we require one more lemma to obtain examples of functionally complete algebras in CP varieties.

Lemma 3.1.19 *Let* **A** *be a simple algebra in a congruence permutable variety and let* $\theta \in$ Con$(\mathbf{A} \times \mathbf{A})$ *be other than either* $0_{A \times A}$ *or* $1_{A \times A}$, *or one of the projection kernels. Then a* θ-*block* C *is a subuniverse of* **A** × **A** *only if* C *is the graph of an automorphism of* **A**.

PROOF Let C be a θ-block, let p_i, $i = 1, 2$, be the projection homomorphisms of **A** × **A** onto **A**, and let π_i be their kernels. Since Con$(\mathbf{A} \times \mathbf{A})$ is modular it follows from the simplicity of **A** that Con$(\mathbf{A} \times \mathbf{A})$ has height 2 so that θ is a complement of each π_i; this means that

$$\theta \cap \pi_i = 0_A \quad \text{and} \quad \theta \circ \pi_i = 1_A, \quad i = 1, 2.$$

From the first of these equalities we have

$$(x, y_1) \equiv_\theta (x, y_2) \implies y_1 = y_2$$
$$\text{and} \quad (x_1, y) \equiv_\theta (x_2, y) \implies x_1 = x_2$$

which means that each of the projections, restricted to C, is 1-1.

From the second equality, for any $x, y, z, w \in A$,

$$(x, y) \equiv (z, w) \; (\theta \circ \pi_i), \quad i = 1, 2.$$

Hence for some $u_1, u_2 \in A$,

$$(x, y) \equiv_\theta (z, u_1) \quad \text{and} \quad (x, y) \equiv_\theta (u_2, w),$$

and this means that each projection, restricted to C, is onto A. Therefore, if p'_i is the restriction of p_i to C, then $\phi = p'_1 {p'_2}^{-1}$ is a bijection on A and $C = \{(x, \phi x) : x \in A\}$ is the graph of ϕ. If C is a subuniverse then ϕ is an automorphism. This proves the lemma. Notice that we have actually shown that each of the θ blocks has the same cardinality as A and the map $y \mapsto (x, y)/\theta$ is 1-1 from A onto $(A \times A)/\theta$.   •

Example 3.1.20 *Functionally complete groups and rings.*

Let $\mathbf{G}$ be a simple group. If θ is any congruence of $\mathbf{G} \times \mathbf{G}$, which is not $0_{G \times G}$, $1_{G \times G}$, or a projection kernel, then one of the θ-blocks is a normal subgroup $N = \{(x, \phi x) : x \in G\}$ of $\mathbf{G} \times \mathbf{G}$, where ϕ is some automorphism of $\mathbf{G}$. If e is the group identity and $y \in A$, then normality implies

$$(y, e)(x, \phi x)(y, e)^{-1} \in N \quad \text{for all} \; x \in A.$$

This means that $(yxy^{-1}, \phi x) \in N$ for all $x \in A$ and hence $yxy^{-1} = x$ for all $x \in A$ so that $\mathbf{G}$ is abelian. By the last lemma and Theorem 3.1.18 we conclude that each finite simple group is either functionally complete (iff it is nonabelian) or is cyclic of prime order. This characterization of functionally complete groups is due to W. Maurer and J. Rhodes (who gave a completely different proof in [Maurer, Rhodes, 1965]). Remarkably, G. Bergman has shown that there is actually a single term in the language of groups which, by only changing the constants, produces the discriminator in any finite simple nonabelian group ([G. Bergman, 1988]).

Likewise if $\mathbf{R}$ is any (not necessarily associative) ring, and θ is neither $0_{R \times R}$, $1_{R \times R}$, nor the kernel of a projection, then some θ-block $N = \{(x, \phi x) : x \in R\}$ is a two sided ideal of $\mathbf{R} \times \mathbf{R}$ where ϕ is an automorphism. Hence if 0 is the zero element of $\mathbf{R}$ and $r \in R$, then

$$(r, 0)(x, \phi x) = (rx, 0) \in N \quad \text{for all} \; x \in R.$$

Therefore $\phi(rx) = 0$; so $rx = 0$ for all $x \in R$. We conclude that every finite simple ring is either functionally complete or is a zero ring.

Finally we discuss a much sharper version of Theorem 3.1.18 which shows the significance of skew congruences. First, following [McKenzie, 1976] we define the concept of an affine algebra: let $\mathbf{A} = \langle A; F \rangle$ be an algebra and suppose there is a binary operation $+$ on A such that $\mathbf{G} = \langle A; + \rangle$ is an abelian group and such that for all operations $f \in F$, if f is k-ary and $\mathbf{x}, \mathbf{y} \in A^k$, then

$$f(\mathbf{x}) + f(\mathbf{y}) = f(\mathbf{x} + \mathbf{y}) + f(\mathbf{0})$$

where $\mathbf{0} = (0, \ldots, 0)$ and 0 is the identity of $\mathbf{G}$. In this case we say that $\mathbf{A}$ is **affine over** $\mathbf{G}$. In [McKenzie, 1976] the following theorem is proved:

Theorem 3.1.21 *Every finite simple algebra in a congruence permutable variety is either functionally complete or is affine over an elementary abelian p-group.*

This theorem was first used by McKenzie to show that every minimal locally finite CP variety has a finite base for its identities. More important, it was one of the early steps in the development of commutator theory as discussed in Section 1.2.6. (See also [McKenzie, 1978].) In the development of the commutator it was shown, for example, that in congruence modular varieties an algebra is abelian iff it is affine. While we are not interested in pursuit of these matters in this book, we sketch below the admirably brief proof of Theorem 3.1.21 obtained by Á. Szendrei ([Szendrei, 1981]). The proof is an excellent example of the power of Rosenberg's primality criterion (Theorem 3.1.9).

PROOF (Sketch) Szendrei's key observation is that it is an easy exercise to show that an algebra $\mathbf{A}$ is affine just exactly in case $\text{Pol}\,\mathbf{A} \subseteq clo(\{r\})$ for some *prime affine* relation (i.e., type (4) in Rosenberg's theorem).

Then to complete the proof suppose $\mathbf{A}$ is a functionally incomplete finite simple algebra having the term m as a Mal'cev operation. Then by Rosenberg's criterion $\text{Pol}\,\mathbf{A} \subseteq clo(\{r\})$ for at least one of the six types in Rosenberg's theorem. First observe that $clo(\{r\})$ with r

of type (2) (permutations with all cycles of the same prime length) or a unary central relation (unary of type (5)) fails to contain all constants. Hence Pol $\mathbf{A} \subseteq clo(\{r\})$ cannot hold for such a relation r. Nor can it hold for a nontrivial equivalence relation (type (3)) since $\mathbf{A}$ is simple. In order to show that Pol $\mathbf{A} \not\subseteq clo(\{r\})$ for all at least binary relations of types (1) (bounded partial orders), (5) (central relations), or (6) (determined by a k-regular family of equivalence relations), Szendrei exhibits three $3 \times k$ matrices, M, representing k-ary r of each of these three types, i.e., whose rows are in r, but for which the values of $m(M)$ are not in r. From this it follows that the only remaining possibility is that Pol $\mathbf{A}$ is necessarily contained in $clo(\{r\})$ for some prime affine relation; hence $\mathbf{A}$ is affine over an elementary abelian p-group. •

In Chapter 5 we shall discuss generalizations of the results and examples of this section to infinite algebras. For example, in that setting an important extension of Theorem 3.1.21 ([Gumm, 1979]) will still be in force where "functional completeness" is replaced by "each finite partial function has an interpolating polynomial".

3.2 Near unanimity varieties

In this section we begin our development of material needed to profitably generalize primality and thus lead to Chapter 4.

A variety V is a **near unanimity variety** if V has a near unanimity term. We have already proved, and in Chapter 2 used, Theorem 1.2.11, which asserts that a near unanimity variety is CD. In order to develop most generalizations of primality we need to examine near unanimity varieties more systematically.

3.2.1 Characterizations

Near unanimity varieties were first studied by A. Huhn ([Huhn, 1972]) who proved the following theorem.

Theorem 3.2.1 *For a variety V and an integer $n \geq 2$ the following conditions are equivalent:*

(0) *For all algebras* $\mathbf{A} \in V$ *and congruences* $\theta_0, \ldots, \theta_n$ *of* $\mathbf{A}$,

$$\theta_0 \circ \bigwedge_{i=1}^{n} \theta_i = \bigwedge_{j=1}^{n} \left(\theta_0 \circ \bigwedge_{\substack{i=1 \\ i \neq j}}^{n} \theta_i\right).$$

(1) *V has an* $(n+1)$-*ary near unanimity term* $u(x_0, \ldots, x_n)$.

(2) *Every algebra* $\mathbf{A}$ *in* V *satisfies the following weak Chinese remainder condition: for any integer* $r \geq n$, *any* $a_1, \ldots, a_r \in A$, *and any congruence relations* $\theta_1, \ldots, \theta_r$ *of* $\mathbf{A}$, *if the congruences*

$$x \equiv a_i \ (\theta_i), \quad 1 \leq i \leq r,$$

are solvable n *at a time, then they are simultaneously solvable.*

PROOF We shall prove the theorem only for the case $n = 2$; only the notation is more complex for the general case. We proceed cyclically.

(0) $\Rightarrow$ (1). For $n = 2$, (0) reads

$$\theta_0 \circ (\theta_1 \wedge \theta_2) = (\theta_0 \circ \theta_2) \wedge (\theta_0 \circ \theta_1).$$

Let $\mathbf{F}$ be the V-free algebra with free generators x, y, z. Then the pair (x, z) is in both $\mathrm{Cg}(x, y) \circ \mathrm{Cg}(y, z)$ and $\mathrm{Cg}(x, y) \circ \mathrm{Cg}(x, z)$. Hence by (0) (x, z) is in $\mathrm{Cg}(x, y) \circ (\mathrm{Cg}(y, z) \wedge \mathrm{Cg}(x, z))$ so there exists a term $u(x, y, z)$ (an element of $\mathbf{F}$) such that

$$x \equiv u(x, y, z) \ (\mathrm{Cg}(x, y)),$$
$$z \equiv u(x, y, z) \ (\mathrm{Cg}(y, z) \wedge \mathrm{Cg}(x, z)).$$

By the same argument as in the proof of Theorem 1.2.2 we conclude that

$$u(x, x, z) \approx x, \quad u(x, z, z) \approx z, \quad u(z, y, z) \approx z$$

are identities of V.

(1) $\Rightarrow$ (2). Assume that V has a 3-ary near unanimity term u and that the congruences $x \equiv a_i \ (\theta_i)$, $i = 1, 2, 3$, have pairwise solutions a_{12}, a_{13}, a_{23}. Therefore $u(a_{12}, a_{13}, a_{23}) \equiv_{\theta_1} u(a_1, a_1, a_{23}) = a_1$. Likewise $u(a_{12}, a_{13}, a_{23})$ solves the other two congruences. If we are given four pairwise solvable congruences we first obtain solutions a_{ijk} for each triple and then observe that, for example, $u(a_{123}, a_{124}, a_{134})$

solves all four. Continuing, by induction it is clear that once we have obtained solutions of r congruences $r - 1$ at a time, then if s_1, s_2, s_3 are any three of the $(r - 1)$-wise solutions, $u(s_1, s_2, s_3)$ solves all r. (We are using the fact that for a set of $r > 3$ elements, any three subsets, each having at least $r - 1$ elements, have an element in common.)

    $(2) \Rightarrow (0)$. Note that the left side of the congruence equation of (0) is contained in the right; consequently we need only show that the right side is contained in the left. To this end suppose the pair (a, c) is contained in $(\theta_0 \circ \theta_2) \wedge (\theta_0 \circ \theta_1)$. Then we have elements b_{01}, b_{02} satisfying the congruences:

$$b_{02} \equiv a \ (\theta_0), \quad b_{02} \equiv c \ (\theta_2), \quad b_{01} \equiv a \ (\theta_0), \quad b_{01} \equiv c \ (\theta_1). \qquad (3.1)$$

Now consider the system of congruences

$$z \equiv a \ (\theta_0),$$
$$z \equiv c \ (\theta_2),$$
$$z \equiv c \ (\theta_1),$$

with z as unknown. Clearly c solves the last two congruences. It follows from (3.1) that b_{02} solves the first two congruences and b_{01} solves the first and the third congruence. Consequently, by (2) the system has a simultaneous solution $b \in A$. But then $a \equiv b \ (\theta_0)$ and $b \equiv c \ (\theta_1 \wedge \theta_2)$ so that the pair (a, c) is in the left side of the congruence equation. •

    For our subsequent purposes, the following theorem provides three more conditions, each equivalent to each of the conditions of Theorem 3.2.1, and thus equivalent to the variety having a near unanimity term. These appear in [Baker, Pixley, 1975] and [Pixley, 1979].

    Recall that if X is a finite subset of A^m, a function $f : X \to A$ is a *finite partial operation* on A. A partial operation f is **interpolated** by a function g provided the domain of g contains the domain of f and both functions agree wherever they are both defined.

Theorem 3.2.2 *For a variety V and an integer $n \geq 2$, each of the following conditions is equivalent to the existence of an $(n + 1)$-ary near unanimity term for V (and hence is equivalent to each of the conditions (0), (1), (2) of Theorem 3.2.1). Each of the conditions is for any algebra $\mathbf{A}$ in V, integer $m \geq 1$, and m-ary finite partial operation f on A.*

(3) *If the restrictions of f to each subset of its domain of n or fewer elements have interpolating term functions, then f has an interpolating term function.*

(4) *f has an interpolating term function iff all subuniverses of $\mathbf{A}^n$ are closed under f (where defined).*

(5) *f has an interpolating polynomial function iff all diagonal subuniverses of $\mathbf{A}^n$ are closed under f (where defined).*

PROOF $(1) \Rightarrow (3)$. The proof is by induction on the cardinality $|dom\ f|$, of the domain of f. The cases $|dom\ f| \leq n$ are trivial. Suppose the assertion holds for $n \leq |dom\ f| < r$, and consider the case $|dom\ f| = r$. By the induction hypothesis, for each integer $i = 1, \ldots, r$ there exists an m-ary term function t_i which agrees with f on each of the $(r-1)$ domain elements other than the i-th. Let $t = u(t_1, \ldots, t_{n+1})$, be the composition of the near unanimity term with $t_1, \ldots, t_{n+1}$. By the near unanimity property of u, t must agree with f on *every* domain element, as desired. (If $r > n+1$ the terms $t_{n+2}, \ldots, t_r$ are superfluous and in fact we could have obtained t as a composition of u with any $n+1$ of the t_i, i.e., in general there are many choices for the interpolating term.)

$(3) \Leftrightarrow (4)$ Recall from Section 1.1.3 that the "closure" of a subuniverse S of $\mathbf{A}^n$ under an m-ary f (where defined) means that for any $m \times n$ matrix M whose rows are n-tuples in S and whose columns are m-tuples in $dom f$, S contains the n-tuple $f(M)$ obtained by applying f to each column of M. (S is a diagonal subuniverse if it contains the "diagonal" Δ of all constant n-tuples, $\{(a, \ldots, a) : a \in A\}$.)

Hence the closure condition of (4) is equivalent to the condition: for any $m \times n$ matrix M with entries in A and columns in $dom f$, the row vector $f(M)$ is in the subuniverse of $\mathbf{A}^n$ generated by the rows of M, and this is equivalent to the condition that $f(M) = t(M)$ for some m-ary term t. But this means precisely that the term function t interpolates f on the n elements of $dom f$ consisting of the columns of M. (If fewer than n domain elements of f were given then M could be filled out by repeating columns.) Thus statements (3) and (4) are just different ways of stating the same condition.

$(4) \Rightarrow (1)$. Let $\mathbf{F}$ be the free algebra of V with free generators x and y. Consider the finite partial $(n+1)$-ary function f on F whose domain consists of $(n + 1)$-tuples

$$(x, \ldots, x, y), (x, \ldots, x, y, x), \ldots, (y, x, \ldots, x),$$

and which assigns to each of them the value x. Now, if M is an arbitrary $(n + 1) \times n$ matrix with all columns in $dom\ f$ then clearly M contains a row $(x, \ldots, x)$ and also $f(M) = (x, \ldots, x)$. This means that f preserves all subuniverses of $\mathbf{F}^n$. Consequently, condition (4) yields the existence of an $(n + 1)$-ary term u such that $u(\mathbf{b}) = x$ for every $\mathbf{b} \in dom\ f$. Since x and y are free generators, this implies that u is a near unanimity term for V.

(1) $\Rightarrow$ (5). Let $\mathbf{A} \in V$ and f be an m-ary finite partial function on A which preserves all diagonal subuniverses of $\mathbf{A}^n$. If V has an $(n + 1)$-ary near unanimity term then so does the variety $V(\mathbf{A}^+)$, for every $\mathbf{A} \in V$. Since f preserves all diagonal subuniverses of $\mathbf{A}^n$, it preserves all subuniverses of $(\mathbf{A}^+)^n$. Applying the implication (1) $\Rightarrow$ (4) to the variety $V(\mathbf{A}^+)$, we see that f must be interpolated on its domain by a term function of $\mathbf{A}^+$. The latter, however, is a polynomial function of $\mathbf{A}$. Thus V satisfies condition (5).

(5) $\Rightarrow$ (1). The idea of the proof is the same as that of the proof of implication (4) $\Rightarrow$ (1). The difference is that here we need more free generators. Let $\mathbf{F}$ be the V-free term algebra with $n + 1$ free generators $x_0, \ldots, x_n$ (whose elements are $(n + 1)$-ary terms $t(x_0, \ldots, x_n)$). Consider the $(n + 1)$-ary finite partial operation f on F as follows: $dom\ f$ consists of the following $n + 1$ elements of F^{n+1}:

$$(x_0, x_1, \ldots, x_1),$$
$$(x_0, x_1, x_0, \ldots, x_0),$$
$$(x_0, x_0, x_2, x_0, \ldots, x_0),$$
$$\cdots\cdots\cdots$$
$$(x_0, \ldots, x_0, x_n);$$

f has value x_1 on the first of these $(n + 1)$-tuples and value x_0 on the remaining n of them. Again, if we have an $(n + 1) \times n$ matrix M with columns from $dom\ f$ then $f(M)$ always coincides with one of the rows of M.

Therefore f can be interpolated by a polynomial function of $\mathbf{F}$; this means that for some $k \geq 0$ there is an $(k + n + 1)$-ary term t of V and elements (terms)

$$t_1(x_0, \ldots, x_n), \ldots, t_m(x_0, \ldots, x_n)$$

of **F** such that

$$t(t_1(x_0, \ldots, x_n), \ldots, t_k(x_0, \ldots, x_n), x_0, x_1, x_1, x_1, \ldots, x_1) = x_1,$$
$$t(t_1(x_0, \ldots, x_n), \ldots, t_k(x_0, \ldots, x_n), x_0, x_1, x_0, x_0, \ldots, x_0) = x_0,$$
$$t(t_1(x_0, \ldots, x_n), \ldots, t_k(x_0, \ldots, x_n), x_0, x_0, x_2, x_0, \ldots, x_0) = x_0,$$
$$\cdots \cdots \cdots \cdots \cdots \quad (3.2)$$
$$t(t_1(x_0, \ldots, x_n), \ldots, t_k(x_0, \ldots, x_n), x_0, x_0, x_0, \ldots, x_0, x_n) = x_0$$

are identities of V. Now put

$$u(x_0, \ldots, x_n) = t(t_1(x_0, \ldots, x_n), \ldots, t_k(x_0, \ldots, x_n), x_0, x_1, \ldots, x_n).$$

Then the near unanimity identities required by statement (1) are particular instances of equations (3.2) and hence are also equations of V. Therefore (5) implies (1). •

3.2.2 More properties of near unanimity varieties

In paragraphs 1–4 below we discuss some additional properties of near unanimity varieties.

1. The combination of Theorems 3.2.1 and 3.2.2 shows that for a variety V and integer n the algebraic conditions in (0)–(5) are equivalent and each constitutes a "strong" Mal'cev condition for V. Other equivalent Mal'cev conditions are known. For example, for a given n another interesting equivalence is

> *If **A** is a subalgebra of a direct product $\mathbf{C}_1 \times \cdots \times \mathbf{C}_r$, $r \geq n$, $\mathbf{C}_i \in V$, then **A** can be uniquely determined from the knowledge of its n-fold projections, i.e., from its images under projection into the products $\mathbf{C}_{i_1} \times \cdots \times \mathbf{C}_{i_n}$, $i_1 < \cdots < i_n$.*

See [Baker, Pixley, 1975] for the proof of this equivalence. In addition G. Bergman has shown (under the same hypotheses as above) that this "uniqueness" result is complemented by a corresponding "existence" result ([G. Bergman, 1977]:

> *Suppose that $\mathbf{C}_1, \ldots, \mathbf{C}_r \in V$ and that for every n-tuple $(i_1, \ldots, i_n)$ we are given a subalgebra of $\mathbf{C}_{i_1} \times \cdots \times \mathbf{C}_{i_n}$. Then these data come (as projections, described above)*

from a subalgebra of $\mathbf{C}_1 \times \cdots \times \mathbf{C}_r$ *iff they satisfy certain natural consistency conditions.*

For the case $n = 2$ the consistency conditions are: if for the 2-tuple (i, j) the given subalgebra of $\mathbf{C}_i \times \mathbf{C}_j$ has universe $C(i, j)$ then it is required that for all $i, j, k \leq r$,

$$C(i, k) \subseteq C(i, j) \circ C(j, k), \ \ C(i, j) = C(j, i)^{\smile}, \ \ C(i, i) \subseteq \Delta,$$

where $\smile$ denotes converse. Bergman also discusses conditions under which his existence result implies the conditions of Theorems 3.2.1 and 3.2.2.

2.   The equivalence of statements (1), (4), and (5) of Theorems 3.2.1 and 3.2.2 has the following important consequence for finite algebras.

Corollary 3.2.3 *Let* $\mathbf{A}$ *be any finite algebra having an* $(n + 1)$-*ary near unanimity term. Then for any positive integer* m, *an* m-*ary total operation* f *on* A *is a term function of* $\mathbf{A}$ *iff every subuniverse of* $\mathbf{A}^n$ *is closed under* f. *The operation* f *is a polynomial iff every diagonal subuniverse of* $\mathbf{A}^n$ *is closed under* f.

Indeed, it suffices to simply regard f as a finite partial function.

Recall from Section 1.1.4 that for any algebra $\mathbf{A}$ the $V(\mathbf{A})$-free algebra with κ free generators is a subalgebra of $\mathbf{A}^{|A|^{\kappa}}$. For this reason $f : A^m \to A$ is a term function of $\mathbf{A}$ iff all subuniverses of $\mathbf{A}^{|A|^m}$ are closed under f. If $\mathbf{A}$ is finite this can be checked by a finite number of tests for any given m, but no upper bound applies to all m. In contrast, if $\mathbf{A}$ has an $(n + 1)$-ary near unanimity term, then for *all* m, only subuniverses of $\mathbf{A}^n$ need be checked. For example, for lattices, which have a ternary majority term, a function f (of any arity) on a finite lattice $\mathbf{L}$ will be a lattice term function iff all sublattices of $\mathbf{L} \times \mathbf{L}$ are closed under f.

The actual construction of a term function on a finite algebra with $(n + 1)$-ary near unanimity term u is accomplished by the interpolation process given in statement (3). If $dom\ f = \{\mathbf{a}^1, \ldots, \mathbf{a}^r\}$ and if each of the terms t_i, $i = 1, \ldots, n + 1$, interpolates f on all of $\{\mathbf{a}^1, \ldots, \mathbf{a}^{n+1}\}$ except $\mathbf{a}^i$, then $s = u(t_1, \ldots, t_{n+1})$ interpolates f on all of $\{\mathbf{a}^1, \ldots, \mathbf{a}^{n+1}\}$. (The t_i are m-ary if f is m-ary.) We put

$s_{n+2} = s$ and continue, computing $s_1, s_2, \ldots, s_{n+1}$ so that each of the s_i interpolates f on all of $\{\mathbf{a}^1, \ldots, \mathbf{a}^{n+2}\}$ except $\mathbf{a}^i$; so u composed with any $n+1$ of the s_i interpolates f on all of $\{\mathbf{a}^1, \ldots, \mathbf{a}^{n+2}\}$. Hence interpolating on $n+1$ points of the domain requires 1 use of u, on $n+2$ points requires $1 + (n+1)$ uses, on $n+3$ points, $1 + (n+1) + (n+1)^2$ uses, etc., so that altogether

$$1 + (n+1) + \cdots + (n+1)^{r-(n+1)} = [(n+1)^{r-n} - 1]/n$$

uses of u are required. The "depth" of the composition is evidently $r - n$.

As an example of the potential complexity of the interpolating term, if $\mathbf{L}$ is a 10 element lattice and f is a ternary function on L (which preserves all sublattices of $\mathbf{L} \times \mathbf{L}$), then $|dom\ f| = 1000$ and f is represented by a ternary term which is a composition of a majority term u for lattices. The depth of the composition is 998 and involves $(3^{998} - 1)/2$ uses of u.

As a practical matter it is entirely possible, at least in some special situations, that the interpolation of a finite partial operation may be accomplished by a composition involving fewer uses of the near unanimity term than in the computation above, if some other strategy is devised. There seems to be little information known on this subject.

3. Another important consequence of the equivalence of statements (1) and (4) is for the theory of clones. If A is a finite set then it is known that a clone on A need not be finitely generated and in general, for a given clone, there seems to be no known criterion for determining precisely when it is finitely generated. Hence the following result is interesting since it provides a simple sufficient condition for finite generation. Corollary 3.2.5 will be of great importance in Chapter 4.

Theorem 3.2.4 *If a clone on a finite set contains an $(n+1)$-ary near unanimity function, then it is finitely generated.*

PROOF If C is a clone on a finite set A, then C is the set of term functions of the algebra $\mathbf{A} = \langle A; C \rangle$. Since A is finite, there are only finitely many subsets in A^n and hence only finitely many which are *not* subuniverses of $\mathbf{A}^n$. For each such nonsubuniverse B, pick an

$f \in C$ such that B is not closed under f. These f together with the near unanimity function of C will then generate a clone $D \subseteq C$. Then the algebra $\langle A; D \rangle^n$ has exactly the same subuniverses as $\mathbf{A}^n$; so by the implication (1) $\Rightarrow$ (4) we have $C = D$.   •

Corollary 3.2.5 *If* $\mathbf{A}$ *is a finite algebra which has a near unanimity term, then* $\mathbf{A}$ *is term equivalent to an algebra of finite type.*

In Chapter 4 we shall make an important application of this corollary to algebras which, as presented, may have infinite type.

4. The criterion of Corollary 3.2.3 for expressibility as a term function does not generally extend to infinite algebras. For an explicit counterexample (for lattices) see [Baker, Pixley, 1975]. On the other hand we do have the following general condition for locally finite varieties.

Theorem 3.2.6 *If* V *is a locally finite variety having an* $(n+1)$*-ary near unanimity term, then for any* $\mathbf{A}$ *in* V*, a finitary function* f *on* A *is a term function of* $\mathbf{A}$ *iff every subuniverse of* $\mathbf{A}^n$ *is closed under* f.

PROOF Let $f : A^m \to A$ and suppose every subuniverse of $\mathbf{A}^n$ is closed under f. Let $\mathbf{F}$ be the free algebra of the variety $V(\mathbf{A})$, with m free generators. Since V is locally finite, so is $V(\mathbf{A})$ and therefore $\mathbf{F}$ is finite. Hence $\mathbf{F}$ can be embedded in a suitable direct power $\mathbf{A}^k$. Let $\mathbf{b}_i = (b_i^1, \ldots, b_i^k)$, $i = 1, \ldots, m$, be the k-tuples in A^k which are free generators of the subalgebra they generate in $\mathbf{A}^k$. Denote $\mathbf{b}^j = (b_1^j, \ldots, b_m^j)$, $j = 1, \ldots, k$, and let $G = \{\mathbf{b}^1, \ldots, \mathbf{b}^k\}$. Since G is finite, by the implication (1) $\Rightarrow$ (4) of Theorem 3.2.2, f is interpolated by some term t on G. If $\mathbf{c}$ is any other element in A^m then f is also interpolated by some term s on $G \cup \{\mathbf{c}\}$. Then s and t agree on G which implies

$$s(\mathbf{b}_1, \ldots, \mathbf{b}_m) = (s(\mathbf{b}^1), \ldots, s(\mathbf{b}^k))$$
$$= (t(\mathbf{b}^1), \ldots, t(\mathbf{b}^k)) = t(\mathbf{b}_1, \ldots, \mathbf{b}_m).$$

We see that s and t agree on free generators of a free algebra of $V(\mathbf{A})$; hence $s \approx t$ is an identity of $V(\mathbf{A})$. In particular, $t(\mathbf{c}) = s(\mathbf{c}) = f(\mathbf{c})$. Since $\mathbf{c} \in A^m$ is arbitrary, we conclude that f is a term function.   •

This proof actually shows that for *any* locally finite variety, finite interpolation by term functions implies global term function interpolation. Notice that if all diagonal subuniverses of $\mathbf{A}^n$ are closed under f the above proof does *not* allow us to conclude that f is a polynomial: agreement of two polynomials on G does not imply their agreement everywhere.

Theorem 3.2.6 applies, in particular, to finitely generated varieties. For example if $\mathbf{L}$ is any distributive lattice and all subuniverses of $\mathbf{L} \times \mathbf{L}$ are closed under f, then f is a lattice term function.

3.3 Arithmetical varieties

The six statements (0)-(5) of Theorems 3.2.1 and 3.2.2 combined give several important conditions characterizing varieties with near unanimity terms. In this section we specialize these theorems to obtain parallel characterizations of arithmetical varieties (Theorems 3.3.1 and 3.3.3 combined). For these and additional equivalences refer to [Pixley, 1979].

We have noticed in the discussion following Theorem 1.2.3 that if a variety is congruence permutable then it is congruence distributive iff it has a ternary majority term and this is the strongest form of Theorem 1.2.3. Since the existence of a near unanimity term for a variety implies congruence distributivity it follows that in the presence of congruence permutability if a variety has any near unanimity term it has a ternary majority term. Thus the following companion to Theorem 3.2.1 is not surprising; as noted earlier, what is surprising is that the three equivalent statements are formally the same as for finite equivalence lattices (Theorem 2.2.6).

Theorem 3.3.1 *For a variety V the following are equivalent:*

(0) *V is arithmetical, i.e., for all $\mathbf{A} \in V$ and congruences $\theta_0, \theta_1, \theta_2$ of $\mathbf{A}$*

$$\theta_0 \circ (\theta_1 \wedge \theta_2) = (\theta_1 \circ \theta_0) \wedge (\theta_2 \circ \theta_0);$$

(1) *V has a Pixley term;*

(2) *For each $\mathbf{A} \in V$, $\mathbf{Con}\,\mathbf{A}$ satisfies the strong Chinese remainder condition (SCRC).*

PROOF Conditions (0) and (2) are equivalent by Theorem 2.2.1; the equivalence of these with (1) follows from Theorem 1.2.2.   •

Now we formulate a theorem characterizing arithmetical varieties and which is analogous to Theorem 3.2.2. In that theorem we saw that conditions (3) and (4) were just restatements of the same condition; that is, n-wise interpolation is the same thing as closure of subuniverses of $\mathbf{A}^n$. The same is true in the context of arithmeticity; hence we state only two conditions in our analogous theorem (Theorem 3.3.3 below).

As in the case of Theorem 3.3.1 we anticipate a version of our analog for single algebras. Indeed we shall establish our analog by first proving Theorem 3.3.2 below. This theorem can also be obtained as a corollary from more general results in [Hagemann, Herrmann, 1982] but the proof we give below is much more direct and, in fact, is almost immediate from Lemma 2.2.3 of Chapter 2. Recall that Lemma 2.2.3 asserts:

> *If* $\mathbf{L}$ *is any equivalence lattice on a set* A, *then for each set of the form*
>
> $$X = \{(x,y,x),(x,y,y),(z,y,z),(y,y,z)\} \subseteq A^3,$$
>
> *the finite partial operation* f *defined on* X *by*
>
> $$f(x,y,x) = f(x,y,y) = x \text{ and } f(z,y,z) = f(y,y,z) = z$$
>
> *is* $\mathbf{L}$*-compatible. If such* f *can always be* $\mathbf{L}$*-compatibly extended to* $X \cup \{(x,y,z)\}$, *then* $\mathbf{L}$ *satisfies the (SCRC).*

Now if $\mathbf{A}$ is an algebra then the partial operation f is certainly $\mathbf{Con\,A}$-compatible. Hence if each $\mathbf{Con\,A}$-compatible operation can be interpolated by a polynomial, it follows that $\mathbf{Con\,A}$ satisfies the (SCRC), i.e., $\mathbf{A}$ is arithmetical. Thus we have established statement (a) of the following theorem.

Theorem 3.3.2   (a) *If an algebra* $\mathbf{A}$ *satisfies the condition*

> *For any finite partial operation* f *on* A, f *has an interpolating polynomial iff* f *is* $\mathbf{Con\,A}$*-compatible (where defined),*

then **A** *is arithmetical.*

(b) *If* **A** *satisfies the condition*

> *For any finite partial operation f on A, f has an interpolating term function iff f is compatible with all rectangular subuniverses in $A \times A$ (i.e., all rectangular subuniverses in $A \times A$ are closed under f, where defined),*

then all subalgebras of **A** *are arithmetical.*

PROOF of (b). Let **B** < **A** and let $X \subseteq B^3$ be the same as in Lemma 2.2.3. We first verify that the function f of the lemma—which we know is **Con B**-compatible—is actually compatible with all rectangular subuniverses of **A** × **A**. This verification is only a slight variation of the proof of Lemma 2.2.3: recall that in that proof, to verify the **L**-compatibility of f, only the single case where $\mathbf{a} = (x, y, y)$ and $\mathbf{b} = (y, y, z)$ required any particular attention; namely, we used the transitivity of an equivalence relation to conclude that $(x, y), (y, y), (y, z) \in L$ implies $(x, z) \in L$. But also in the present situation only this case requires attention and here any rectangular subuniverse of **A** × **A** which contains $(x, y), (y, y), (y, z)$ must also contain (x, z). Hence f is compatible with all rectangular subuniverses of **A** × **A** and therefore, by the condition of (b), has an interpolating term function. This term function is a **Con B**-compatible extension of f to $X \cup \{(x, y, z)\}$; so by Lemma 2.2.3 **B** is arithmetical. ●

We notice that the converses of the statements of Theorem 3.3.2 are false: for example let **A** be the two element algebra with no operations, which is obviously arithmetical, but the conditions of both (a) and (b) are not satisfied. For varieties we do have the full equivalence; in particular here is our analog of Theorem 3.2.2. It provides the means for obtaining the most common generalizations of primality.

Theorem 3.3.3 *For a variety V each of the following conditions is equivalent to arithmeticity (and hence is equivalent to each of the conditions of Theorem 3.3.1). Each of the conditions is for every* *algebra* **A** *in V, integer $m \geq 1$, and m-ary finite partial operation f on A.*

(3) *f has an interpolating term function iff all rectangular subuniverses of* **A** × **A** *are closed under f (where defined);*

(4) *f has an interpolating polynomial function iff f is* **Con A***-compatible (where defined).*

PROOF Each of the statements (3) and (4) implies that V is arithmetical by Theorem 3.3.2. If V is arithmetical then by Fleischer's theorem (Theorem 1.2.13) all subuniverses of **A** × **A** are rectangular; so (3) follows from the corresponding statement of Theorem 3.2.2 with $n = 2$.

To prove that (3) implies (4), suppose f is **Con A**-compatible for **A** $\in V$. Since (3) implies arithmeticity, it follows that the variety generated by $\mathbf{A}^+$ is arithmetical. Also the subuniverses of $\mathbf{A}^+ \times \mathbf{A}^+$ are just the diagonal subuniverses of **A** × **A** and, by Theorem 1.2.13 again, the rectangular subuniverses of $\mathbf{A}^+ \times \mathbf{A}^+$ are just the congruences of **A**. Hence f has an interpolating term function of $\mathbf{A}^+$ and thus an interpolating polynomial of **A**.   •

Condition (4) of the preceding theorem is one of the most important characterizations of arithmetical varieties; certainly this is true in the context of the present book, for reasons we shall see shortly. For locally finite varieties Theorem 3.2.6 (for varieties with near unanimity term) directly specializes:

Theorem 3.3.4 *If V is a locally finite arithmetical variety, then for any* **A** *in V, $f : A^m \to A$ is a term function of* **A** *iff every rectangular subuniverse of* **A** × **A** *is closed under f.*

For example, as we observed in Section 1.2.1, any variety of arithmetical rings is finitely generated ([Michler, Wille, 1970]); hence this test applies in that setting.

3.4 Generalizations of primality

3.4.1 Generalizations involving term function characterizations

For an algebra **A** recall that we have the following important clones on A:

$$o(A) \subseteq \text{Clo } \mathbf{A} \subseteq \text{Pol } \mathbf{A} \subseteq \text{Comp } \mathbf{A} \subseteq O(A).$$

Now Theorem 3.3.3 asserts that for arithmetical varieties the test for an operation to be a term function of a finite algebra is that it preserve all (rectangular) subuniverses of $\mathbf{A} \times \mathbf{A}$. Hence the most obvious generalizations of primality are obtained by restricting ourselves to arithmetical varieties and requiring that the term functions be characterized by some simple preservation properties; specifically this means that we require, for a finite algebra **A**, that all of the functions in $clo(R)$ be representable by functions in Clo **A** where R is some "natural" set of subuniverses of $\mathbf{A} \times \mathbf{A}$. (Thus **A** is primal iff $\text{Clo}_n \, \mathbf{A} = clo_n(R)$, $n > 0$, where R is the set consisting of only the diagonal $\Delta = \{(a, a) : a \in A\}$.) One can then hope to obtain reasonable interesting algebraic characterizations of such algebras by applying Theorem 3.3.3.

For example, let us give a short proof of Theorem 3.1.2, that is, that the four conditions of that theorem imply primality. Since $V(\mathbf{A})$ is arithmetical and **A** is finite, the term functions of **A** are just those functions compatible with all graph subalgebras (= rectangular subalgebras) of $\mathbf{A} \times \mathbf{A}$. But since **A** is simple and has no proper subalgebras, these are just $\mathbf{A} \times \mathbf{A}$ itself and the graphs of automorphisms, and the only automorphism graph is Δ. Hence **A** is primal.

To generalize primality, the most obvious procedure, at least within arithmetical varieties, is to relax some of the three remaining conditions, stating these conditions in terms of subuniverses of $\mathbf{A} \times \mathbf{A}$. An obvious first choice is to allow **A** to have subalgebras, i.e., require Clo **A** be the subalgebra preserving functions on A; this means we take R to be the set of all subuniverses of $\mathbf{A} \times \mathbf{A}$ contained in Δ. Such an algebra will be called **subalgebra primal**. (In earlier literature the terminology *semiprimal algebra* was often used.) Notice that any subalgebra **S** of a subalgebra primal algebra **A** is also subalgebra primal, for if $f : S^m \to S$ preserves subuniverses, then we can extend f to $\mathbf{x} \in A^m \setminus S^m$ by setting $f(\mathbf{x}) = x_1$, and obtain a term function of $\mathbf{A}$. It is easy to verify the following characterization:

Theorem 3.4.1 *A finite algebra* $\mathbf{A}$ *is subalgebra primal iff the following conditions are satisfied:*

(i) *the nontrivial subalgebras of* $\mathbf{A}$ *are simple;*

(ii) *any isomorphism between nontrivial subalgebras is the identity function (i.e., subalgebras have only the identity automorphism and any two distinct nontrivial subalgebras are nonisomorphic);*

(iii) $V(\mathbf{A})$ *is arithmetical.*

PROOF Indeed, if $\mathbf{A}$ is subalgebra primal, the discriminator preserves subuniverses so that conditions (i) and (iii) are satisfied.

For (ii) let $\mathbf{S}$ and $\mathbf{T}$ be subalgebras of $\mathbf{A}$ and let $\alpha : \mathbf{S} \to \mathbf{T}$ be an isomorphism which is not the identity function. Then there is an $a \in S$ such that $\alpha(a) \neq a$. Let b be any element of S and define $f : A^2 \to A$ by $f(x,y) = y$ if $(x,y) \neq (a,b)$ and $f(a,b) = a$. Then certainly f preserves subalgebras and hence is a term function. Thus

$$\alpha(a) = \alpha f(a,b) = f(\alpha(a), \alpha(b)) = \alpha(b)$$

since $(\alpha(a), \alpha(b)) \neq (a,b)$. Hence $a = b$ and therefore $\mathbf{S}$ and $\mathbf{T}$ must be a trivial.

Conversely, if conditions (i)–(iii) hold for $\mathbf{A}$ then Theorem 3.3.3 applies; and conditions (i) and (ii) clearly restrict the rectangular subuniverses of $\mathbf{A} \times \mathbf{A}$ to the subuniverses contained in Δ. Thus $\mathbf{A}$ is subalgebra primal.   ●

Notice further that in relaxing the definition of primality to allow subalgebras the other three conditions characterizing primality still hold; thus a primal algebra is simply a subalgebra primal algebra having no proper subalgebras.

Another generalization of primality is to allow $\mathbf{A}$ to be nonsimple which suggests taking R to be the set of congruences of $\mathbf{A}$. These algebras were called **congruence primal** in Section 1.1.3 and were formally defined by requiring $\text{Clo}_n \mathbf{A} = \text{Comp}_n \mathbf{A}$, $n > 0$. (In earlier literature the name *hemiprimal* has been widely used.) Since any constant function is in Comp **A** it follows that a congruence primal algebra has among its term functions constant term functions of all possible arities. Hence a congruence primal algebra can have neither proper subalgebras nor automorphisms. Also observe that the variety generated by **A** need no longer be arithmetical. Indeed *every* algebra is a reduct of a congruence primal algebra having the same congruence lattice (just add all operations in Comp **A**); thus nothing can be said about the congruence lattice of a congruence primal algebra and no characterization of general congruence primal algebras is known. On the other hand if we restrict ourselves to finite arithmetical algebras then we have a quite simple characterization:

Theorem 3.4.2 *A finite arithmetical algebra* **A** *is congruence primal iff the following conditions are satisfied:*

(i) **A** *has no proper subalgebras;*

(ii) *all rectangular subuniverses of* **A** × **A** *are congruences of* **A***;*

(iii) $V(\mathbf{A})$ *is arithmetical.*

Notice that statement (ii) is equivalent to the requirement that for all $\theta, \phi \in$ Con **A**, any isomorphism $\alpha : \mathbf{A}/\theta \to \mathbf{A}/\phi$ is the identity function (so in particular $\theta = \phi$).

PROOF If **A** is arithmetical congruence primal then **Con A** supports a Pixley function (Theorem 2.2.6) which must be a term function; hence condition (iii). Since each constant function is in Comp **A** each is a term function and this implies (i). Let S be a rectangular subuniverse of **A** × **A**. Then, using (i), there are congruences ϕ and ψ of **A** and an isomorphism $g : \mathbf{A}/\phi \to \mathbf{A}/\psi$ with graph S. Let t be a unary term function with constant value $a \in A$; then for any $x \in A$

$$g(a/\phi) = g(t(x)/\phi) = g(t(x/\phi)) = t(g(x/\phi)) = a/\psi.$$

Hence $a/\phi = b/\phi$ implies $a/\psi = b/\psi$ and, conversely, so $\phi = \psi$ and it follows that g is the identity. Therefore S is the congruence $\phi = \psi$; hence condition (ii) holds.

Conversely, if (i)–(iii) hold, then by (iii) and Theorem 3.3.3, Clo **A** consists of functions preserving the rectangular subuniverses of **A** × **A** and, by (i) and (ii), these are just the congruences of **A**.   •

As we noted above (ii) simply means that if $g : \mathbf{A}/\phi \rightarrow \mathbf{A}/\psi$ is an isomorphism, then g is the identity function. Hence (ii) implies that $\mathbf{A}$ has no nontrivial automorphisms. Thus for finite arithmetical congruence primal algebras three out of the four characterizing conditions for primality continue to hold. Since a primal algebra is just a simple congruence primal algebra, and observing that any simple algebra is arithmetical, we see that Theorem 3.4.2 is a natural generalization of Theorem 3.1.2.

To provide a natural example of a nonsimple congruence primal algebra it is convenient to first introduce the concept of independence. A finite set of varieties $V_1, \ldots, V_n$, of the same type, is said to be **independent** if there is a term $t(x_1, \ldots, x_n)$ of the type such that for each $i = 1, \ldots, n$, $t(x_1, \ldots, x_n) \approx x_i$ is an identity of V_i. Algebras $\mathbf{A}_1, \ldots, \mathbf{A}_n$, of the same type, are said to be independent if the varieties they generate are independent.

An example of a nonsimple arithmetical congruence primal algebra is the direct product $\mathbf{A}_1 \times \cdots \times \mathbf{A}_n$ of nonisomorphic independent primal algebras $\mathbf{A}_i$. In this case independence obviously implies that if for each $i = 1, \ldots, n$, $p_i(x, y, z)$ is a Pixley term for $V(\mathbf{A}_i)$ then the composition $t(p_1(x, y, z), \ldots, p_n(x, y, z))$ is a Pixley term for $V(\mathbf{A}_1, \ldots, \mathbf{A}_n)$. Moreover, for nonisomorphic primal algebras $\mathbf{A}_i$ the converse is also true: $V(\mathbf{A}_1, \ldots, \mathbf{A}_n)$ arithmetical implies that the $\mathbf{A}_i$ are independent ([Pixley, 1971]). Using this criterion it is easily seen, using Theorem 3.4.2, that $\mathbf{A}_1 \times \cdots \times \mathbf{A}_n$ is congruence primal. For more information on independence and its properties see the exercises at the end of this section.

Note that in general we need not expect a quotient of a congruence primal algebra to again be congruence primal. However, using Theorem 3.4.2 above one can easily show that for finite arithmetical congruence primals, it is true that quotients are again congruence primal.

Finally let us allow $\mathbf{A}$ to have proper automorphisms: define $\mathbf{A}$ to be **automorphism primal** if $Clo_n \mathbf{A} = clo_n (\text{Aut } \mathbf{A})$ for > 0. (The latter condition means that the term functions represent all functions which are compatible with the graphs of the automorphisms of $\mathbf{A}$.) Since the discriminator is a pattern function, it follows directly that if $\mathbf{A}$ is automorphism primal, the discriminator is a term function. From this it follows as before that $V(\mathbf{A})$ is arithmetical and $\mathbf{A}$ is

simple. The difference now is that **A** may very well have proper subalgebras (though, by virtue of the discriminator, they must be simple). For example the set of fixed points of an automorphism of any
algebra is a subalgebra (the intersection of the graph of the given automorphism with the graph Δ of the identity automorphism). Thus
it is not clear how to characterize automorphism primal algebras algebraically by some analog of the preceding two theorems. For this
reason Quackenbush introduced "demiprimal" algebras: finite automorphism primal algebras having no proper subalgebras. Since this
terminology is not very suggestive, we shall simply call them **automorphism primal algebras with no subalgebras**. Such algebras
are easily characterized as follows:

Theorem 3.4.3 *A finite algebra* **A** *is automorphism primal with no
subalgebras iff the following conditions are satisfied:*

(i) **A** *has no proper subalgebras;*

(ii) **A** *is simple;*

(iii) $V(\mathbf{A})$ *is arithmetical.*

PROOF Clearly the conditions hold for **A** automorphism primal
with no subalgebras. Conversely, if the conditions hold then, apply-
ing Theorem 3.3.3 again, we see that the only subuniverses of $\mathbf{A} \times \mathbf{A}$
are the graphs of automorphisms. •

See [Quackenbush, 1971] for more in this direction, including
other generalizations.

Despite the fact that we shall otherwise seldom need to refer to
this class of algebras, they do turn out to play an important role in
the theory of affine complete varieties (Chapter 4, Theorem 4.4.13).

Obviously there are other ways in which primality may be generalized along the lines just discussed. However, within the context of
arithmetical varieties, the most important generalization of primality has been *quasiprimality* which was defined earlier. Statement (3)
of the following theorem characterizes quasiprimal algebras by describing Clo **A**. This statement was actually the original definition of
quasiprimality ([Pixley, 1971]). Notice, as is apparent from either of
statements (1) or (3), subalgebra primal and automorphism primal
algebras are special kinds of quasiprimal algebras.

Theorem 3.4.4 *For a finite algebra* **A** *the following statements are equivalent:*

(1) **A** *is quasiprimal (i.e., the discriminator is a term function).*

(2) *All nontrivial subalgebras of* **A** *are simple and* $V(\mathbf{A})$ *is arithmetical.*

(3) Clo **A** *represents the set of all operations on A which preserve all* **inner isomorphisms** *of* **A**, *that is, all isomorphisms between (not necessarily distinct) subalgebras of* **A**.

Inner isomorphisms are also often called internal isomorphisms or (less accurately) inner automorphisms.

PROOF We have already observed that (1) implies (2). Condition (3) follows from (2) and Theorem 3.3.4, for if all subalgebras are simple then the only graph subalgebras of **A** × **A** are graphs of isomorphisms between subalgebras of **A**. Since the discriminator is a pattern function, we again see that it preserves isomorphisms between subalgebras and hence (3) implies (1).   •

Let us denote the set of all inner isomorphisms of **A** by Inn **A**. We emphasize the role of an inner isomorphism as a relation (its graph). Notice that for inner isomorphisms f and g, the relation product $f \circ g$ may be the empty function. With this in mind, Inn **A** is closed under relation product $\circ$ and inverses (converses) $^{-1}$, and hence the algebra **Inn A** $= \langle$ Inn **A**; $\circ$, $^{-1} \rangle$ is an inverse semigroup. Condition (3) simply asserts that Clo **A** represents $clo(\text{Inn } \mathbf{A})$. Notice that the subuniverses of **A** correspond precisely with the idempotent members of Inn **A**.

Because of condition (3) it would be difficult to find a name for quasiprimal algebras which would be more suggestive of their characterization in terms of term functions. Of course we can always refer to them as "finite discriminator algebras" and often do, but the name quasiprimal is so embedded in the literature that it makes sense to continue using it.

In Chapter 4, when studying affine complete varieties, finite algebras with no proper subalgebras turn out to play a special role. Therefore it is important to make explicit the fact that in this class of algebras three notions considered in the present chapter coincide.

Theorem 3.4.5 *If **A** is a finite algebra with no proper subalgebras, then the following conditions are equivalent:*

(1) **A** *is functionally complete;*

(2) **A** *is quasiprimal;*

(3) **A** *is automorphism primal.*

PROOF (1) $\Rightarrow$ (2). If **A** is functionally complete then it is simple and by Theorem 3.1.11 the variety $V(\mathbf{A})$ is arithmetical. Hence by Theorem 3.4.4 **A** is quasiprimal.

(2) $\Rightarrow$ (3). If **A** is quasiprimal then by Theorem 3.4.4 **A** is simple and $V(\mathbf{A})$ is arithmetical. Hence by Theorem 3.4.3 **A** is automorphism primal.

(3) $\Rightarrow$ (1). If **A** is automorphism primal then the discriminator is a term function of **A**. Then it follows from Theorem 3.1.4 that **A** is functionally complete.   •

Quasiprimal algebras have only simple (but possibly isomorphic) subalgebras while congruence primal algebras have no subalgebras and no automorphisms between quotients. In seeking ways to further extend primality in such a way to capture both of these concepts it is then natural to allow isomorphisms between quotients of subalgebras and to ask for a characterization of finite algebras for which the term functions are precisely those functions preserving isomorphisms of quotients of subalgebras. But (graphs of) isomorphisms of quotients of subalgebras are precisely rectangular (i.e., graph) subuniverses. Thus we define a finite algebra **A** to be **rectangular primal** if Clo **A** represents all operations on A which preserve all rectangular subuniverses of $\mathbf{A} \times \mathbf{A}$. From Theorem 3.3.4 we see that if **A** is finite and $V(\mathbf{A})$ is arithmetical, then **A** is rectangular primal. In Theorem 3.4.6 below we shall show that if a finite algebra **A** has only arithmetical subalgebras, then the converse is true, i.e., if **A** is rectangular primal then it generates an arithmetical variety.

Theorem 3.4.6 says that we can regard rectangular primality as a kind of ultimate generalization of primality, at least within the context of algebras generating arithmetical varieties. Observe that the hypothesis of the theorem requiring that each subalgebra be arithmetical is essential; otherwise, if **A** were an algebra having no proper subalgebras and no nontrivial isomorphisms of quotients (i.e., the only rectangular subuniverses of $\mathbf{A} \times \mathbf{A}$ are congruences), then we could conclude that $\mathbf{A}$ is congruence primal iff $V(\mathbf{A})$ is arithmetical, which we know is not true. Also this requirement is not really a new additional hypothesis; it was explicitly present in Theorem 3.4.2 and, implicitly, in Theorem 3.4.4 since simple subalgebras are trivially arithmetical.

While in the sense described above, Theorem 3.4.6 is not so surprising, it is worthwhile to observe its significant content. We already know from Theorem 3.3.3 that if $\mathbf{A}$ is a finite algebra and we wish to determine if $V(\mathbf{A})$ is arithmetical then we need to know for *each member* of $V(\mathbf{A})$ that each graph subuniverse preserving *partial* operation can be interpolated by a term function. Theorem 3.4.6 says that if we know that $\mathbf{A}$ has only arithmetical subalgebras then we need to know this information only for the generating algebra $\mathbf{A}$, and moreover, we need only know that each graph subuniverse preserving *total* operation of $\mathbf{A}$ is a term function. This is a considerable reduction in the amount of information needed in the general situation. More important for our purposes, Theorem 3.4.6 will be important in Section 4.4.3.

Rectangular primal algebras were called *polynomial complete* in [Pixley, 1972a] where they were introduced. This was partly because at that time what are now called terms were then commonly called polynomials. But even revising this to *term complete* does not seem to yield a sufficiently descriptive name (and might even confuse them with primal algebras). Also "rectangular primal" is consistent with the terminology "subalgebra primal", "congruence primal", etc., already introduced. Also, of course, we now are using "polynomial completeness" to describe the entire range of completeness properties.

Theorem 3.4.6 *If $\mathbf{A}$ is a finite algebra having all subalgebras arithmetical, then $\mathbf{A}$ is rectangular primal (i.e., $\text{Clo}\,\mathbf{A}$ represents the set of all operations on A which preserve all rectangular subuniverses of $\mathbf{A} \times \mathbf{A}$) if and only if $V(\mathbf{A})$ is arithmetical.*

PROOF As noted we only need to prove the "only if" direction and to do this we need to prove the following:

If a finite algebra $\mathbf{A}$ has only arithmetical subalgebras,

then there is a Pixley function f on A which preserves
all graph subuniverses of **A** × **A**.

The idea of the proof is to proceed in somewhat the same way as
we did in adapting the proof of Theorem 2.2.6 to Section 2.2.4 (Para-
graph 3, on prescribed automorphisms). The potential obstruction
in the process described there was that we required f to be principal.
Without this requirement we can proceed successfully in the present
case but with the complication that we must consider proper subalgebras. However, since the sought Pixley function is 3-ary we need
only require that it be compatible with graph subalgebras which are
graphs of isomorphisms of quotients of subalgebras of **A** of no more
than 3 generators. For this reason, in what follows, for each ordered
triple (x, y, z) of (not necessarily distinct) elements of A, for brevity
let $[x, y, z]$ denote the subalgebra generated by the elements of the
triple. Then to prove the statement above, and hence the theorem,
it will be sufficient to prove the following proposition.

Proposition 3.4.7 *If* **A** *is a finite algebra having only arithmetical
subalgebras, then there is a Pixley function f on A which preserves
all subalgebras* $[x, y, z]$, *induces a congruence compatible function on
each of these subalgebras, and in addition satisfies the following condition*

(A) *for each pair of triples* $(x, y, z), (u, v, w) \in A^3$, *and congruences*
θ, σ *of* $[x, y, z]$ *and* $[u, v, w]$ *respectively, if the statement*

there exists an isomorphism

$$\alpha : [x, y, z]/\theta \to [u, v, w]/\sigma \tag{3.3}$$

such that $\alpha(x/\theta) = u/\sigma$, $\alpha(y/\theta) = v/\sigma$, $\alpha(z/\theta) = w/\sigma$

holds, then

$$\alpha f_\theta(x/\theta, y/\theta, z/\theta) = f_\sigma(u/\sigma, v/\sigma, w/\sigma). \tag{3.4}$$

PROOF If the desired function f is constructed then we automatically have a system of elements

$$f_\theta(x/\theta, y/\theta, z/\theta) \in [x, y, z]/\theta,$$

for all quotient algebras $[x, y, z]/\theta$ where $x, y, z \in A$, which satisfies
condition (A) and the following conditions as well:

(B) For every $x/\phi, y/\phi, z/\phi$

$$f_\phi(x/\phi, x/\phi, z/\phi) = z/\phi,$$
$$f_\phi(x/\phi, y/\phi, x/\phi) = f_\phi(x/\phi, y/\phi, y/\phi) = x/\phi.$$

(C) If $\phi, \psi \in \mathrm{Con}\,[x, y, z]$ and $\phi \leq \psi$ then

$$f_\phi(x/\phi, y/\phi, z/\phi) \subseteq f_\psi(x/\psi, y/\psi, z/\psi).$$

Note that the condition (C) follows from the compatibility of f and means that the function f_ϕ must induce f_ψ.

On the other hand any such system of elements $f_\theta(x/\theta, y/\theta, z/\theta)$ uniquely determines f because $f(x, y, z) = f_0(x/0, y/0, z/0)$ where 0 is the zero congruence of $[x, y, z]$. Our proof is based on this observation.

Since the congruence lattice of each $[x, y, z]$ is distributive, each congruence has a well defined height. Let us measure the height of a congruence from the top of the congruence lattice, assigning height 0 to the diversity congruence. We use induction to prove that the system of elements $f_\theta(x/\theta, y/\theta, z/\theta)$ satisfying (A), (B), and (C) exists. In the proof, at the m-th step we will consider, simultaneously, all congruences θ of height m occurring in any $[x, y, z]$.

If θ has height 0, i.e., θ is the diversity congruence, then the elements $f_\theta(x/\theta, y/\theta, z/\theta)$ are defined trivially. If θ has height 1, θ is maximal; in this case we define

$$f_\theta(x/\theta, y/\theta, z/\theta) = z/\theta$$

if $x/\theta = y/\theta$ and $= x/\theta$ otherwise (i.e., as the discriminator). Then the properties (B) and (C) are clearly satisfied. If (3.3) holds for some (u, v, w), $\sigma \in \mathrm{Con}[u, v, w]$, and an isomorphism α then (3.4) also holds since $f_\sigma(u/\sigma, v/\sigma, w/\sigma)$ must also be defined to be the discriminator. Hence (A) holds too.

Now assume that the elements $f_\theta(x/\theta, y/\theta, z/\theta)$ satisfying (A), (B), and (C) are defined for all $x, y, z \in A$, and all congruences $\theta \in \mathrm{Con}\,[x, y, z]$ of height less than m. We show how to construct the appropriate elements for congruences θ of height m.

We first note that if θ has more than one cover in $\mathrm{Con}\,[x, y, z]$ then condition (C) uniquely determines $f_\theta(x/\theta, y/\theta, z/\theta)$ exactly as in Case 1 in the proof of Theorem 2.2.6. Moreover, the same proof takes care of condition (B). We only have to check that condition (A) is satisfied as well.

Suppose that for some triples $(x, y, z), (u, v, w) \in A^3$ and congruences θ and σ of subalgebras $[x, y, z]$ and $[u, v, w]$, respectively, (3.3) holds, with isomorphism α. Then certainly θ and σ have the same number of covers in $\mathrm{Con}\,[x, y, z]$ and $\mathrm{Con}\,[u, v, w]$, respectively. Let two of these (distinct) covers of θ be θ_1 and θ_2. We may assume that σ_1 and σ_2 are (distinct) covers of σ such that the isomorphisms

$$\alpha_i : [x, y, z]/\theta_i \rightarrow [u, v, w]/\sigma_i$$

induced by α also satisfy (3.3).

First, it is easily verified that for any $a \in [x, y, z]$, we have

$$\alpha(a/\theta_1 \cap a/\theta_2) = \alpha(a/\theta_1) \cap \alpha(a/\theta_2) \tag{3.5}$$

(where we are using the same symbol α to denote the appropriate induced isomorphisms). Then, as in (2.6) of Case 1 of the proof of Theorem 2.2.6, we have

$$f_\theta(x/\theta, y/\theta, z/\theta) = f_{\theta_1}(x/\theta_1, y/\theta_1, z/\theta_1) \cap f_{\theta_2}(x/\theta_2, y/\theta_2, z/\theta_2)$$

and

$$f_\sigma(u/\sigma, v/\sigma, w/\sigma) = f_{\sigma_1}(u/\sigma_1, v/\sigma_1, w/\sigma_1) \cap f_{\sigma_2}(u/\sigma_2, v/\sigma_2, w/\sigma_2).$$

Now the equality (3.4) follows, using (3.5), from the fact that by the induction hypothesis (3.4) holds for the values of the $f_{\theta_i}, f_{\sigma_i}$ occurring on the right side of these two equalities under the induced isomorphisms. This proves (A) and concludes the case where θ has at least two covers.

It remains to consider the congruences θ of $[x, y, z]$ having a single cover θ_1. Now both conditions (B) and (C) will clearly be satisfied if we choose $f_\theta(x/\theta, y/\theta, z/\theta)$ to be any element in $f_{\theta_1}(x/\theta_1, y/\theta_1, z/\theta_1)$.

To establish (A), before making any such choices, we partition the set of all pairs $((x, y, z), \theta)$, where $\theta \in \mathrm{Con}\,[x, y, z]$ has a single cover, into equivalence classes where $((x, y, z), \theta)$ and $((u, v, w), \sigma)$ are placed in the same class provided (3.3) holds for these pairs. Within each of these classes, for some member $((x, y, z), \theta)$ make a choice and then for all other members of the class define

$$f_\sigma(u/\sigma, v/\sigma, w/\sigma) = \alpha f_\theta(x/\theta, y/\theta, z/\theta).$$

For each $(x, y, z) \in A^3$ the construction of f_θ will terminate when
$\theta = 0$, the least element of $\mathbf{Con}[x, y, z]$. Then we define $f(x, y, z)$
to be the single element in $f_0(x/0, y/0, z/0)$. The process continues
for all remaining triples until the number of steps taken equals the
maximum height of any $\mathbf{Con}[x, y, z]$. At the end we obtain a function $f : A^3 \rightarrow A$ meeting the two assertions of Proposition 3.4.7.   •

Related generalizations of primality The generalized primal algebras discussed in this section so far have been algebras which (a)
are definable by the condition that Clo $\mathbf{A}$ represents the operations
in $clo(R)$ for a suitable collection R of subuniverses of $\mathbf{A} \times \mathbf{A}$, and (b)
generate arithmetical varieties (the case of general congruence primal
algebras being an exception). We briefly mention two generalizations
of primality (actually of quasiprimality) which do not satisfy both
(a) and (b).

The first of these are **paraprimal** algebras: a finite algebra $\mathbf{A}$ is
paraprimal if all subalgebras of $\mathbf{A}$ are simple, and the variety generated by $\mathbf{A}$ is CP. From Theorem 3.4.4 $\mathbf{A}$ is quasiprimal iff it is
paraprimal and $V(\mathbf{A})$ is CD. In some ways paraprimal algebras generalize groups of prime order. These algebras have a few interesting
properties akin to primality and are discussed in the survey article
on Primality by Quackenbush (Appendix 5 of [Grätzer, 1979]). However there is no term function characterization for them (Exercise 10,
below). This is apparently due to the fact that these algebras have
no near unanimity term and this, as we shall see in Chapter 4, sets
them quite apart from the study of affine complete algebras. See
[McKenzie, 1978] for further information and the exercises at the
end of this section for interesting connections between paraprimal
and quasiprimal algebras. Paraprimal algebras were introduced in
[Clark, Krauss, 1969].

The second class of algebras which generalize quasiprimal algebras are the finite dual discriminator algebras.

The dual discriminator function on a set A was defined in Section
3.1.3 by $d(x, y, z) = x$ if $x = y$ and $d(x, y, z) = z$ if $x \neq y$, and
in this sense, at least, is dual to the discriminator. A variety is
a dual discriminator variety if it has a ternary term which induces
the dual discriminator on each subdirectly irreducible member of
the variety. Study of dual discriminator varieties was initiated in
[Fried, Pixley, 1979]. Obviously the dual discriminator is a ternary

majority function so that a dual discriminator variety is CD. The dual discriminator can be constructed from the discriminator,

$$d(x, y, z) = t(x, t(x, y, z), z)$$

where t is the discriminator, but, as we have seen earlier from the example of a lattice majority term, not the other way around.

We have observed that as a consequence of Theorem 3.1.7 that if the finite algebra **A** has more than two elements then it is functionally complete iff the dual discriminator is a polynomial of **A**. Combining this fact with Theorems 3.1.11 and 3.4.3 we have the following consequence.

Corollary 3.4.8 *If a finite dual discriminator algebra* **A** *has more than two elements and has no proper subalgebras, then* **A** *is automorphism primal.*

Dual discriminator varieties have many deeper properties, which are dual to corresponding properties of discriminator varieties, other than just the simple formal duality displayed in the definitions of the two discriminators. Also, finite dual discriminator algebras can be characterized by the property Clo **A** $= clo(R)$ for an appropriate collection of subuniverses R of **A** $\times$ **A**. This collection of subuniverses properly contains the graphs of "isomorphisms between subalgebras" which according to Theorem 3.4.4 is the characterizing collection of subuniverses for the quasiprimal algebras. These and other properties are presented in detail in [Fried, Pixley, 1979].

Still another generalization of quasiprimal algebras and, more specifically, of automorphism primal algebras is the class of **endoprimal** algebras: the clone of term functions consists of those operations which preserve all endomorphisms. Not surprisingly, these algebras generally do not generate arithmetical varieties, and include, for example, vector spaces of dimension greater than 1. Endoprimal algebras have been of interest primarily in connection with duality issues; see [Clark, Davey, 1998] for references.

Algebras with prescribed related structures In Section 2.2.4 (Theorem 2.2.12) we showed how to construct an arithmetical algebra with a prescribed congruence lattice and discussed the problem of prescribing the automorphism group as well. In light of this our discussion of rectangular primal algebras suggests the following question:

Problem 3.4.9 *Given a finite set A and a set R of rectangular subsets in A × A, under what conditions can we find a rectangular primal algebra* $\mathbf{A} = \langle A; F \rangle$ *having only arithmetical subalgebras, and with R as the collection of rectangular subuniverses of* $\mathbf{A} \times \mathbf{A}$?

To qualify as the set of rectangular subuniverses the set R must satisfy certain consistency conditions. For example it must be closed under the operations of intersection, converse, and relation product (since each nonempty member of R must be the subuniverse of a graph subalgebra). The set must also include $A \times A$ and the diagonal. In Section 3.5 we will see evidence that this is enough.

This problem is a special case of the general *concrete representation problem* for "related structures" of an algebra (i.e., congruence lattice, automorphism group, subalgebra lattice, endomorphism semigroup, etc.). A survey of this problem and solutions can be found in [Pöschel, 1979]. Quackenbush, in his survey article on Primality (Appendix 5 of [Grätzer, 1979]), describes a positive solution to the related structures problem appropriate in the special case of a quasiprimal algebra. Specifically Quackenbush (specializing a general result of M. G. Stone, [Stone, 1972]) gives the following prescription:

Let S be a collection of subsets of a finite set A and let I be a set of bijections with domains and ranges being members of S. Suppose that S and I satisfy the following closure conditions: (i) S is closed under intersection, (ii) I is closed under composition, taking of inverses, and restriction, meaning that if $\alpha : A_1 \to A_2$ is in I, $A_3 \in S$, and $A_3 \subseteq A_1$, then the restriction of α to A_3 is in I. Also we require the nonempty members of S to be precisely the sets of fixed points of members of I, and I to contain all maps between one element members of S. Then it can easily be shown that $\mathbf{A} = \langle A; F \rangle$, with $F = clo(I)$ is quasiprimal with S and I, respectively, the sets of prescribed subuniverses and inner isomorphisms. (If the empty set is in S then no nullary operations will be in F.) The algebra $\langle I; \circ, ^{-1} \rangle$ is, of course, the inverse semigroup **Inn A** of inner isomorphisms of **A**.

If I is any group of permutations on A and S consists of all of the fixed point sets of the members of I, then $\mathbf{A}$ will be automorphism primal with $\mathrm{Aut}\,\mathbf{A} = I$. If no permutation has a fixed point then $\mathbf{A}$ will be automorphism primal with no subalgebras.

On the other hand, if S consists of any collection of subsets of A closed under intersection, and I consists of all maps between one element subsets in S together with the identity maps on each subset in S, then $\mathbf{A}$ will be subalgebra primal with the system S of subuniverses.

If $\mathbf{L}$ is any arithmetical 0-1 equivalence lattice on the finite set A and f is a principal Pixley function for $\mathbf{L}$, then Theorem 2.2.12 asserts that the algebra $\mathbf{A} = \langle A; f \rangle$ has $\mathbf{Con}\,\mathbf{A} = \mathbf{L}$. Then by Theorem 3.4.2 $\mathbf{A}^+$ is congruence primal. If $\mathbf{B} = \langle A; F \rangle$ and $\mathbf{Con}\,\mathbf{B} = \mathbf{L}$, then $\mathbf{B}$ is congruence primal iff $\mathbf{B}$ and $\mathbf{A}^+$ are term equivalent. More generally, any finite arithmetical algebra is congruence primal iff it is weakly isomorphic to an algebra of the form $\mathbf{A}^+$ above.

Hence if we partition the class of all finite arithmetical congruence primal algebras by the relation of weak isomorphism, then there is a canonical algebra of the form $\mathbf{A}^+$, $\mathbf{A}$ as above, in each class.

Let us indicate that algebras $\mathbf{A}$ and $\mathbf{B}$ are term equivalent by writing $\mathbf{A} \equiv_t \mathbf{B}$. Also we designate the set of all rectangular subuniverses of $\mathbf{A} \times \mathbf{A}$ by $\mathrm{Rect}\,\mathbf{A}$. Then it is suggestive to summarize some of the concepts discussed in this section as follows:

Let $\mathbf{A} = \langle A; F \rangle$ and $\mathbf{B} = \langle A; F' \rangle$ be finite algebras with the same universe A. If $\mathbf{A}$ has any of the properties:

(i) *subalgebra primal,* (ii) *automorphism primal,* (iii) *quasiprimal,*
(iv) *congruence primal,* (v) *rectangular primal,*

then $\mathbf{A} \equiv_t \mathbf{B}$ if and only if $\mathbf{B}$ has the same property and in the corresponding cases the following hold:

(i) $\mathrm{Sub}\,\mathbf{A} = \mathrm{Sub}\,\mathbf{B}$, (ii) $\mathrm{Aut}\,\mathbf{A} = \mathrm{Aut}\,\mathbf{B}$, (iii) $\mathrm{Inn}\,\mathbf{A} = \mathrm{Inn}\,\mathbf{B}$,
(iv) $\mathbf{Con}\,\mathbf{A} = \mathbf{Con}\,\mathbf{B}$, (v) $\mathrm{Rect}\,\mathbf{A} = \mathrm{Rect}\,\mathbf{B}$.

It is easy to formulate a paraphrase of this summary in which term equivalence is replaced by the coarser partition of weak isomorphism. For example, in the case of congruence primality, we replace the condition, (iv) $\mathbf{Con}\,\mathbf{A} = \mathbf{Con}\,\mathbf{B}$, by the requirement that there be a bijection $\beta : A \to B$ and a corresponding isomorphism $\alpha : \mathbf{Con}\,\mathbf{A} \to \mathbf{Con}\,\mathbf{B}$ such that for each $\theta \in \mathbf{Con}\,\mathbf{A}$,

$$(a,b) \in \theta \quad \Longleftrightarrow \quad (\beta(a), \beta(b)) \in \alpha(\theta).$$

In Section 3.5 we discuss the formulation of the summary which can be established when term equivalence is replaced by categorical equivalence, a very much coarser partition.

Exercises

1. Let **A** be an algebra of cardinality no larger than 4 and suppose that $V(\mathbf{A})$ is semisimple and arithmetical. Show that **A** is quasiprimal. Show that 4 is the least integer for which this statement is true.

2. Prove that a variety is a discriminator variety iff it is arithmetical and has the property (PCC) that each principal congruence of each of its members **A** has a complement in **Con A**. ([Fried, Kiss, 1983])

3. Show that a variety V is a dual discriminator variety iff V has the properties: a) CD, b) PCC, and c) if a principal congruence $\mathrm{Cg}(a,b)$ of an algebra **A** in V has complement θ in **Con A**, then for all $x \in A$
$$a/\mathrm{Cg}(a,b) \cap x/\theta \neq \emptyset$$

   i.e., the block of $\mathrm{Cg}(a,b)$ containing a intersects every θ block. (Note that this property is implied by congruence permutability.) (A. F. Pixley, Studia Sci. Math. Hungarica, **19**, 1984.)

4. **Definition** The **product** $V_1 \times \cdots \times V_k$ **of** k **varieties** $V_1, \ldots, V_k$ of the same type is the class of all algebras which are isomorphic to the direct product of members from each of the V_i.

   Prove that if $V_1, \ldots, V_k$ are independent varieties then their product is a variety (equal to their join in the lattice of varieties of the given type).

5. Let $\mathbf{A}_1, \ldots, \mathbf{A}_k$ be nonisomorphic (finite) primal algebras of the same type. Show that they are independent iff the variety they generate is arithmetical.

6. Show that in Exercise 5 "arithmetical" can be replaced by "congruence permutable". (J. Froemke)

7. **Definition** Varieties $V_1, \ldots, V_k$ of the same type are **weakly independent** if for each set $t_1, \ldots, t_k$ of terms such that $t_i \approx t_j$

is an identity of $V_i \cap V_j$ for all $1 \leq i < j \leq k$, there is a term t such that $t \approx t_i$ is an identity of V_i for all i. Algebras are weakly independent if they generate weakly independent varieties.

Prove that finite algebras $\mathbf{A}_1, \ldots, \mathbf{A}_k$ are weakly independent quasiprimal algebras iff each $\mathbf{A}_i$ and each of its nontrivial subalgebras is simple and $V(\mathbf{A}_1, \ldots, \mathbf{A}_k)$ is arithmetical. ([Pixley, 1971])

8. Generalize Exercise 6 by showing that a finite set of quasiprimal algebras of the same type is weakly independent iff the algebras generate a CP variety. ([Pixley, 1972b])

9. Show that a variety V generated by a paraprimal algebra $\mathbf{A}$ is CD (and hence $\mathbf{A}$ is quasiprimal) iff V contains no nontrivial affine algebra. Show that a finite algebra $\mathbf{A}$ is quasiprimal iff it is paraprimal and contains no nontrivial affine subalgebras. (See Theorem 3.1.21 and [McKenzie, 1978].)

10. Show that there is no term function characterization for paraprimal algebras. (E. Kiss)

11. Show that the existence of an $(n+2)$-ary near unanimity term for a variety does not imply the existence of an $(n+1)$-ary near unanimity term for the variety. (A. P. Huhn)

3.4.2 Affine complete algebras

We have seen that functional completeness and primality are somewhat parallel concepts and congruence primal algebras were introduced to generalize primality to nonsimple algebras. The primary concept in this book—affine completeness—involves a parallel extension of functional completeness; recall from Section 1.1.3 that an algebra $\mathbf{A}$ is affine complete if each congruence compatible operation on A is a polynomial function of $\mathbf{A}$, that is, if $\mathrm{Pol}\,\mathbf{A} = \mathrm{Comp}\,\mathbf{A}$. Thus an algebra is functionally complete iff it is finite, simple, and affine complete.

The most common examples of nonsimple affine complete algebras are all Boolean algebras. This fact is due to G. Grätzer ([Grätzer, 1962]) and provides a starting point of much of what is included in this book. A variety V is **affine complete** if each algebra of V is affine complete. Thus the variety of Boolean algebras is the most important example of an affine complete variety. Later we shall see that, more generally, any finite arithmetical congruence primal algebra generates an affine complete variety. In particular if a finite set of primal algebras generates an arithmetical variety, then this variety is an affine complete variety. (See Exercise 5 of the last section and also [Hu, 1971].)

As noted earlier congruence primal algebras do not occur in classical algebra but the concept will occasionally be useful for technical reasons, primarily because $\mathbf{A}$ is affine complete iff $\mathbf{A}^+$ is congruence primal. Also we should notice that there is no such thing as a nontrivial congruence primal variety; for if V is a congruence primal variety and κ the cardinality of the free term algebra $\mathbf{F}_V(\omega)$, then for any $\mathbf{A} \in V$, $\mathbf{A}^\kappa$ is in V and hence its distinct elements correspond to distinct elements in F so that $|A^\kappa| \leq \kappa$ which is possible only if $\mathbf{A}$ is trivial.

It is worthwhile to appreciate the origin of the terminology *affine complete*. In his book [Wille, 1970], R. Wille established a connection between universal algebras and certain geometries (defined by an appropriate variation of the usual axioms for affine geometries). For a universal algebra $\mathbf{A}$ the associated geometry is called a congruence class geometry (Kongruenzklassengeometrie); in this geometry the points are just the elements of A while the congruence classes (blocks) of the congruence relations form the geometric entities (comparable with lines and planes, etc.) formed from sets of points. For a given congruence relation the various congruence classes enjoy a weak form of parallelism.

The classic motivating example is that of vector spaces regarded as universal algebras $\mathbf{A} = \langle A; F \rangle$ where the elements of A are the vectors, the operations F are vector addition, and for each element λ of the scalar field, a unary operation of left scalar multiplication of a vector by λ. The resulting geometry is then an affine coordinatizable geometry in the familiar sense; each subspace N determines a congruence relation of the algebra and the blocks of the congruence are just the various translations $a + N$ of N by $a \in A$, all parallel. In E. Artin's book *Geometric Algebra* ([Artin, 1957]) it is shown, in particular, how the affine plane is coordinatized by taking the points of the plane as the universe of an algebra and obtaining the basic operations of the algebra from the dilations. The dilations are the transformations carrying lines into parallel lines and in any vector space **A** over a field K, if f is a dilation of the affine geometry of a vector space, then $f(x) = \lambda x + a$ for some $\lambda \in K$ and $a \in A$, which means f is a polynomial of the vector space considered as a universal algebra.

In the more general setting considered by Wille, a geometry which is the congruence class geometry of a suitable algebra is called an affine coordinatizable geometry and a central question then is "which geometries are affine coordinatizable?" The answer is provided by Wille in a way which can be seen as a generalization of the method of Artin noted above: in Wille's development it turns out that an essential property of an affine coordinatizable geometry is that it possess "sufficiently many dilations" in a sense made precise in [Wille, 1970] (Satz 3.5) and where "dilations" are those transformations which are appropriate generalizations of the members of the traditional affine group, that is, certain transformations which carry congruence classes into other classes of the same congruence relation. Moreover, one obtains the coordinatizing algebra for an affine coordinatizable geometry by choosing the dilations as basic operations of the algebra. Conversely, the dilations of the congruence class geometry of an algebra **A** are exactly the congruence compatible operations on **A**.

For this reason, from the geometrical standpoint it turns out that of particular interest are those algebras in which all **Con A**-compatible functions are algebraically expressible, which is the case when all compatible functions are polynomials. For this reason algebras with this property are called affine complete. These algebras were first studied by Wille's student H. Werner in his (Darmstadt) Ph.D. dissertation; references can be found in [Werner, 1970]. In Chapter 5 we will establish the affine completeness of many algebras. In view of the present remarks it will be no surprise to find (Theorem 5.2.9) that vector spaces are affine complete iff their dimension is not 1, a result first established by Werner. (Note that dimension 1 is excluded since the multiplication of pairs of elements of the coordinatizing field, though certainly compatible with the congruence relations of the vector space (regarded as a universal algebra), is not a polynomial of the vector space; this will be implicit in the proof of Theorem 5.2.9.)

The name "affine complete" has persisted in the literature although there has been little, if any, subsequent development of the geometric implications of the now highly developed theory of affine

complete algebras and this gap may very well be a fruitful area of research. A basis for this can be found in Chapter 4 which is devoted to
affine complete varieties and in Chapter 5, much of which is devoted
to the study of affine complete algebras in particular varieties.

Aside from the remarks above, in the present section we shall
make only two simple but important observations about *arithmetical*
affine complete algebras which are consequences of earlier results
of the present chapter and which will be needed later. These two
observations represent the simplest example of the merging of the
geometrical and logical (Boolean) sources of affine completeness.

First, from Theorem 3.3.3 we have immediately:

Corollary 3.4.10 *Every finite algebra in an arithmetical variety is
affine complete.*

Later we shall use this result to describe all arithmetical affine
complete varieties of finite type.

Werner's Theorem 3.1.4 asserts that a finite algebra is functionally complete iff it is both simple and has a Pixley polynomial. Since
a simple algebra is certainly arithmetical the following consequence
of Theorem 3.3.3 generalizes Werner's theorem.

Theorem 3.4.11 *A finite arithmetical algebra* **A** *is affine complete
iff some Pixley function is a polynomial of* **A**.

PROOF Since **A** is arithmetical there is a **Con A**-compatible Pixley function (Theorem 2.2.6) which is a polynomial if **A** is affine
complete. Conversely, if **A** has a Pixley polynomial then $V(\mathbf{A}^+)$ is
arithmetical so $\mathbf{A}^+$ is congruence primal and hence **A** is affine complete.   •

For any 0-1 arithmetical equivalence lattice **L** on a finite set A
we have seen that Theorem 2.2.12 asserts that if f is a principal
Pixley function for **L**, then $\mathbf{A} = \langle A; f \rangle$ has congruence lattice **L**. By
Theorem 3.4.11 **A** is also affine complete. Moreover, if **B** is any alge-
bra with the same universe A as **A**, then **A** and **B** are polynomially
equivalent iff **B** is affine complete and **Con B**=**L**. Hence the algebras $\langle A; f \rangle$ are canonical representatives in the class of all arithmetical affine complete algebras on A under the relation of polynomial

equivalence.

Finally, recall from Theorem 2.2.15 of Chapter 2 that every finite arithmetical equivalence lattice satisfies the compatible choice function condition. It follows therefore that for any finite arithmetical affine complete algebra **A**, for each $a \in A$ and $\theta \in \mathrm{Con}\,\mathbf{A}$, there is a *polynomial* choice function modulo θ having a as a fixed point, i.e., membership in each congruence block is definable by a polynomial.

Exercises

1. Let **A** be an affine complete arithmetical algebra with a finite congruence lattice. Show that *any* partial operation on A, which is **Con A**-compatible where defined, has an interpolating polynomial function. ([Kaarli, 1983])

2. Let **A** be an affine complete algebra and suppose either that **A** is countable or Con **A** is finite. Show that **A** is arithmetical iff every finite partial operation on A which is Con **A** compatible, where defined, has an interpolating polynomial. ([Kaarli, 1983])

3.4.3 Polynomial interpolation

In Theorems 3.2.2 and 3.3.3 we showed that near unanimity and arithmetical varieties have useful characterizations in terms of interpolation. The most important of these, since it corresponds to classical polynomial interpolation, is the characterization of arithmeticity; that is, a variety V is arithmetical iff for every algebra **A** in V has the property:

> *Every* **Con A**-*compatible finite partial function can be interpolated by a polynomial function.*

For this reason, the present section is devoted to presenting some definitions of other related interpolation properties for an isolated algebra **A**. We shall return to this topic in Chapter 5 where we will examine it in the context of several particular varieties.

In the literature, for example in [Pixley, 1982], algebras with this particular interpolation property have often been referred to as *locally affine complete*. On the other hand, more recently it has been observed that if we change "partial function" to "total function" in this property, and require the interpolation at every finite subset of its domain, we obtain another interesting and formally larger class of algebras which has also been called locally affine complete, for example in the papers [Kaarli, Márki, Schmidt 1983], [Kaarli, 1985]. To resolve the conflict in terminology we introduce the following definition.

Definition A *total* m-ary function on an algebra **A** is said to be a **local polynomial function** of **A** if it can be interpolated by a polynomial on every finite subset of A^m. An algebra **A** is called **locally affine complete** if every **Con A**-compatible *total* function on A is a local polynomial function of **A**. An algebra **A** is called **strictly locally affine complete** if any **Con A**-compatible *finite partial* function on A can be interpolated on its domain by a polynomial. Varieties, all of whose members have either of these properties, are called by the corresponding name.

The new terminology allows us to restate some results we obtained earlier. For example Theorem 3.3.2 (a) asserts that

> *Every strictly locally affine complete algebra is arithmetical,*

and Theorem 3.3.3 says, in particular, that

> *A variety V is arithmetical if and only if it is strictly locally affine complete.*

It should be emphasized that while both affine completeness and strict local affine completeness imply local affine completeness, there is, except for finite or simple algebras, generally no comparison between the two stronger properties. Later in Chapter 5 we see that there exist nonarithmetical affine complete algebras and clearly they cannot be strictly locally affine complete. On the other hand, infinite members of arithmetical varieties (lattice ordered groups for example) are strictly locally affine complete but need not be affine complete.

The following scheme (Figure 3.1) visualizes the relationship between four classes of algebras. (The classes above are larger than those below.)

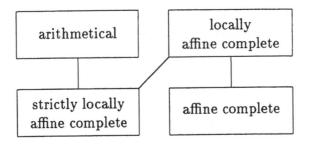

Figure 3.1

In the case of finite algebras we get a different picture since obviously in that case both local versions of affine completeness imply affine completeness. So affine completeness and local affine completeness coincide for finite algebras. However, finite affine complete algebras need not be strictly locally affine complete. The reason is that partial compatible functions need not have a total compatible extension. Thus Figure 3.2 below describes the relationship between properties in the case of finite algebras.

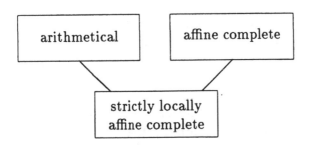

Figure 3.2

We mentioned already that locally affine complete (even affine complete) algebras need not be strictly locally affine complete. On the other hand we have the following important special case where local affine completeness implies strict local affine completeness.

Theorem 3.4.12 *An arithmetical algebra which is at most countable is strictly locally affine complete iff it is locally affine complete.*

PROOF We only need to prove the sufficiency. Let **A** be an at most countable arithmetical locally affine complete algebra and f

be a finite partial compatible function on A with domain $X \subseteq A^m$. Since A is countable, so is A^m and we may arrange its elements in the form of a sequence starting with m-tuples belonging to X. By Theorem 2.2.2 the lattice **Con A** satisfies the compatible function extension property. Applying this property at most countably many times, we eventually come to a compatible m-ary function g on **A** which extends f. Since **A** is locally affine complete, g can be interpolated by a polynomial function on X.   •

Problem 3.4.13 *Does there exist an uncountable arithmetical algebra which is locally affine complete but not strictly locally affine complete?*

Also we observe that the class of locally affine complete varieties contains both the class of arithmetical varieties and the class of affine complete varieties. While we shall exhibit numerous locally affine complete algebras, we do not know, aside from affine complete and arithmetical varieties, if any other locally affine complete varieties exist. Thus the following problem:

Problem 3.4.14 *Find a locally affine complete variety which is neither arithmetical nor affine complete. Does there exist a characterization of such varieties in terms of some kind of polynomial completeness properties?*

If we omit the phrase "**Con A**-compatible" in the above definitions, which amounts to assuming that **A** is simple, then we come to two concepts: "locally functionally complete" and "strictly locally functionally complete". However, these two collapse into one because simplicity implies that finite partial operations can always be extended to **Con A**-compatible total operations. While algebras with this property have been called **locally functionally complete**, it has become more common to cut through the thicket of terminology and simply say that the algebra has **the interpolation property**.

The interpolation properties just described deal with polynomial interpolation. We can of course make parallel definitions for interpolation by term functions, but just as congruence primal finite algebras do not appear in classical algebra, the same is true for the extensions via term function interpolation. An exceptional case which has been studied is that of **locally primal algebras**, algebras of arbitrary cardinality for which each operation can be interpolated by a term function on each finite subset of its domain. (The distinction between interpolation of partial and total operations is again unnecessary because of simplicity.) In particular in [Rosenberg, Szabó, 1984] extensions of Rosenberg's theorem and primality criterion (Theorems 1.1.5 and 3.1.9) are obtained. (In their paper what we here call locally primal algebras are called *locally complete.*) We shall briefly discuss these results and applications in Section 5.1.2.

To conclude the section, we draw the reader's attention to an open general problem. Affine completeness of an algebra **A** means that all finitary **Con A**-compatible functions must be polynomials. Practically, for all varieties where affine complete members have been described (see Chapter 5) it has turned out that an algebra **A** is affine complete iff all binary **Con A**-compatible functions are polynomials. Also, it follows from Theorem 3.1.6 that a finite algebra **A** is functionally complete iff all binary functions on A are polynomial. Hence we have the following problem.

Problem 3.4.15 *Does there exist a natural number n such that, for arbitrary algebra* **A***, the algebra is affine complete iff all n-ary* **Con A***-compatible functions on A are polynomials? Is this true at least for finite algebras?*

Note that the similar problem for strict local affine completeness has been solved. It follows from [Hagemann, Herrmann, 1982] that an algebra **A** is strictly locally affine complete iff every ternary finite partial **Con A**-compatible function on A can be interpolated by a polynomial function on its domain. Recently E. Aichinger ([Aichinger, 2000]) has improved this result by showing that ternary functions can be replaced by binary functions. Since strictly locally affine complete algebras are arithmetical and arithmetical algebras have the compatible function extension property, it follows that a finite arithmetical algebra is affine complete iff all of its binary compatible functions are polynomials.

3.5 Categorical equivalence

In Section 3.1 we stated that T. K. Hu directly generalized the Stone duality theorem for Boolean algebras by showing that if **P** is a primal algebra then a variety W is categorically equivalent to $V(\mathbf{P})$ iff W is the variety generated by some primal algebra. In the present section we shall briefly summarize some recent results of C. Bergman, J. Berman, and G. Gierz, which extend this result to several generalizations of primal algebras. Their papers, [C. Bergman, 1998], [Bergman, Berman, 1996], [Gierz, 1994], from which our discussion is summarized, should be consulted for further details. Categorical equivalence of algebras is a rapidly developing field and even as we complete this book new results have emerged, e.g., [Snow, 1998]. For this reason we have kept our summary brief and present just some of the recent results. The minimal amount of category theory we require may be found in a few pages of [MacLane, 1997].

First we recall the concepts of equivalence and isomorphism of categories. Two categories C and D are said to be **equivalent** if there are functors $F : C \to D$ and $G : D \to C$ such that the composite functors $F \circ G$ and $G \circ F$ are naturally isomorphic to the identities on D and C respectively. Any variety of algebras forms a category in which the morphisms are taken to be all homomorphisms between algebras. A classical example of categorical equivalence of varieties of algebras is Morita's Theorem which provides necessary and sufficient conditions on two rings with unity in order for their varieties of unitary modules to be categorically equivalent.

A related but generally much stronger notion is that of isomorphism of categories. An **isomorphism of categories** C and D is a functor $F : C \to D$ such that for some functor $G : D \to C$, the composite functors $F \circ G$ and $G \circ F$ are identity functors (and not just naturally isomorphic to them). The simplest example of the distinction between isomorphism and equivalence of categories is the following: for a given category C partition the objects into equivalence classes, each class consisting of isomorphic objects. Pick one object from each class. The resulting category (a skeleton of C) is then equivalent to C but not isomorphic. We will require the concept of isomorphism of categories later in our discussion.

As is noted by Bergman and Berman ([Bergman, Berman, 1996]), a surprising number of "algebraic" properties have been shown to be preserved under categorical equivalence of varieties. They observe that some of these, such as "cartesian products" and "homomorphic kernels", are familiar to anyone who has worked with categories of algebras; others are somewhat unexpected, including "surjective homomorphism" and "finite algebra".

The main tools for working with categorical equivalence of varieties have been, first of all, the representation theory of algebras by sections in sheaves, the method used by Gierz and which is presented for quasiprimal algebras in [Keimel, Werner, 1974]. Most past results on categorical equivalence have been obtained by this method. A general description appears in [Davey, Werner, 1983] and more recently and completely in [Clark, Davey, 1998]. Second, a purely algebraic characterization of equivalence is due to R. McKenzie (the method used by Bergman and Berman). This characterization is expressed in terms of matrix powers of algebras and idempotent invertible terms, concepts which we do not need to introduce here. The characterization is completely general and gives some new insights into the strength of categorical equivalence. For example, from it one can see that "homogeneous Mal'cev conditions" are preserved by categorical equivalence. (Homogeneous Mal'cev conditions are Mal'cev conditions in which no compositions of terms occurs, as for example is the case with congruence permutability, distributivity, and arithmeticity.)

The third method, which is based on McKenzie's method, is due to O. Lüders and appears in [Denecke, Lüders, 2000]. Since application of this method by C. Bergman ([C. Bergman, 1998]) has yielded results which will be of the greatest interest in Chapter 4 we explain it in more detail below. Also most of the results we shall survey are summarized from the paper [C. Bergman, 1998].

First recall from Section 1.1.3 (Theorem 1.1.6) that for finite sets the Galois connection *inv-clo* establishes a dual isomorphism between the lattice of clones on A and the lattice of coclones on A as subuniverses of the algebra $\mathbf{R} = \langle R(A); \zeta, \tau, \nu, \pi, \sqcap, \delta \rangle$. In particular, for a given algebra $\mathbf{A} = \langle A; F \rangle$, $inv(F)$ is a subuniverse of $\mathbf{R}$; it determines a subalgebra of $\mathbf{R}$ which we call the **coclone algebra of A** and denote it by $\mathbf{C}(\mathbf{A})$. Conversely (since A is finite), every coclone on A is of the form $inv(F)$ for an algebra with basic operations F, i.e., every coclone algebra on A is the coclone algebra of an algebra on A.

Our interest in the present discussion is in characterizing, up to categorical equivalence, the varieties generated by generalizations of primal algebras. For example Hu's result for primal algebras actually asserts that if $\mathbf{P}$ is primal then the class of all algebras of the form $F(\mathbf{P})$ as F ranges through all equivalences between varieties is exactly the class of all primal algebras. We shall describe below how this result can be generalized. In order to be precise we have the following definition due to Bergman and Berman:

Definition Let $\mathbf{A}$ and $\mathbf{B}$ be algebras, not necessarily of the same type. We say that $\mathbf{A}$ and $\mathbf{B}$ are **categorically equivalent** if there is a functor F which is an equivalence from the category $V(\mathbf{A})$ to $V(\mathbf{B})$ such that $F(\mathbf{A}) = \mathbf{B}$. We write $\mathbf{A} \equiv_c \mathbf{B}$ to indicate this relationship.

It is not hard to show that if F is an equivalence between two varieties, then for any algebra $\mathbf{A}$, F restricts to an equivalence between the varieties $V(\mathbf{A})$ and $V(F(\mathbf{A}))$. Thus our definition exactly captures the concept discussed above. It is also obvious that $\equiv_c$ is an equivalence relation on algebras. (Hu's result can be stated as: an algebra $\mathbf{P}$ is primal iff $\mathbf{P}/\equiv_c$ is the class of primal algebras.)

Before proceeding we must make one additional assumption. We shall certainly want term equivalent algebras to be categorically equivalent. But since one of two term equivalent algebras may have nullary operations and the other not, the empty set will not be a subuniverse of the first but will be of the second, and hence the varieties generated by the two could not be equivalent as categories, since the "empty algebra", which is usually admissible as an object in a category, occurs in the second but not the first. As before we avoid this difficulty by simply disallowing nullary operations for the remainder of this discussion. This is no problem since, as noted earlier, any nullary operation can (with a change of type the only penalty) always be represented as a unary operation. For example for any algebra $\mathbf{A}$ we can hereafter redefine the algebra $\mathbf{A}^+$ to be the result of adding unary, rather than nullary, constant operations for each element of A. Thus all of the algebras considered will have the empty set as a subuniverse.

Finally, with this convention, the definition of categorical equivalence and the definition of $\mathbf{C}(\mathbf{A})$ given above, Lüders' theorem can be now be stated as follows:

Theorem 3.5.1 *For finite algebras* $\mathbf{A}$ *and* $\mathbf{B}$, $\mathbf{A} \equiv_c \mathbf{B}$ *iff* $\mathbf{C}(\mathbf{A})$ *is isomorphic to* $\mathbf{C}(\mathbf{B})$.

Given the nature of $C(A)$, from this theorem it is clear, for example, that term equivalence or weak isomorphism of finite algebras implies their categorical equivalence. To apply Lüders' theorem to generalizations of primality as discussed in the preceding sections we can take advantage of the existence of a ternary majority term. By Corollary 3.2.3 this implies that $\text{Clo } A = clo(\text{Sub}(A \times A))$ and this means that for our purposes it will be sufficient to consider only a modified version of the algebra $C(A)$ which results from restricting to binary relations and hence is finite. Thus to apply Lüders' theorem to our present situation we introduce one more concept from [C. Bergman, 1998]: for an algebra A the algebra

$$S_2(A) = \langle \text{Sub}(A^2); \cap, \circ, \smile, 0_A, 1_A \rangle$$

is called the S_2-*structure* on A. Thus the elements of $S_2(A)$ are the subuniverses of $A \times A$ considered as binary relations. The operations are the binary operations of intersection and relation product, the unary relation of converse, and the nullary operations 0_A and 1_A. Bergman then applies Theorem 3.5.1 to prove his main result:

Theorem 3.5.2 *Let A be a finite algebra with a ternary near unanimity (i.e., majority) term. Then $A \equiv_c B$ iff B is finite, has a majority term, and $S_2(A) \cong S_2(B)$.*

With this background we can present some generalizations of Hu's original result for primality cited above.

First consider subalgebra primal algebras (Theorem 3.4.1). Here we have the following result which appears in both [Gierz, 1994] and [Bergman, Berman, 1996]:

Theorem 3.5.3 *Let A be any finite subalgebra primal algebra and let B be any algebra. Then $A \equiv_c B$ iff B is a finite subalgebra primal algebra and there is a lattice isomorphism between the subalgebra lattices of A and B which is a bijection between the one element subalgebras of A and B.*

Thus a complete set of invariants for the relation of categorical equivalence among subalgebra primal algebras is the (abstract) subalgebra lattice with labels attached to those elements corresponding to one element subalgebras.

For the case of general finite automorphism primal algebras Bergman and Berman also establish necessary and sufficient conditions for categorical equivalence. The conditions are complicated by the existence of proper subalgebras. In Chapter 4 we shall see that only the case of automorphism primal algebras with no proper subalgebras are those relevant to the study of affine complete varieties: we shall show that they are precisely the generators of the minimal affine complete varieties. Hence we restrict our attention to this case. Here we have the following easily stated result established in both [Bergman, Berman, 1996] and [Gierz, 1994].

Theorem 3.5.4 *Let* **A** *be a finite automorphism primal algebra with no proper subalgebras and let* **B** *be any algebra. Then* $\mathbf{A} \equiv_c \mathbf{B}$ *if and only if* $\operatorname{Aut} \mathbf{A} \cong \operatorname{Aut} \mathbf{B}$ *and* **B** *is automorphism primal with no proper subalgebras.*

Next we consider quasiprimal algebras. In this case [Gierz, 1994] presents the following description of a complete set of invariants. For an algebra **A**, as before, let **Inn A** denote the inverse semigroup of all inner isomorphisms of **A**. If **A** and **B** are two quasiprimal algebras whose inverse semigroups **Inn A** and **Inn B** are isomorphic under an isomorphism α, then α will map idempotents to idempotents, and therefore induce an isomorphism α^* between the lattices **Sub A** and **Sub B** of subuniverses. We say that α *preserves one element subuniverses* iff α^* maps the one element subuniverses of **A** onto the one-element subuniverses of **B**. According to the following theorem **Inn A** with labels corresponding to the one-element subuniverses constitutes a complete set of invariants of **A** relative to categorical equivalence. The proof appears both in [Gierz, 1994] (using the sheaf approach) and in [C. Bergman, 1998] (using Theorem 3.5.2).

Theorem 3.5.5 *Let* **A** *be a quasiprimal algebra and let* **B** *be any algebra. Then* $\mathbf{A} \equiv_c \mathbf{B}$ *iff the inverse semigroups* **Inn A** *and* **Inn B** *are isomorphic under an isomorphism which preserves one-element subuniverses.*

If **A** is a subalgebra primal algebra then, as we noted earlier, no two distinct subalgebras are isomorphic and no subalgebra has a nontrivial automorphism, i.e., the inner isomorphisms are just the identity maps on the subuniverses. Hence every element of **Inn A** is an idempotent so that Theorem 3.5.3 is a corollary of Theorem 3.5.5. Likewise if **A** is automorphism primal with no subalgebras then the inner isomorphisms are exactly the automorphisms so that Theorem 3.5.4 is also a corollary of Theorem 3.5.5. Also, by restricting each of the theorems above to simple rigid algebras with no proper subalgebras, i.e., to primal algebras, we have Hu's result, **A** $\equiv_c$ **B** if **A** and **B** are any two primal algebras. (The algebra **A** is primal if **Inn A** consists of just the identity map on A.)

For the case of finite arithmetical congruence primal algebras the abstract congruence lattice is a complete set of invariants. This is established both in [Bergman, Berman, 1996] and more simply using Theorem 3.5.2 in [C. Bergman, 1998]:

Theorem 3.5.6 *Let* **A** *be a finite arithmetical congruence primal algebra and let* **B** *be any algebra. Then* **A** $\equiv_c$ **B** *iff* **B** *is finite arithmetical and congruence primal, and* **Con A** $\cong$ **Con B**.

Comparing these results with Theorem 3.4.6 and the summary at the end of Section 3.4.1 suggests the following problem.

Problem 3.5.7 *Find a complete set of invariants relative to categorical equivalence for rectangular primal algebras having only arithmetical subalgebras.*

This problem has been solved by C. Bergman for the case in which **A** has no proper subalgebras. We will describe his result at the end of Chapter 4 where it is of special relevance. Of course the summary at the end of Section 3.4.1 suggests that Rect **A** should in some way describe the categorical equivalence class of **A** and this turns out to be the case in Bergman's result (Theorem 4.4.24), but the more general case has not yet been settled.

We should note here that the complete results of Bergman and Berman are actually stronger than the statements of the theorems above in that they also construct minimal members of the categorical equivalence classes for both congruence primal algebras and automorphism primal algebras with no subalgebras, and also for general finite automorphism primal algebras. For finite arithmetical congruence primal algebras they also exhibited a canonical member of each class, the dual Heyting algebra in the class. Recall that at the end of

Section 3.4.2 we also showed that the algebras $\mathbf{A}^+$ where $\mathbf{A} = \langle A; f \rangle$, f a principal Pixley function, are canonical representatives relative to weak isomorphism.

Finally, in [Davey, Werner, 1983] it was shown that for any finite algebra having a ternary majority term there is a natural way to extend the Stone topological duality theory. This result then applies, in particular, to each of the generalizations of primality which we have considered in the present section. The book [Clark, Davey, 1998] provides a full discussion of this and many other consequences for topological duality theory of results of this chapter. We shall return to this topic at the end of Chapter 4.

Chapter 4

Affine Complete Varieties

4.1 Introduction and instructive examples

In Section 3.4.2 we briefly introduced affine complete algebras, defining them as algebras for which the polynomials coincide with the congruence compatible functions. A variety was then defined to be affine complete if all of its members were affine complete. We also observed that the variety of Boolean algebras is the most important example of an affine complete variety, a discovery due to G. Grätzer [Grätzer, 1962]. T. K. Hu [Hu, 1971] extended Grätzer's result by showing that varieties generated by finite sets of independent primal algebras are also affine complete. Almost simultaneously with Hu, Iskander [Iskander, 1972] characterized affine complete p-rings (i.e., associative rings satisfying the identity $x^p \approx x$ for a given prime number p). Iskander's result implies that the variety of p-rings with a distinguished identity element (i.e., with the identity element as a nullary operation) is affine complete. It is known that the latter variety is generated by the prime field $GF(p)$ with a distinguished identity element and the latter algebra is primal. Hence Iskander's result restricted to rings with an identity element is covered by Hu's result. On the other hand, Iskander's result includes Grätzer's theorem because the variety of Boolean algebras is term equivalent to the variety of 2-rings with identity element. It is interesting that Grätzer's and Iskander's proofs used specific properties of Boolean algebras and p-rings, while Hu's proof depended on his extension of the Boolean Stone duality theory to varieties generated by independent primal algebras.

The existence of these important natural examples was motivation for the systematic study of affine complete varieties. The hope was to discover a result somewhat similar to the known description of strictly locally affine complete varieties (see Theorem 3.3.3). Eventually it was discovered that affine complete varieties form a subclass of CD varieties that enjoy certain rather easily described additional properties. Though the problem of describing all affine complete varieties is not solved yet, there is a reasonably satisfactory characterization of locally finite affine complete varieties. The purpose of the present chapter is to systematically present these results.

We complete this introductory section by extending the list of varieties that are affine complete. These results originally appeared in [Kaarli, Pixley, 1987]. A key observation is that each of the varieties considered by Hu is arithmetical and generated by a finite algebra having no proper subalgebras. Indeed recall from Section 3.4 that we observed that the direct product of finitely many non-isomorphic primal algebras is congruence primal and hence has no proper subalgebras.

The following theorem considerably sharpens Hu's theorem. Unlike Hu's proof (depending on Stone duality extended to primal algebras), the proof below is completely elementary, depending only on basic properties of CD varieties. Later (Theorem 4.4.6) we shall obtain a simple characterization of locally finite affine complete varieties which is in the form of a greatly strengthened version of the following.

Theorem 4.1.1 *If a CD variety V is generated by a finite algebra having no proper subalgebras and all finite members of V are affine complete, then V is affine complete.*

PROOF Suppose $V = V(\mathbf{A})$ where $\mathbf{A} = \langle A; F \rangle$ is finite having no proper subalgebras. By congruence distributivity $V = IP_S H(\mathbf{A})$ so that if $\mathbf{B} \in V$, we may suppose

$$\mathbf{B} \leq \prod (\mathbf{A}_i : i \in I)$$

where each $\mathbf{A}_i$ is $\mathbf{A}/\theta$ for some $\theta \in \operatorname{Con} \mathbf{A}$. Let f be a m-ary compatible function on $\mathbf{B}$. Then the algebra $\mathbf{B}^* = \langle B; F \cup \{f\} \rangle$ has the same universe and congruence lattice as $\mathbf{B}$. Also, by compatibility, f naturally induces various functions on the subdirect factors of $\mathbf{B}$, say f_i is the function induced on A_i. It follows from this that $\mathbf{B}^*$ is a subdirect product

$$\mathbf{B}^* \leq \prod(\mathbf{A}_i^* : i \in I),$$

where $\mathbf{A}_i^*$ is obtained by adding f_i to the operations of $\mathbf{A}_i$. Since A is finite there are only finitely many nonisomorphic $\mathbf{A}_i^*$ and the variety generated by all of them is locally finite. Hence if we pick any $a \in B$, then the subalgebra $\mathbf{C}^*$ it generates in $\mathbf{B}^*$ is finite and therefore isomorphic to a subdirect product in $\prod(\mathbf{A}_i^* : i \in I_0)$ for some finite subset $I_0 \subseteq I$. Then the reduct $\mathbf{C}$ obtained from $\mathbf{C}^*$ by discarding the operation f is a finite subalgebra of $\mathbf{B}$ and $f(C^m) \subseteq C$. Since V is CD, the algebra $\mathbf{C}$ has no skew congruences with respect to the above subdirect decomposition. Thus the function $f|_C$ defined componentwise with respect to this decomposition is in $\mathrm{Comp}\mathbf{C}$.

Since by our hypothesis the algebra $\mathbf{C}$ is affine complete, the function $f|_C$ is a polynomial of $\mathbf{C}$. Hence for some $(m+k)$-ary term t and $\mathbf{c} \in C^k$, $f(\mathbf{x}) = t(\mathbf{x}, \mathbf{c})$ for all $\mathbf{x} \in C^m$. But since $\mathbf{A}$ has no subalgebras, the projections of C are onto each of the factors A_i, for *all* $i \in I$. Hence for any $\mathbf{x} \in B^m$ and $i \in I$, choose $\mathbf{y} \in C^m$ with $\mathbf{y}_i = \mathbf{x}_i$. Then

$$f(\mathbf{x})_i = f_i(\mathbf{x}_i) = f_i(\mathbf{y}_i) = f(\mathbf{y})_i$$
$$= t(\mathbf{y}, \mathbf{a})_i = t(\mathbf{y}_i, \mathbf{a}_i) = t(\mathbf{x}_i, \mathbf{a}_i) = t(\mathbf{x}, \mathbf{a})_i$$

and hence f is a polynomial of $\mathbf{B}$. Thus V is affine complete. ●

The next corollary is immediate from Corollary 3.4.10.

Corollary 4.1.2 *Every arithmetical variety generated by a finite algebra having no proper subalgebras is affine complete.*

This corollary shows that arithmetical affine complete varieties are easy to construct. One has only to take a finite algebra having a Pixley function among its fundamental operations and, if necessary, to add more fundamental operations in order to get rid of proper subalgebras. The algebra we get in this way generates an affine complete variety. In particular, any finite arithmetical congruence primal algebra generates an affine complete variety.

The corollary also raises the question of the existence of nonarithmetical affine complete varieties. Later in the chapter we shall see

that while all affine complete varieties are CD they need not be CP
and it is worthwhile at this point to see how this can happen. The
following series of examples was constructed in [Kaarli, 1990].

Example 4.1.3 *A nonarithmetical affine complete variety.*

Our example can be most easily constructed by referring to Example
2.3.1 in Section 2.3 where we constructed a class of Boolean equivalence lattices on a finite set which had the (CFL) property but which
were not arithmetical. Referring to the notation and details of that
construction, we only have to choose the set $A \subseteq A_1 \times \cdots \times A_n$ so
that some of the kernels π_i of the projections onto the A_i do not
permute. If we let **L** be an equivalence lattice on a set A as described in that Example, and let $F = clo(L)$ then it is possible, with
some computation, to show that the algebra $\mathbf{A} = \langle A; F \rangle$ generates
a nonarithmetical affine complete variety. On the other hand this
will also follow directly from a later result of the present chapter
(Corollary 4.4.9) and the fact that from the definition **A** is seen to
be both congruence primal and to have all homomorphic images also
congruence primal (i.e., in the terminology used in Corollary 4.4.9
A is *hereditarily congruence primal*). Hence we will not go through
all of the details of the verification that **A** generates an affine complete variety, but instead leave this as an instructive exercise; rather
we note below some interesting features of the example. (The full
details appear in [Kaarli, 1990]. A very instructive special case with
$n = |A_1| = |A_2| = 2$ is presented in [Kaarli, Pixley, 1987].)

First notice that all of the operations $f \in F$ are of the form
$f = (f_1, \ldots, f_n)$ where each f_i can be taken arbitrarily on A_i so long
as A is closed under f. Since it is clear that for an arbitrary f_i on
A_i, A is closed under $f = (c_1, \ldots, c_{i-1}, f_i, c_{i+1}, \ldots, c_n)$, where each
c_j is considered as a constant function, it follows that the quotients
$\mathbf{A}_1/\pi_1, \ldots, \mathbf{A}_n/\pi_n$ are primal algebras. As was indicated in the discussion following Theorem 3.4.2 they would be independent only if
they generated an arithmetical variety, which they clearly do not.

Also, directly from Theorem 2.3.9, we see that among the operations of **A** there must be a near unanimity function and hence that
$V(\mathbf{A})$ is CD (by Theorem 1.2.11). Also, by Theorem 3.2.4, we see
that although $F = clo(L)$ is infinite, in fact it is finitely generated so
that $V(\mathbf{A})$ is equivalent to a variety of finite type. We will see in the

present chapter that all finitely generated affine complete varieties will have these properties.

Finally, since it is instructive, let us show that actually, in our case, the near unanimity function in F is $(n + 1)$-ary (which is in general of lesser rank than the number $|L|$ given by Theorem 2.3.9). To do this, and referring to the details of Example 2.3.1, let u_i be the near unanimity function on A_i which has the value c_i if it is not the case that n of its arguments agree. Then $u = (u_1, \ldots, u_n)$ is obviously a near unanimity function on A. To show that $u \in F$ consider an $(n + 1) \times n$ matrix M (with i-th column entries from A_i) and suppose for some $k \leq n$, M has an $(n + 1) \times k$ submatrix in which each of the columns has n identical entries. Then by the pigeon-hole principle, application of the u_i to this submatrix will yield one of its $n + 1$ rows while application of the appropriate u_i to the other columns of M will yield c_i. Hence, by the way in which A is chosen, $u \in F$.

Exercises

1. Complete the detailed proof from first principles that the variety generated by the algebra **A** of Example 4.1.3 is affine complete.

2. Construct a pair of primal algebras of the same type which generate a CD but not affine complete variety.

3. Show that no nontrivial variety can have the property that for each of its members *every* congruence compatible partial operation has an interpolating polynomial function. Exhibit an infinite algebra having this property.

4.2 General properties

In this section we establish three important properties enjoyed by all affine complete varieties. The results are from [Kaarli, Pixley, 1987] and [Kaarli, McKenzie, 1997].

The first of the three is residual finiteness. It is easy to observe that an infinite SI algebra **A** of countable type cannot be affine complete. Indeed, if μ is the monolith of **A** and K is a μ-block of size at least 2, then all functions $A \to K$ are compatible. The cardinality of the set of such functions is at least $2^{|A|}$ while the cardinality of the set of polynomial functions is only $|A|$. Now we show that an affine complete variety cannot contain an infinite SI, no matter what the type of the variety is. The proof requires some model theory.

Theorem 4.2.1 *Every affine complete variety is residually finite.*

PROOF Let **A** be an infinite SI member of an affine complete variety V with monolith μ and suppose a_1, a_2 are distinct elements in the same μ-block. Define $f : A^2 \to A$ for all $u, v \in A$ by

$$f(u, v) = \begin{cases} a_1 & \text{if } u = v, \\ a_2 & \text{otherwise.} \end{cases}$$

Then f is compatible since its values are in the same monolith class. Hence for some $(k + 2)$-ary term $t(x_1, x_2, \mathbf{y})$ and $\mathbf{c} \in A^k$,

$$f(u, v) = t(u, v, \mathbf{c})$$

for all $u, v \in A$. Now let Φ be the sentence

$$(\exists x_1, x_2)\, [x_1 \not\approx x_2 \wedge$$
$$(\exists \mathbf{y})(\forall u, v)(t(u, u, \mathbf{y}) \approx x_1 \wedge (u \not\approx v \to t(u, v, \mathbf{y}) \approx x_2))].$$

By the definition of f we have $\mathbf{A} \models \Phi$. Since A is infinite, by the Löwenheim-Skolem theorem (Theorem 1.1.2) there is a $\mathbf{B} \in V$ such that $|B| > |F| + \omega$ and $\mathbf{B} \models \Phi$. Hence there exist distinct $b_1, b_2 \in B$ and $\mathbf{d} \in B^k$ such that for any distinct $u, v \in B$,

$$b_1 = t(u, u, \mathbf{d}) \equiv_{\mathrm{Cg}(u,v)} t(u, v, \mathbf{d}) = b_2\,.$$

This means that $\mathrm{Cg}(b_1, b_2)$ must be the monolith congruence of **B**; that is, **B** is SI. Then, however, **B** has at least $2^{|B|} > |B|$ polynomials which contradicts the assumption $|B| > |F| + \omega$.   •

Notice that the proof does not say that an infinite SI affine complete algebra does not exist (unless the type is at most countable); rather, it can be interpreted to say that in a variety containing an infinite SI affine complete algebra there is always another algebra which is SI but not affine complete.

The next important property of affine complete varieties is an easy consequence of results of Chapter 2.

Theorem 4.2.2 *Every affine complete variety is congruence distributive.*

PROOF Let V be an affine complete variety, $\mathbf{A} \in V$, and $\mathbf{L} = \mathbf{Con}\,\mathbf{A}$. Then by Theorem 4.2.1 the equivalence lattice $\mathbf{L}$ is residually finite (each element is the meet of elements of finite index). Moreover, since all quotient algebras of $\mathbf{A}$ are affine complete, $\mathbf{L}$ has the (CFL) property. Hence, by Theorem 2.3.11 the lattice $\mathbf{L}$ is distributive. $\bullet$

The proof of the special case of Theorem 4.2.2 for locally finite varieties was an early and straightforward application of tame congruence theory made by R. McKenzie in 1986. It has not been published.

The third important general property of affine complete varieties is that none of their SIs has proper subalgebras. In order to establish this result we first prove that whenever $\mathbf{B}$ is a subalgebra of some algebra $\mathbf{A}$ contained in an affine complete variety then each compatible function on $\mathbf{B}$ has a unique compatible extension on $\mathbf{A}$. Note that since a compatible function on $\mathbf{B}$ is a polynomial the existence of such extension is obvious. Thus the only new content of the following theorem is the uniqueness of the extension.

Theorem 4.2.3 *If V is an affine complete variety, $\mathbf{A}$ an algebra of V, and $\mathbf{B} < \mathbf{A}$, then any compatible function on $\mathbf{B}$ has exactly one compatible extension on $\mathbf{A}$.*

PROOF We suppose the theorem is false, namely that f has extensions g and h and that $g(u) \neq h(u)$ for some $u \in A \setminus B$. Let

$$D = \{\mathbf{a} = (a_1, a_2, \ldots) \in A^\omega : a_i \text{ are eventually constant and in } B\}.$$

Hence $(a_1, a_2, \ldots) \in D$ iff there exist an integer n and $b \in B$ such that $a_i = b$ for all $i \geq n$. Obviously D is a subuniverse of $\mathbf{A}^\omega$ and $\mathbf{D} \in V$. Define $c : D \to D$ by

$$c(\mathbf{x})_i = \begin{cases} g(x_i) & \text{if } i \text{ is odd}, \\ h(x_i) & \text{if } i \text{ is even}. \end{cases}$$

Since $\mathbf{x}$ is eventually constant in B, so is $c(\mathbf{x})$ and thus $c : D \to D$.

Next we show that $c \in \operatorname{Comp} \mathbf{D}$. This requires that we show that

$$(c(\mathbf{x}), c(\mathbf{y})) \in \operatorname{Cg}(\mathbf{x}, \mathbf{y})$$

for all $\mathbf{x}, \mathbf{y} \in D$. Now for some integer n and elements $a, b \in B$, we have

$$\mathbf{x} = (x_1, x_2, \ldots, x_n, \mathbf{a})$$

and

$$\mathbf{y} = (y_1, y_2, \ldots, y_n, \mathbf{b})$$

where $\mathbf{a} = (a, a, a, \ldots)$ and $\mathbf{b} = (b, b, b, \ldots)$ are constants. But

$$\mathbf{D} = \mathbf{A}^n \times \mathbf{D}_n, \quad \mathbf{D}_n \cong \mathbf{D},$$

where D_n results from erasing everything before the $(n+1)$-st components of the members of D. In view of Theorem 4.2.2 $\mathbf{D}$ is CD and therefore has no skew congruences. Since c is defined componentwise with respect to the direct decomposition $\mathbf{D} = \mathbf{A}^n \times \mathbf{D}_n$, it is compatible. Therefore for some, say $(k+1)$-ary term, t,

$$c(\mathbf{x}) = t(\mathbf{x}, \mathbf{d}^1, \ldots, \mathbf{d}^k)$$

where each $\mathbf{d}^j = (d_1^j, d_2^j, \ldots)$ is eventually constant, say d^j in B. Hence for n_0 sufficiently large and $i \geq n_0$

$$c(\mathbf{x})_i = t(x_i, d^1, \ldots, d^k).$$

For such an i select $\mathbf{x} \in D$ with $x_i = x_{i+1} = u$. Then

$$\begin{aligned} g(u) = c(x_i) = t(x_i, d^1, \ldots, d^k) = \\ t(x_{i+1}, d^1, \ldots, d^k) = c(x_{i+1}) = h(u), \end{aligned}$$

a contradiction.   •

Theorem 4.2.3 has important consequences which are among the most striking features of affine complete varieties.

Corollary 4.2.4 *If V is an affine complete variety and $\mathbf{B} < \mathbf{A} \in V$ then $V(\mathbf{B}) = V(\mathbf{A})$.*

PROOF If p and q are terms and $p \approx q$ is one of the defining equations of $V(\mathbf{B})$, then by Theorem 4.2.3 $p \approx q$ is also an equation of $\mathbf{A}$. Hence $V(\mathbf{B}) = V(\mathbf{A})$. •

Notice that in Theorem 4.2.3 it is possible for the subalgebra $\mathbf{B}$ of $\mathbf{A}$ to have only a single element. Hence we have

Corollary 4.2.5 *If $\mathbf{A}$ is a nontrivial algebra in an affine complete variety, then $\mathbf{A}$ has no one element (i.e., trivial) subalgebras.*

Corollary 4.2.6 *Subdirectly irreducible algebras in an affine complete variety have no proper subalgebras.*

PROOF Suppose $\mathbf{B} < \mathbf{A} \in V$ where $\mathbf{A}$ is SI. Corollary 4.2.4 asserts that $\mathbf{A} \in V(\mathbf{B}) = IP_S HS(\mathbf{B})$ by the finiteness of $\mathbf{B}$. Hence the subdirect irreducibility of $\mathbf{A}$ yields $\mathbf{A} \in HS(\mathbf{B})$ so $\mathbf{B} = \mathbf{A}$. •

Corollary 4.2.7 *An affine complete variety is locally finite if and only if it is generated by a finite algebra with no proper subalgebras.*

PROOF If V is a locally finite affine complete variety then its free algebra $\mathbf{F}_1$ in one generator is finite. Since by Corollary 4.2.6 the SI members of V have no proper subalgebras, they all are isomorphic to quotient algebras of $\mathbf{F}_1$. Obviously $\mathbf{F}_1$ has only finitely many different quotient algebras; hence V has only finitely many SIs implying that V has a finite generator $\mathbf{A}$. However, the algebra $\mathbf{A}$ has a subalgebra with no proper subalgebras which by Corollary 4.2.4 generates V as well. •

In Section 3.4.3 we visualized the relationship between certain properties of algebras related to affine completeness. The following scheme does the same on the level of varieties.

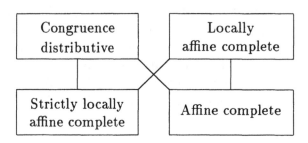

Figure 4.1

Recall that strictly locally affine complete varieties are precisely the arithmetical varieties and so they form a subclass of CD varieties.

4.3   Varieties with a finite residual bound

A variety V has a **finite residual bound** if there is a finite upper bound on the sizes of the SI members of V. In this section we shall show, most importantly, that affine complete varieties of countable type satisfy a condition (Property (S)) which guarantees a finite residual bound. We begin with some categorical properties which will result from this fact. All results of this section are from [Kaarli, McKenzie, 1997]. Under the assumption of congruence distributivity (not known at that time to be superfluous) the corollaries of Theorem 4.3.6 appeared in [Kaarli, Pixley, 1987]

4.3.1   Some categorical properties

We prove in this subsection that all affine complete varieties with finite residual bound have some important categorical properties. Namely, they have enough injectives and consequently they also have the congruence extension property and the amalgamation property. (See [Kiss, Márki, Pröhle, Tholen, 1983].) Actually, these properties follow already from congruence distributivity and the absence of proper subalgebras in SIs.

Recall that an algebra $\mathbf{E}$ in a variety V is said to be **injective in V** if for every $\mathbf{A} \in V$, every homomorphism $\alpha : \mathbf{A} \rightarrow \mathbf{E}$ can be extended to every $\mathbf{B} \in V$ that contains $\mathbf{A}$ as a subalgebra. A variety V **has enough injectives** iff every $\mathbf{A} \in V$ is a subalgebra of some injective member of V. An algebra $\mathbf{A}$ is said to have the **congruence extension property** if for any subalgebra $\mathbf{B} < \mathbf{A}$, the restriction map $\operatorname{Con} \mathbf{A} \rightarrow \operatorname{Con} \mathbf{B}$ is surjective. A class of algebras has the congruence extension property if so do all its members. A class K of algebras has the **amalgamation property** if for every $\mathbf{A}, \mathbf{B}, \mathbf{C} \in K$ and embeddings $\beta : \mathbf{A} \rightarrow \mathbf{B}$, $\gamma : \mathbf{A} \rightarrow \mathbf{C}$, there exists an algebra $\mathbf{D} \in K$ and embeddings $\beta_1 : \mathbf{B} \rightarrow \mathbf{D}$, $\gamma_1 : \mathbf{C} \rightarrow \mathbf{D}$ such that $\beta_1 \beta = \gamma_1 \gamma$.

Lemma 4.3.1 *Let V be a congruence distributive variety whose subdirectly irreducible algebras have no proper subalgebras. Assume that*

A, **B** *and* **S** *are finite algebras in* V, $\mathbf{A} < \mathbf{B}$, **S** *is subdirectly irreducible, and* $\alpha : \mathbf{A} \to \mathbf{S}$ *is a homomorphism. Then there exists a homomorphism* $\beta : \mathbf{B} \to \mathbf{S}$ *which extends* α.

PROOF Let **B** be subdirect in $\mathbf{S}_1 \times \cdots \times \mathbf{S}_n$ where all the $\mathbf{S}_i$ are SI. Then **A** is also subdirect in $\mathbf{S}_1 \times \cdots \times \mathbf{S}_n$. Since **S** has no proper subalgebra, α is surjective. Let ρ be the kernel of α and π_i be the kernel of the projection $p_i : \mathbf{B} \to \mathbf{S}_i$. Note that p_i is surjective and $p_i|_A$ is likewise surjective, because $\mathbf{S}_i$ has no proper subalgebras. Because of congruence distributivity and the fact that ρ is meet irreducible, there is an i with $\pi_i|_A \le \rho$. This gives a homomorphism $\gamma : \mathbf{S}_i \to \mathbf{S}$ for which $\gamma p_i(a) = \alpha(a)$ for all $a \in A$. Thus $\gamma p_i : \mathbf{B} \to \mathbf{S}$ extends α. ●

Theorem 4.3.2 *Suppose that V is a congruence distributive variety with a finite residual bound and no subdirectly irreducible algebra in V possesses a proper subalgebra. Then V has enough injectives; hence V has the congruence extension property and the amalgamation property.*

PROOF Obviously, every algebra in V can be embedded in a product of SIs in V, and it is well known (and easily verified) that the class of injective algebras in V is closed under direct products. Hence all statements of this theorem will follow if we can prove that every SI algebra in V is injective in V.

To prove this, let $\mathbf{A} < \mathbf{B} \in V$ and let $\alpha : \mathbf{A} \to \mathbf{S}$ be a homomorphism where **S** is a SI algebra in V. We can assume that **B** is subdirect in $\prod(\mathbf{S}_i : i \in I)$ where each $\mathbf{S}_i$ is SI. Then **A** is also subdirect in $\prod(\mathbf{S}_i : i \in I)$. We denote by ρ the kernel of α and remark that α is surjective, so that ρ is a completely meet irreducible congruence of **A**.

By Lemma 1.2.20, we can choose an ultrafilter $\mathcal{U}$ on I such that $\theta_{\mathcal{U}}|_A \le \rho$. Let $\nu : \mathbf{B} \to \prod(\mathbf{S}_i/\theta_{\mathcal{U}} : i \in I)$ be the natural map, whose kernel is $\theta_{\mathcal{U}}|_B$. Since $\theta_{\mathcal{U}}|_A \le \rho$, there is a homomorphism $\gamma : \mathbf{A}_1 \to \mathbf{S}$, where $\mathbf{A}_1 = \nu(\mathbf{A})$, such that $\gamma\nu(a) = \alpha(a)$ for all $a \in A$. Now the algebra $\mathbf{B}_1 = \nu(\mathbf{B}) < \prod(\mathbf{S}_i/\theta_{\mathcal{U}} : i \in I)$ is finite, since the $\mathbf{S}_i$ are of bounded finite size. By Lemma 4.3.1, there is a homomorphism $\delta : \mathbf{B}_1 \to \mathbf{S}$ such that $\delta(\pi(a)) = \gamma(\nu(a)) = \alpha(a)$ for all $a \in A$. Thus $\delta\nu|_B$ extends α to **B**, as was desired. ●

4.3.2   Property (S)

In this subsection we consider a class of affine complete varieties that lies between finitely generated varieties and those with finite residual bound.

That an algebra **A** has no proper subalgebras is equivalent to the condition that the set of unary term functions acts transitively on A, i.e., for every $a, b \in A$ there exists a unary term t such that $t(a) = b$. If **A** is finite, this condition takes the form $\mathbf{A} \models \Phi(T)$ where $T = \{t_1, \ldots, t_n\}$ is some finite set of unary terms and $\Phi(T)$ is the sentence

$$(\forall x, y)\, [t_1(x) \approx y \vee \cdots \vee t_n(x) \approx y] \,.$$

If we have a finite collection of algebras $\mathbf{A}_i$ of the same type and each of them models a certain sentence $\Phi(T_i)$ then obviously every $\mathbf{A}_i$ models the sentence $\Phi(T)$ where T is the union of all sets T_i. This situation occurs if the $\mathbf{A}_i$ are all of the SIs of a finitely generated affine complete variety. Indeed, by results of Section 4.2, all $\mathbf{A}_i$ are finite with no proper subalgebras, and because of congruence distributivity, their number is finite. Summarizing, every finitely generated affine complete variety V satisfies the following condition.

Definition A variety V has **property** (S) if there exists a finite set T of unary terms such that all subdirectly irreducible members of V satisfy the sentence $\Phi(T)$.

Obviously, property (S) implies that all SI members of the variety have at most $|T|$ elements (hence the variety has a finite residual bound) and all SI members have no proper subalgebras. However, it does not imply that the variety is finitely generated. We shall show that under certain natural restrictions on the variety, property (S) is either sufficient or necessary for affine completeness.

Theorem 4.3.3 *Every arithmetical variety with property* (S) *is affine complete.*

PROOF Assume that the unary terms $t_1, \ldots, t_n$ witness property (S) for an arithmetical affine complete variety V. Let f be an m-ary compatible function for the algebra $\mathbf{A} \in V$. We can assume that **A** is subdirect in $\prod(\mathbf{S}_i : i \in I)$ where the $\mathbf{S}_i$ are SI. Let $c \in A$ and put $C = \{t_1(c), \ldots, t_n(c)\}$.

Because V is arithmetical and C is finite, there is an m-ary polynomial g of $\mathbf{A}$ which agrees with f on C. Because of condition (S), the set C projects onto every S_i. Then, since both g and f are compatible for $\mathbf{A}$, it easily follows that $g(x_1, \ldots, x_m) = f(x_1, \ldots, x_m)$ for all $x_1, \ldots, x_m \in A$. •

Now it is natural to ask whether property (S) is satisfied in all affine complete varieties. As we shall see later, this is not the case. However, in some important classes of affine complete varieties, the property does hold.

Theorem 4.3.4 *Suppose that V is an affine complete variety whose class of subdirectly irreducible members is first order definable. Then V has property (S). In particular an affine complete variety with definable principal congruences has property (S).*

PROOF Assume that Ψ is a first order sentence in the language of V so that an algebra in V is SI iff it satisfies Ψ. Let T_i, $i \in I$, be the collection of all finite sets of unary terms in the language of V and consider the set of first order sentences

$$Eq(V) \cup \{\Psi\} \cup \{\neg\Phi(T_i) : i \in I\}.$$

This set is inconsistent. Indeed, if it were not then it would have a model $\mathbf{A}$, which would be an SI algebra in V that fails to satisfy any of the sentences $\Phi(T_i)$. This would mean that $\mathbf{A}$ has at least one proper subalgebra contradicting Corollary 4.2.6. Hence by the compactness theorem (Theorem 1.1.1) there is a finite set $\{i_1, \ldots, i_k\}$ such that the set

$$Eq(V) \cup \{\Psi, \neg\Phi(T_{i_1}), \ldots, \neg\Phi(T_{i_k})\}$$

is inconsistent. This means that any SI algebra $\mathbf{A} \in V$ satisfies at least one of the sentences $\Phi(T_{i_j})$ where $1 \leq j \leq k$. Obviously then, every SI in V satisfies the sentence $\Phi(T_{i_1} \cup \cdots \cup T_{i_k})$. •

Corollary 4.3.5 *A discriminator variety is affine complete iff it has property (S). More generally, any arithmetical variety with definable principal congruences is affine complete iff it has property (S).*

PROOF. Recall that discriminator varieties are those having a term $t(x, y, z)$ so that an algebra in V is SI iff it satisfies the first order sentence

$$(\forall x, y, z) \left[t(x, x, z) \approx z \,\&\, (x \not\approx y \longrightarrow t(x, y, z) \approx x) \right].$$

Now the necessity is immediate from Theorem 4.3.4. Since the function $t(x, y, z)$ is a Pixley function, discriminator varieties are arithmetical and the sufficiency follows from Theorem 4.3.3.   •

Theorem 4.3.6 *Every affine complete variety of countable type has property* (S).

PROOF Let V be an affine complete variety of countable type and t_n, $n \in \omega$, be a list of all the unary terms in the language of V. First assume, to obtain a contradiction, that for every $i \in \omega$ there is an SI algebra $\mathbf{A}_i \in V$ and an element $a_i \in A_i$ for which the set $\{t_0(a_i), \ldots, t_i(a_i)\}$ does not intersect any of the nontrivial blocks of the monolith congruence μ_i of $\mathbf{A}_i$.

We choose $b_i, c_i \in A_i$ so that $b_i \neq c_i$ and $b_i \equiv c_i \,(\mu_i)$. Let

$$\mathbf{P} = \prod (\mathbf{A}_i \times \mathbf{A}_i : i \in \omega)$$

and a be the element $(a_0, a_0, a_1, a_1, \ldots)$ of P. A general element $x \in P$ will be denoted as $(x(0), x(1), x(2), \ldots)$ so that for example $a(2i) = a(2i{+}1) = a_i$ for every $i \in \omega$. We take $\mathbf{D}$ to be the subalgebra of $\mathbf{P}$ generated by a. Clearly it consists of all the elements $t_i^{\mathbf{P}}(a)$ for $i \in \omega$. Then we take $\mathbf{D}'$ to be the subalgebra of $\mathbf{P}$ whose universe consists of all elements which for some $x \in D$ differ only at finitely many places from x. For every i, let f_i be the identity function on A_i and take g_i to be the function that differs from f_i only at c_i and has $g_i(c_i) = b_i$. Both f_i and g_i are compatible functions on $\mathbf{A}_i$. Let

$$f = (f_0, g_0, f_1, g_1, \ldots, f_i, g_i, \cdots),$$

a function mapping D' into P. Since for every $x \in D'$ there is an n such that for all sufficiently large i,

$$x(2i) = x(2i + 1) = t_n(a_i) \neq c_i,$$

it follows that $f(x)$ eventually agrees with x and so $f(x) \in D'$.

The function f is a compatible function of $\mathbf{D}'$. Indeed, suppose $x, y \in D'$ and choose m large enough so that $f(x)$ agrees with x and $f(y)$ agrees with y at $2m$ and at all later places. Identifying $\mathbf{D}'$ with a subdirect product of $\mathbf{A}_0 \times \mathbf{A}_0 \times \cdots \times \mathbf{A}_{m-1} \times \mathbf{A}_{m-1}$ and a certain subalgebra $\mathbf{D}_m$ of $\mathbf{A}_m \times \mathbf{A}_m \times \cdots$, we have that

$$x = (x_0, \ldots, x_{2m-1}, u), \quad f(x) = (x_0, g_0(x_1), \ldots, g_{m-1}(x_{2m-1}), u),$$

$$y = (y_0, \ldots, y_{2m-1}, v), \quad f(y) = (y_0, g_0(y_1), \ldots, g_{m-1}(y_{2m-1}), v).$$

This subdirect decomposition of $\mathbf{D}'$ into $2m + 1$ factors has no skew congruences (since the variety V is CD). From this and the fact that each of the coordinate functions f_i, g_i of f is compatible, it is clear that f is a compatible function of $\mathbf{D}'$. However, if we choose $p(x) = t(x, d^1, \ldots, d^r)$ to be any polynomial function of $\mathbf{D}'$ where t is a term and $\{d^1, \ldots, d^r\} \subseteq D'$, and if we choose m large enough so that for $k \geq m$ we have

$$(d^1(2k), \ldots, d^r(2k)) = (d^1(2k+1), \ldots, d^r(2k+1))$$

then the coordinate functions of $p(x)$ at the $2k$ and $2k+1$ coordinates will be equal when $k \geq m$. Since $f(x)$ does not have this property, it follows that $p(x)$ and $f(x)$ cannot coincide on D'. This contradicts the fact that the algebra $\mathbf{D}'$ is affine complete.

The contradiction shows that there exists an integer m such that whenever a belongs to an SI algebra $\mathbf{A} \in V$ then among the values $t_0(a), \ldots, t_m(a)$ there is an element which lies in a nontrivial class of the monolith of $\mathbf{A}$. All that remains is to show that there is an integer n so that whenever a lies in a nontrivial monolith class for an SI algebra $\mathbf{A} \in V$ then $A = \{t_0(a), \ldots, t_n(a)\}$, for then the terms $t_i(t_j(x))$ ($i \leq n$, $j \leq m$) will satisfy the requirement of the theorem.

Assume to the contrary that for all i there is an SI algebra $\mathbf{A}_i$ in V and distinct elements a_i, b_i congruent modulo the monolith of $\mathbf{A}_i$ and an element $c_i \in A_n \setminus \{t_0(a_n), \ldots, t_n(a_n)\}$. We form $\mathbf{P}$, $\mathbf{D}$ and $\mathbf{D}'$ just as before. Here we take f_i and g_i to be the compatible functions of $\mathbf{A}_i$ such that $f_i(x) = a_i$ is constant and $g_i(x) = a_i$ for all x except $x = c_i$ where $g_n(c_i) = b_i$. Exactly the same argument as before yields precisely the same contradiction. ●

Corollary 4.3.7 *Each affine complete variety of countable type has a finite residual bound.*

Corollary 4.3.8 *Each affine complete variety of finite type is finitely generated.*

PROOF Let V be an affine complete variety of finite type. Then by Theorem 4.3.6 V has a finite bound for the sizes of its SIs. The latter together with the finite type assumption implies that V has only finitely many SIs up to isomorphism. Because all of the SIs are finite their direct product is therefore also a finite algebra which generates V.   ●

4.3.3 Two counter-examples

We present here two examples which will show that Theorem 4.3.6 and Corollary 4.3.7 cannot be extended to varieties of arbitrary type. These examples first appeared in [Kaarli, McKenzie, 1997]. Both of them rely on the following lemma.

Lemma 4.3.9 *There exists an uncountable family S of infinite subsets of ω with the property that every two members of S have finite intersection.*

PROOF For example we may construct such a family as follows. For every real number r pick a rational sequence $(a_i : i \in \omega)$ converging to r and put $S_r = \{\{a_0, \ldots, a_i\} : i \in \omega\}$. Then the set of all S_r forms an uncountable family of subsets in the countable set of all finite rational sequences and two distinct such sets can have only finitely many members in common.   ●

Example 4.3.10 *An arithmetical affine complete variety having uncountable type which is residually ≤ 4 and fails to have property* (S).

By Lemma 4.3.9 there exists an uncountable family S of subsets of ω such that every two distinct members of S have a finite intersection. Applying Zorn's Lemma, we can assume that S is maximal, that is that every infinite subset of ω has an infinite intersection with some member of S. Also, if $U_0, \ldots, U_n$ are finitely many members of S then $\omega \setminus (U_0 \cup \cdots \cup U_n)$ is infinite, because it contains all but finitely many of the elements in any set $U \in S$ different from $U_0, \ldots, U_n$.

We construct a variety V of type

$$F = \{p, 0, 1\} \cup \{t_n, c_n, d_n : n \in \omega\} \cup \{e_U : U \in S\}$$

where p and t_n are ternary operation symbols and all remaining operation symbols are nullary. In what follows we shall need the subtypes $H = \{p, 0, 1\}$ and $G = H \cup \{t_n, c_n, d_n : n \in \omega\}$ of F. Our variety will be generated by a two element algebra $\mathbf{T}$ and a countable series of four element algebras $\mathbf{S}_n$, $n \in \omega$, which are defined as follows.

The universe of the algebra $\mathbf{T}$ is $T = \{0, 1\}$ where the nullary operation symbols 0 and 1 are interpreted by themselves. The $p(x, y, z)$ is interpreted in $\mathbf{T}$ as the ternary discriminator. For all $n \in \omega$ and $U \in S$, we put $t_n(x, y, z) = x$ and $c_n = d_n = e_U = 1$ in $\mathbf{T}$.

The reduct of every $\mathbf{S}_n$, $n \in \omega$, to the type H is $\mathbf{T} \times \mathbf{T}$. The remaining operations are interpreted in $\mathbf{S}_n$ as follows:

$$t_k(x, y, z) = \begin{cases} x \text{ if } k < n, \\ \text{the ternary discriminator on } S_n \text{ if } k \geq n, \end{cases}$$

$$c_k = \begin{cases} (1, 1) & \text{if } k < n, \\ (0, 1) & \text{if } k \geq n, \end{cases} \quad d_k = \begin{cases} (1, 1) & \text{if } k < n, \\ (1, 0) & \text{if } k \geq n, \end{cases}$$

$$e_U = \begin{cases} (0, 1) & \text{if } n \in U, \\ (1, 1) & \text{if } n \notin U. \end{cases}$$

Note that each of the algebras $\mathbf{T}$, $\mathbf{S}_n$ has a ternary discriminator among its term functions; so they all are quasiprimal, hence simple. Obviously $\mathbf{T}$ has no proper subalgebras. Since 0, 1, c_n and d_n are interpreted as (0,0), (1,1), (0,1) and (1,0), correspondingly, in $\mathbf{S}_n$, the algebras $\mathbf{S}_n$ have no proper subalgebras as well. Hence by Theorem 1.2.22 each of the algebras $\mathbf{T}$, $\mathbf{S}_n$, but also $\langle T; G \rangle$, is the only SI of the variety it generates. Furthermore, the variety V is arithmetical because $p(x, y, z)$ is a Pixley term for V.

Our first task is to find all the SIs of V. Let $\mathbf{S}$ be any SI in V. Since V is CD, by Theorem 1.2.22 $\mathbf{S} \in HSP_U(\mathbf{T}, \{\mathbf{S}_n : n \in \omega\})$. That is, we can write $\mathbf{S}$ as isomorphic to $\mathbf{D}/\rho$ where

$$\mathbf{D} < \mathbf{E} = \prod(\mathbf{B}_i : i \in I)/\theta_{\mathcal{U}},$$

each $\mathbf{B}_i$ is one of our generating SIs, and $\mathcal{U}$ is an ultrafilter on I. For $n \in \omega$ let $I_n = \{i \in I : \mathbf{B}_i = \mathbf{S}_n\}$. If the ultrafilter $\mathcal{U}$ contains one of the sets I_n, then $\mathbf{E}$ satisfies all the equations of $\mathbf{S}_n$. Hence $\mathbf{E}$, but then $\mathbf{D}$ and $\mathbf{S}$ as well, are members of $V(\mathbf{S}_n)$. Since $\mathbf{S}_n$ is the only SI of the variety it generates, $\mathbf{S} \cong \mathbf{S}_n$.

Now assume that the ultrafilter $\mathcal{U}$ contains none of the sets I_n. Then it must contain all the complements $I \setminus I_n$ and therefore also the complements of all unions $I_0 \cup I_1 \cup \cdots \cup I_n$, $n \in \omega$. This implies that the equations

$$t_k(x, y, z) \approx x, \quad c_k \approx d_k \approx 1$$

hold in $V(\mathbf{E})$. Thus the reduct of algebra $\mathbf{E}$ to G belongs to the variety generated by $\langle T; G \rangle$. Since $\langle T; G \rangle$ is the only SI of that variety, we have the isomorphism $\langle S; G \rangle \cong \langle T; G \rangle$.

Thus, the list of the SIs of V is the following:

1. the four element algebras $\mathbf{S}_n$, $n \in \omega$,

2. the two element algebras whose reducts to type G are $\langle T; G \rangle$ and that differ with interpretation of operations e_U in them.

Now observe that V does not satisfy (S). Let F' be any finite subset F. Then choose $n \in \omega$ to be larger than all k such that c_k or d_k or t_k occurs in F' and to lie outside of every set $U \in S$ such that $e_U \in F'$. Note that the latter is possible due to the maximality of the family S. The algebra $\langle S_n; F' \rangle$ is then term equivalent to $\mathbf{T} \times \mathbf{T}$ and therefore has no unary term t satisfying $t((1,1)) = (0,1)$.

It remains to show that V is affine complete. Let f be an m-ary compatible function on the algebra $\mathbf{A} \in V$. We assume that $\mathbf{A}$ is subdirect in $\prod(\mathbf{B}_i : i \in I)$ where every $\mathbf{B}_i$ is a SI from the above list. Again, let $I_n = \{i \in I : \mathbf{B}_i = \mathbf{S}_n\}$ and let f_i denote the function which f induces on $\mathbf{B}_i$.

First, suppose that there are infinitely many $n \in \omega$ for which some f_i, $i \in I_n$, is not compatible for the algebra $\langle B_i; H \rangle$, i.e., does not preserve both kernels of the projections of $B_i = S_n$ onto T. Then, because of maximality of S, the set of those n must have an infinite intersection with some $X \in S$. That is, there exists a set $X \in S$ and an infinite set $Y \subseteq X$, and for every $n \in Y$ an $i_n \in I_n$ such that f_{i_n} fails to be compatible for $\langle S_n; H \rangle$.

Since $n \in Y \subseteq X$, the operation e_X is interpreted in $\mathbf{S}_n$ as $(0,1)$. Therefore the subset $M = \{0, 1, e_X, p(0, e_X, 1)\} \subseteq A$ projects onto B_{i_n} at every i_n, $n \in Y$. Since f is compatible and V is arithmetical, there is a polynomial $t(\mathbf{x}, \mathbf{a})$ of $\mathbf{A}$, where t is a term of type F and $\mathbf{a} \in A^k$, which agrees with f on M. Since M projects onto B_{i_n}, we have that $t(\mathbf{x}, \mathbf{a})$ induces the function f_{i_n} on B_{i_n}.

Pick $n \in Y$ larger than all the i such that t_i occurs in t. Then there is a term t' involving only p and constants such that the equation $t \approx t'$ holds in $\mathbf{B}_{i_n}$. Hence the polynomial $t'(\mathbf{x}, \mathbf{a})$ induces the function f_{i_n} on B_{i_n}. But this is a contradiction because $t'(\mathbf{x}, \mathbf{a})$ is a compatible function of $\langle B_{i_n} ; H \rangle$.

Thus there is $n_0 \in \omega$ such that f_i is compatible for the algebra $\langle B_i ; H \rangle$ for all $i \in J$ where J is the union of all I_n with $n > n_0$.

Now observe that the variety generated by $\langle T; H \rangle$ is term equivalent to the variety of Boolean algebras with the constants 0 and 1 having their usual meaning. Hence it is arithmetical and $\mathrm{Cg}(0, 1) = 1_A$ holds in $\langle A; H \rangle$. Therefore there exists a polynomial q of $\langle A; H \rangle$ that agrees with f on the finite set $\{0^{\mathbf{A}}, 1^{\mathbf{A}}\}^m$, so the functions q_i and f_i they induce on B_i agree on $\{0^{\mathbf{B}_i}, 1^{\mathbf{B}_i}\}^m$. The latter set, however, projects onto the m-th power of both direct factors of $B_i = T \times T$. Assuming now $i \in J$ we have that both q_i and f_i are compatible functions of $\langle B_i; H \rangle$; so they must induce equal functions on both of those two direct factors. Hence $f_i = q_i$.

Let r be a polynomial of $\mathbf{A}$ which agrees with f on $\{0, 1, c_{n_0}, d_{n_0}\}$. Note that this set projects onto B_i for all $i \in I \setminus J$; hence $r_i = f_i$ for all those i. We put

$$s(\mathbf{x}) = p(p(c_{n_0}, d_{n_0}, q(\mathbf{x})), p(c_{n_0}, d_{n_0}, r(\mathbf{x})), r(\mathbf{x}))$$

and prove that $s(x) = f(x)$ for all $\mathbf{x} \in A^m$. Obviously it is sufficient to prove that $s_i = f_i$ for all $i \in I$; that is, the s and f induce the same functions on all subdirect factors $\mathbf{B}_i$.

First suppose that $i \notin J$. Then

$$
\begin{aligned}
s(\mathbf{x})_i &= p(p((0,1),(1,0),q(\mathbf{x})_i), p((0,1),(1,0),r(\mathbf{x})_i), r(\mathbf{x})_i) \\
&= p((0,1),(0,1),r(\mathbf{x})_i) = r(\mathbf{x})_i = f(\mathbf{x})_i .
\end{aligned}
$$

If, on the other hand, $i \in J$, then c_{n_0} and d_{n_0} are interpreted in $\mathbf{B}_i$ as $(1,1)$ so

$$s(\mathbf{x})_i = p(q(\mathbf{x})_i, r(\mathbf{x})_i, r(\mathbf{x})_i) = q(\mathbf{x})_i = f(\mathbf{x})_i .$$

Example 4.3.11 *An arithmetical affine complete variety which has no finite residual bound.*

This example is a modification of the previous one. First put $T = \{0, 1\}$ and for $n \in \omega$ put $S_n = T^{n+2}$. Let $P = \prod(S_n : n \in \omega)$.

The type of the variety will be

$$F = \{p, 0, 1\} \cup \{t_n : n \in \omega\} \cup \{c_y : y \in P\}$$

where p and t_n are ternary operation symbols and all others are nullary operation symbols. Let

$$H = \{p, 0, 1\} \quad \text{and} \quad G = H \cup \{t_n : n \in \omega\}.$$

The variety V will be generated by algebras $\mathbf{T}$ and $\mathbf{S}_n$, $n \in \omega$, which are defined as follows. The operation p is interpreted as the ternary discriminator in $\mathbf{T}$. The operations 0 and 1 are interpreted as themselves in $\mathbf{T}$. The reduct of each $\mathbf{S}_n$ to G is $\langle T; G \rangle^{n+2}$. Each operation c_y, $y \in P$, is interpreted as 1 in $\mathbf{T}$ and as $y(n)$ in each $\mathbf{S}_n$. The operations t_n are formally defined exactly as in Example 4.3.10.

The same proof used in Example 4.3.10 shows that every SI of V is isomorphic either to some $\mathbf{S}_n$, $n \in \omega$, or is isomorphic to an algebra whose reduct to G is $\langle T; G \rangle$.

The proof that V is affine complete is almost the same as in Example 4.3.10. Let f be an m-ary compatible function on the algebra $\mathbf{A} \in V$. We assume that $\mathbf{A}$ is subdirect in $\prod(\mathbf{B}_i : i \in I)$ where every $\mathbf{B}_i$ either is equal to some $\mathbf{S}_n$ or has $\langle T; G \rangle$ as the reduct to G. We again denote $I_n = \{i \in I : \mathbf{B}_i = \mathbf{S}_n\}$. Let f_i denote the function which f induces on $\mathbf{B}_i$.

The proof is essentially different from that in Example 4.3.10 only in the argument that there exists only finitely many $n \in \omega$ for which some f_i, $i \in I_n$, is not compatible for the algebra $\langle B_i; H \rangle$. Suppose that this fails. Choose an infinite set $Y \subseteq \omega$ and for every $n \in Y$ we choose $i_n \in I_n$ and $\mathbf{a}(n), \mathbf{b}(n) \in S_n^m$ so that $(f_{i_n}(\mathbf{a}(n)), f_{i_n}(\mathbf{b}(n)))$ does not belong to the principal congruence of $\langle S_n; H \rangle$ generated by the pair $(\mathbf{a}(n), \mathbf{b}(n))$. Choose $y^j, z^j \in P$, $j = 1, \ldots, m$, so that for every $n \in Y$,

$$(y^1(n), \ldots, y^m(n)) = \mathbf{a}(n) \quad \text{and} \quad (z^1(n), \ldots, z^m(n)) = \mathbf{b}(n).$$

Since $f \in \mathrm{Comp}\,\mathbf{A}$ and V is arithmetical, there is a polynomial q of $\mathbf{A}$ which agrees with f on the set $\{(c_{y^1}, \ldots, c_{y^m}), (c_{z^1}, \ldots, c_{z^m})\}$. Suppose that $q(\mathbf{x}) = t(\mathbf{x}, \mathbf{d})$ where t is an $(m + l)$-ary term and $\mathbf{d} \in A^l$. Pick $n \in Y$ larger than all the j such that t_j occurs in t and let t' be the term involving only p and constants such that the polynomials $q(\mathbf{x})$ and $t'(\mathbf{x}, \mathbf{d})$ induce the same function at $i = i_n$. This implies that $f_{i_n}(\mathbf{a}(n)) = t'(\mathbf{a}(n), \mathbf{d}_{i_n})$ and $f_{i_n}(\mathbf{b}(n)) = t'(\mathbf{b}(n), \mathbf{d}_{i_n})$. Of course, this is impossible because in the subdirect factor $\mathbf{B}_{i_n}$ the polynomial $t'(\mathbf{x}, \mathbf{d}_{i_n})$ induces a compatible function of $\langle S_n; H \rangle$.

The remainder of the proof that f is a polynomial of $\mathbf{A}$ can be found by making a small modification in the proof of the corresponding part of Example 4.3.10.

Exercise

Construct an affine complete arithmetical variety of countable type which is not finitely generated ([Kaarli, Pixley, 1987]).

4.4 Locally finite affine complete varieties

In the beginning of this section (Subsections 4.4.1 and 4.4.2) we seek a characterization of locally finite affine complete varieties. Though the result is not as transparent as one might desire, it is still quite satisfactory. We know already that a locally finite affine complete variety must be CD and be generated by a finite algebra with no proper subalgebras (Theorem 4.2.2 and Corollary 4.2.7). It will turn out that these necessary conditions plus the condition that all algebras with no proper subalgebras be affine complete are already necessary and sufficient. Note that there are only finitely many algebras with no proper subalgebras in a given locally finite variety. An important tool throughout the section is Theorem 2.3.8. As a matter of fact, the introduction of compatible function systems in [Kaarli, 1997] was inspired by the problem of characterizing locally finite affine complete varieties. The results of these two subsections also were originally presented in [Kaarli, 1997].

4.4.1 Existence of a near unanimity term

We have already established (Corollary 4.2.7) that all locally finite affine complete varieties are finitely generated. A remarkable fact is that one more finiteness condition is equivalent to local finiteness in the case of affine complete varieties. We first observe that these varieties always have a near unanimity term.

Theorem 4.4.1 *Every locally finite affine complete variety is a near unanimity variety.*

PROOF Let V be a locally finite affine complete variety. By Corollary 4.2.7 V is finitely generated; moreover, we may suppose that $V = V(\mathbf{A})$ where $\mathbf{A}$ is a finite algebra with no proper subalgebras. Let $\mathbf{F}_1$ be the free algebra in one generator of V so that we can think of the elements of $\mathbf{F}_1$ as the unary term functions on $\mathbf{A}$. By Theorem 2.3.9, $\mathbf{F}_1$ admits a compatible m-ary near unanimity function ($m \geq 3$). Hence, due to the affine completeness of $\mathbf{F}_1$, there exist an $(m+p)$-ary term t and unary terms $w_1, \ldots, w_p$ such that

$$t(u(x), \ldots, u(x), v(x), u(x), \ldots, u(x), w_1(x), \ldots, w_p(x)) = u(x) \quad (4.1)$$

for every two unary terms u and v and for every position of v. Note that (4.1) is the equality of functions. Thus it holds when replacing the variable x by arbitrary $a \in A$.

We claim that $t(x_1, \ldots, x_m, w_1(x_1), \ldots, w_p(x_1))$ is then a near unanimity term for V. We have to prove the following two equalities for arbitrary $a, b \in A$:

$$t(a, \ldots, a, b, a, \ldots, a, w_1(a), \ldots, w_p(a)) = a \, ; \quad (4.2)$$

$$t(b, a, \ldots, a, w_1(b), \ldots, w_p(b)) = a \, . \quad (4.3)$$

(Note that in (4.2) b is not in the first position.) Since $\mathbf{A}$ has no proper subalgebras, there are unary terms u and v such that $u(a) = a$ and $v(a) = b$. Hence in view of (4.1) we have

$$t(a, \ldots, a, b, a, \ldots, a, w_1(a), \ldots, w_p(a)) =$$
$$t(u(a), \ldots, u(a), v(a), u(a), \ldots, u(a), w_1(a), \ldots, w_p(a)) =$$
$$u(a) = a,$$

proving (4.2). Likewise, choosing the terms u and v so that $u(b) = b$ and $v(b) = a$, we get the equality (4.3). $\bullet$

Corollary 4.4.2 *An affine complete variety is finitely generated if and only if it is term equivalent to a variety of finite type.*

PROOF By Corollary 4.3.8 every affine complete variety of finite type is finitely generated. For the converse, suppose that V is an affine complete variety generated by a finite algebra $\mathbf{A}$. By Theorem 4.4.1 the clone of term functions on $\mathbf{A}$ contains a near unanimity function; hence, by Theorem 3.2.4, it is generated by a finite set F.

Then the algebras **A** and $\langle A; F \rangle$ are term equivalent and therefore so are the varieties they generate. •

Combining Corollary 4.4.2 and Corollary 4.3.8 and (for (4)) applying the remarks following part (b) of Theorem 1.2.22, we summarize with the following theorem which is one of most striking properties of affine complete varieties.

Theorem 4.4.3 *For an affine complete variety V the following are equivalent:*

(1) *V is finitely generated;*

(2) *V is locally finite;*

(3) *V is term equivalent to a variety of finite type;*

(4) *V has only finitely many subvarieties.*

Although we know by virtue of Example 4.3.11 that in general, for uncountable type, even in the arithmetical case, an affine complete variety need not have a finite residual bound, the following simple observation is worth making.

Theorem 4.4.4 *For any affine complete variety the number of subdirectly irreducible members is, up to weak isomorphism, at most countable.*

PROOF For each integer n, by Corollary 4.4.2, each SI algebra of size n is term equivalent to an algebra of finite type and up to weak isomorphism there are at most a countable number of these; so altogether there are at most a countable number. •

4.4.2 Structure theorems

Since locally finite affine complete varieties are finitely generated, a natural question is: can we characterize these varieties in terms of generators? Knowing that affine complete varieties are CD we may put the question as follows: what are the finite algebras **A** of CD varieties that generate affine complete variety? In view of Corollary 4.2.7 it is enough to consider the algebras **A** that have

no proper subalgebras. By Corollary 4.1.2 no more restrictions are
needed if we are in an arithmetical variety. The general situation,
however, is by far more complicated. For example, the variety of
bounded distributive lattices is not affine complete though it is CD
and generated by a two element lattice that has no proper sublattices
containing 0 and 1. In this example the generating algebra itself is
not affine complete because the transposition interchanging 0 and 1
cannot be realized by a polynomial. Indeed, all polynomial functions
of lattices must preserve the order. Hence we should require at least
that the generating algebra **A** be affine complete. We shall see later
that this is still not enough because there does exist a finite algebra
with no proper subalgebras that generates an affine complete variety
but which has a quotient algebra that is not affine complete. The next
hope is that the algebra **A** has to be **hereditarily affine complete**;
that is, all quotient algebras of **A** are affine complete. The next
theorem shows that this requirement certainly brings us closer to
the answer.

Theorem 4.4.5 *Let V be a congruence distributive variety generated by a finite hereditarily affine complete algebra* **A** *which has no
proper subalgebras. Then all algebras* **B** $\in V$ *which have a subalgebra
isomorphic to* **A** *are affine complete.*

PROOF By Theorem 1.2.22, $V(\mathbf{A})$ is contained in $IP_S H(\mathbf{A})$.
Thus, let $\mathbf{B} \in V(\mathbf{A})$ and suppose that **A** is contained in **B** as a
subalgebra. Since **A** has no proper subalgebras, **B** is subdirect in
$\prod(\mathbf{A}_i : i \in I)$ where $\mathbf{A}_i$ are quotient algebras of **A**. Let

$$\mathbf{A}_i = \mathbf{A}/\tau_i\,, \quad \tau_i \in L = \mathrm{Con}\,\mathbf{A}\,, \quad i \in I\,.$$

We denote the members of $\prod(\mathbf{A}_i : i \in I)$ as families $(a_i/\tau_i)_{i \in I}$ where
$a_i \in A$. It is easily seen that the embedding $\mathbf{B} < \prod(\mathbf{A}_i : i \in I)$
can be chosen so that every $a \in A$ regarded as an element of B is
represented as $(a/\tau_i)_{i \in I}$.

Consider an arbitrary compatible m-ary function f on **B**. Because of the compatibility, the function f factors as $f = (f_i)_{i \in I}$
where f_i is an m-ary compatible function on $\mathbf{A}_i\,, i \in I$.

Now we introduce some notation based on an enumeration of $f(A^m)$:

$$
\begin{aligned}
f(A^m) &= \{s_1, \ldots, s_p\}, \\
s_j &= (s_j^i/\tau_i)_{i \in I}, && s_j^i \in A, \quad j = 1, \ldots, p, \\
\mathbf{s}^i &= (s_1^i, \ldots, s_p^i), && i \in I.
\end{aligned}
$$

For every $\rho \in L = \operatorname{Con} \mathbf{A}$ define an $(m+p)$-ary partial function g_ρ on $\mathbf{A}/\rho$ as follows:

$$
\begin{aligned}
\operatorname{dom} g_\rho &= (A/\rho)^m \times \{\mathbf{s}^i/\rho : \tau_i = \rho\}, \\
g_\rho(\mathbf{a}/\rho, \mathbf{s}^i/\rho) &= f_i(\mathbf{a}/\rho).
\end{aligned}
$$

First check that all g_ρ are well defined. Indeed, suppose that $i \neq j$, $\tau_i = \rho = \tau_j$ and $\mathbf{s}^i/\rho = \mathbf{s}^j/\rho$ and let $f(\mathbf{a}) = f(a_1, \ldots, a_m) = s_u$. Then $f_i(\mathbf{a}/\rho) = s_u^i/\rho = s_u^j/\rho = f_j(\mathbf{a}/\rho)$.

We show now that $g = (g_\rho)_{\rho \in L}$ is a compatible function system. Thus we have to check the conditions $(\mathcal{A})$ and $(\mathcal{B})$ of Definition 2.3.2. Let

$$
\mathbf{a}^1 = (a_1^1, \ldots, a_m^1), \quad \mathbf{a}^2 = (a_1^2, \ldots, a_m^2)
$$

be arbitrary elements of A^m, $i, j \in I$ and

$$
\begin{aligned}
g(\mathbf{a}^1/\tau_i, \mathbf{s}^i/\tau_i) &= f_i(\mathbf{a}^1/\tau_i) = b^1/\tau_i, \\
g(\mathbf{a}^2/\tau_j, \mathbf{s}^j/\tau_j) &= f_j(\mathbf{a}^2/\tau_j) = b^2/\tau_j.
\end{aligned}
$$

It follows from the definition of g that $(\mathbf{a}^1/\tau_j, \mathbf{s}^j/\tau_j) \in \operatorname{dom} g$. Let

$$
g(\mathbf{a}^1/\tau_j, \mathbf{s}^j/\rho_j) = f_j(\mathbf{a}^1/\tau_j) = b/\tau_j.
$$

By compatibility of f_j we have immediately

$$
(b^2, b) \in \tau_j \vee \operatorname{Cg}(\mathbf{a}^1, \mathbf{a}^2). \tag{4.4}
$$

On the other hand, if $f(\mathbf{a}^1) = s_u$ then

$$
\begin{aligned}
b^1/\tau_i &= f_i(\mathbf{a}^1/\tau_i) = s_u^i/\tau_i, \tag{4.5} \\
b/\tau_j &= f_j(\mathbf{a}^1/\tau_j) = s_u^j/\tau_j. \tag{4.6}
\end{aligned}
$$

Obviously $(s_u^i, s_u^j) \in \operatorname{Cg}(\mathbf{s}^i, \mathbf{s}^j)$, hence by (4.5) and (4.6),

$$
(b, b^1) \in \tau_i \vee \tau_j \vee \operatorname{Cg}(\mathbf{s}^i, \mathbf{s}^j). \tag{4.7}
$$

Now (4.4) and (4.7) imply

$$(b^1, b^2) \in \tau_i \vee \tau_j \vee Cg(\mathbf{a}^1, \mathbf{a}^2) \vee Cg(\mathbf{s}^i, \mathbf{s}^j)$$

meaning that g satisfies $(\mathcal{A})$.

The next step is to prove that g satisfies $(\mathcal{B})$ as well. Let $\rho_l \in L$ and

$$(\mathbf{a}^l/\rho_l, \mathbf{c}^l/\rho_l) \in dom\ g, \quad l = 1, \ldots, r.$$

This means that for every $l \in \{1, \ldots, r\}$ there is $i_l \in I$ such that $\rho_l = \tau_{i_l}$ and $\mathbf{c}^l = \mathbf{s}^{i_l}$. Let

$$g(\mathbf{a}^l/\rho_l, \mathbf{c}^l/\rho_l) = f_{i_l}(\mathbf{a}^l/\rho_l) = b^l/\rho_l, \quad l = 1, \ldots, r.$$

Suppose that there are

$$\mathbf{a} = (a_1, \ldots, a_m) \in A^m, \quad \mathbf{c} = (c_1, \ldots, c_p) \in A^p$$

and $\sigma_l \in L$ such that

$$\rho_l \leq \sigma_l, \quad (\mathbf{a}^l, \mathbf{a}), (\mathbf{c}^l, \mathbf{c}) \in \sigma_l, \quad l = 1, \ldots, r.$$

By the definition of g, all $(\mathbf{a}/\rho_l, \mathbf{c}^l/\rho_l)$ are in $dom\ g$. Let

$$g(\mathbf{a}/\rho_l, \mathbf{c}^l/\rho_l) = d^l/\rho_l, \quad l = 1, \ldots, r.$$

Since f_{i_l} is compatible, $(\mathbf{a}, \mathbf{a}^l) \in \sigma_l$ implies

$$(b^l, d^l) \in \sigma_l, \quad l = 1, \ldots, r. \tag{4.8}$$

On the other hand, if $f(\mathbf{a}) = s_u$ then $d^l/\rho^l = f_{i_l}(\mathbf{a}/\rho_l) = s_u^{i_l}/\rho_l$ and consequently

$$(d_l, s_u^{i_l}) \in \sigma_l, \quad l = 1, \ldots, r. \tag{4.9}$$

Because of $\mathbf{c}^l = \mathbf{s}^{i_l}$ and $(\mathbf{c}^l, \mathbf{c}) \in \sigma_l$, we also have

$$(c_u, s_u^{i_l}) \in \sigma_l, \quad l = 1, \ldots, r. \tag{4.10}$$

Now (4.8)–(4.10) imply $(c_u, b^l) \in \sigma_l$ for every $l = 1, \ldots, r$ and we are done.

It remains to show that f is a polynomial function. By Theorem 2.3.8 g is induced by some $(m + p)$-ary polynomial q of $\mathbf{A}$. We prove that the function $f = f(x_1, \ldots, x_m)$ coincides with the m-ary polynomial function $q(x_1, \ldots, x_m, s_1, \ldots, s_p)$. To do this it is enough to verify that they induce the same function on each $\mathbf{A}_i$, $i \in I$. Take arbitrary $a_1, \ldots, a_m \in A$ and compute:

$$(q(a_1, \ldots, a_m, s_1, \ldots, s_p))_i =$$
$$g_{\tau_i}(a_1/\tau_i, \ldots, a_m/\tau_i, s_1^i/\tau_i, \ldots, s_p^i/\tau_i) =$$
$$f_i(a_1/\tau_i, \ldots, a_m/\tau_i) = (f(a_1, \ldots, a_m))_i.$$

Thus f is a polynomial. ●

Now every algebra $\mathbf{B}$ of a locally finite variety V contains a minimal subalgebra $\mathbf{A}$ which obviously has no proper subalgebras. Moreover, if all subdirectly irreducible members of V have no proper subalgebras then $V(\mathbf{A}) = V(\mathbf{B})$. Thus Theorem 4.4.5 implies the following characterization of locally finite affine complete varieties.

Theorem 4.4.6 *A locally finite variety V is affine complete if and only if the following conditions are satisfied:*

(i) *V is congruence distributive;*

(ii) *V is generated by an algebra which has no proper subalgebras;*

(iii) *all algebras in V which have no proper subalgebras (and which are thus necessarily finite) are affine complete.*

Recall that by Theorem 4.1.1, if V satisfies the conditions (i) and (ii) then it is affine complete whenever (iii'): all of its finite members are. Thus Theorem 4.4.6 significantly sharpens this result: namely it characterizes locally finite affine complete varieties with almost the same conditions, replacing (iii') by the weaker condition (iii), which asserts that we need only know that the finite algebras having no proper subalgebras are affine complete.

Now if we want to check whether a given locally finite CD variety is affine complete, we need to have information about its members which have no proper subalgebras. The next proposition will be useful in this respect.

Proposition 4.4.7 *Let V be a locally finite variety and let $\mathbf{M}$ be any minimal subalgebra of $\mathbf{F}_V(1)$ (the V-free algebra with one free* *generator). Then any* $\mathbf{A} \in V$ *has no proper subalgebras if and only if it is a homomorphic image of* $\mathbf{M}$. *Up to isomorphism* $\mathbf{M}$ *is the only algebra in V having this property.*

Notice that the proposition implies that the algebra $\mathbf{M}$ is the unique largest algebra of V which has no proper subalgebras. While $\mathbf{M}$ could be the trivial algebra, in affine complete varieties Corollary 4.2.5 tells this could occur only if the variety is trivial.

PROOF Let $\mathbf{A}$ be an algebra in V having no proper subalgebras. Then $\mathbf{A}$ is a homomorphic image of $\mathbf{F}_V(1)$ and hence the image of $\mathbf{M}$ under the homomorphism is a subalgebra of $\mathbf{A}$ which must then be $\mathbf{A}$. On the other hand, if $\mathbf{A}$ is a homomorphic image of $\mathbf{M}$, then since $\mathbf{M}$ has no proper subalgebras, neither does $\mathbf{A}$.

If $\mathbf{M}'$ is any other algebra in V with this property, then since it is a homomorphic image of itself it has no proper subalgebras. Then $\mathbf{M}$ and $\mathbf{M}'$ are both finite and are homomorphic images of each other. Hence they are isomorphic.   •

Now suppose that $\mathbf{A}$ is a finite algebra of finite type and let the elements of A be $\{a_1, \ldots, a_n\}$. Then the $V(\mathbf{A})$-free algebra with one free generator is isomorphic to the subalgebra of $\mathbf{A}^n$ generated by $(a_1, \ldots, a_n)$ and is effectively constructible. Moreover, by Proposition 4.4.7, a minimal subalgebra $\mathbf{M} = \mathbf{M}(\mathbf{A})$ is uniquely determined by $\mathbf{A}$ up to isomorphism. It is obviously also constructible if the type is finite, and generates $V(\mathbf{A})$. Hence we may reformulate Theorem 4.4.6 as follows.

Theorem 4.4.8 *A locally finite variety V is affine complete if and only if the following conditions are satisfied:*

  (i) *V is congruence distributive;*

 (ii) *V is generated by a finite algebra* $\mathbf{A}$ *which has no proper subalgebras;*

(iii) *the algebra* $\mathbf{M}(\mathbf{A})$, *uniquely determined by* $\mathbf{A}$, *is hereditarily affine complete.*

It is natural to call an algebra **hereditarily congruence primal** if all its quotient algebras are congruence primal. Obviously an algebra is congruence primal iff it is affine complete and all its constants are terms. Hence an algebra is hereditarily congruence primal iff it is hereditarily affine complete and all its constants can be represented by terms. Let $\mathbf{A}$ be a finite hereditarily congruence primal algebra and $\mathbf{L} = \mathbf{Con}\,\mathbf{A}$. Then $\mathbf{L}$ has the (CFL) property and by Theorem 2.3.9 the algebra $\mathbf{A}$ admits a compatible near unanimity function. By the congruence primality of $\mathbf{A}$, the latter is a term function and thus $V = V(\mathbf{A})$ is a near unanimity variety. Furthermore, every member of V contains a canonical homomorphic image of $\mathbf{A}$ as a subalgebra; that is, the algebra $\mathbf{A}$ itself is already $\mathbf{M}(\mathbf{A})$. Hence, we have the following corollary from Theorem 4.4.8.

Corollary 4.4.9 *Every finite hereditarily congruence primal algebra generates an affine complete variety.*

Theorems 4.4.6 and 4.4.8 characterize the generating algebras of a locally finite affine complete variety. In particular, with reference to Theorem 4.4.8, if the variety is of finite type then each quotient of $\mathbf{M}(\mathbf{A})$ is constructible so the characterization problem effectively reduces to that of determining when a finite algebra of finite type in a CD variety is affine complete. By Theorem 4.4.1 it is enough to consider near unanimity varieties.

In summary we are led to one of the principal open problems in the theory of affine complete varieties.

Problem 4.4.10 *Find an intrinsic characterization of the finite algebras which generate affine complete varieties. In particular find a characterization of the affine complete finite algebras of finite type in a near unanimity variety.*

4.4.3   The arithmetical case

In contrast to Problem 4.4.10 the intrinsic characterization problem for locally finite arithmetical affine complete varieties is relatively easy because every finite algebra belonging to an arithmetical variety and having no proper subalgebras generates an affine complete variety. In this section we discuss this situation in some detail; the material comes largely from [Kaarli, Pixley, 1994].

We begin with the following intrinsic characterization which directly follows from Corollaries 4.1.2 and 4.2.7.

Theorem 4.4.11 *A locally finite arithmetical variety is affine complete iff it is generated by a finite algebra having no proper subalgebras.*

Thus the characterization of locally finite arithmetical affine complete varieties is equivalent to finding all of the finite algebras which have no proper subalgebras and which generate an arithmetical variety. The following theorem which is a direct specialization of Theorem 3.4.6 characterizes these algebras as generalizations of primal algebras.

Theorem 4.4.12 *Let* **A** *be a finite arithmetical algebra having no proper subalgebras. The variety* $V(\mathbf{A})$ *is arithmetical iff* $\mathrm{Clo}\,\mathbf{A}$ *consists of all functions which preserve all isomorphisms between quotients of* **A**.

Indeed, since **A** has no proper subalgebras, the rectangular subuniverses of $\mathbf{A} \times \mathbf{A}$ are exactly the graph subuniverses corresponding to isomorphisms between quotients of **A**.

It may happen that, given a finite arithmetical algebra **A** with no proper subalgebras, the question whether a function f belongs to $\mathrm{Clo}\,\mathbf{A}$ can be tested on a subset of the set of all isomorphisms between quotients. For example, suppose that $f \in \mathrm{Clo}\,\mathbf{A}$ iff f preserves identical automorphisms of all quotients of **A**. It is easy to see that the latter condition is equivalent to saying that f preserves all congruences of **A**, that is, $f \in \mathrm{Comp}\,\mathbf{A}$. Hence the algebra **A** is congruence primal. We see that, in particular, Theorem 4.4.12 implies that every finite arithmetical congruence primal algebra generates an arithmetical variety, a result already known by Theorem 3.4.2.

We get other interesting examples if we consider the algebras **A** whose term functions are precisely the functions that preserve all automorphisms of **A**. Recall that these are the automorphism primal algebras with no subalgebras described in Section 3.4. These algebras perhaps deserve more attention because they are exactly the generators of the minimal affine complete varieties. Note that in an affine complete variety every simple algebra generates a minimal subvariety. Also note that since every subvariety of an affine complete variety is affine complete, the minimal affine complete varieties are in fact minimal varieties. We have the following equivalences:

Theorem 4.4.13 *Let V be a variety. Then the following conditions are equivalent:*

(1) *V is a minimal affine complete variety;*

(2) *V is generated by a functionally complete algebra which has no proper subalgebras;*

(3) *V is generated by a automorphism primal algebra with no subalgebras;*

(4) *V is generated by a quasiprimal algebra which has no proper subalgebras.*

PROOF The last three conditions are equivalent due to Theorem 3.4.5. If V is a minimal affine complete variety, then since V is residually finite, it contains a nontrivial locally finite subvariety. By Corollary 4.2.7 the latter is generated by a finite algebra **A** having no proper subalgebras. Now, if ρ is a maximal congruence of **A**, then $\mathbf{A}/\rho$ is a simple algebra with no proper subalgebras which because of minimality of V also generates V. However, a simple algebra of an affine complete variety is functionally complete. This proves the implication (1) $\Rightarrow$ (2). For the converse, let V be generated by a functionally complete algebra **A** which has no proper subalgebras. By Theorem 3.1.11 V is arithmetical and then by Corollary 4.1.2 it is affine complete. That V is minimal follows from Theorem 1.2.22. By this theorem V is the subdirect closure of **A** and hence every nontrivial member of V generates V. •

In view of Theorem 4.4.13, if a two element algebra generates a minimal affine complete variety, then it must be an automorphism primal algebra with no subalgebras and there are precisely two choices: the rigid algebra which is, up to term equivalence, just the two element Boolean algebra $\mathbf{B} = \mathbf{B_2}$, and the algebra **A** having complementation $0' = 1, 1' = 0$ as an automorphism. These two cases are easily compared by taking the discriminator t and complementation ' as operations of each, and in the Boolean case, the nullary operations 0 and 1 as well:

$$\mathbf{A} = \langle \{0,1\}; t,' \rangle,$$
$$\mathbf{B} = \langle \{0,1\}; t,', 0, 1 \rangle.$$

Consequently we have the interesting characterization of the variety of Boolean algebras, up to term equivalence, as *the unique minimal affine complete variety with a rigid two element generator.* In view of the all important mathematical role of the variety generated by $\mathbf{B}_2$ it is remarkable that the variety generated by $\mathbf{A}$ so far seems to have no particularly significant role.

Returning to the discussion preceding Theorem 4.4.13, it is also possible to consider algebras $\mathbf{A}$ such that $\operatorname{Clo}\mathbf{A}$ consists of all functions on A that preserve both the congruences and the automorphisms of $\mathbf{A}$. Assuming, in addition, that $\mathbf{A}$ has no proper subalgebras, we arrive at a class of algebras which naturally generalizes both finite congruence primal algebras and finite automorphism primal algebras with no subalgebras. We are going to show that this class properly contains all finite weakly diagonal arithmetical affine complete algebras. (Recall from Section 3.1.4 that $\mathbf{A}$ is weakly diagonal if every subuniverse of $\mathbf{A} \times \mathbf{A}$ contains the graph of an automorphism of $\mathbf{A}$.) First we establish an interesting feature of weakly diagonal algebras.

Proposition 4.4.14 *Let $\mathbf{A}$ be a finite algebra having no proper subalgebras. Then the following conditions are equivalent:*

(1) *algebra $\mathbf{A}$ is weakly diagonal;*

(2) *all rectangular subuniverses of $\mathbf{A} \times \mathbf{A}$ have the form $\{(a,b) : \gamma a \equiv_\rho b\}$ where γ is an automorphism of $\mathbf{A}$ and ρ is a congruence of $\mathbf{A}$;*

(3) *all rectangular subuniverses of $\mathbf{A} \times \mathbf{A}$ are isomorphic copies of congruences of $\mathbf{A}$.*

Whenever any of the equivalent conditions of the proposition holds the picture is that all constituent rectangles of a rectangular subuniverse in $\mathbf{A} \times \mathbf{A}$ are "squares".

PROOF Assume that $\mathbf{A}$ is weakly diagonal and consider a rectangular subuniverse R in $\mathbf{A} \times \mathbf{A}$. Let R be determined by an isomorphism $\alpha : \mathbf{A}/\phi_1 \to \mathbf{A}/\phi_2$ where $\phi_1, \phi_2 \in \operatorname{Con}\mathbf{A}$, that is

$$R = \{(a,b) \in A \times A : \alpha(a/\phi_1) = b/\phi_2\}.$$

Since **A** is weakly diagonal, there exists an automorphism γ of **A** whose graph is contained in R. Hence $\alpha(a/\phi_1) = \gamma a/\phi_2$ for any $a \in A$ and we see that γ in fact induces α. Moreover, we can write $R = \{(a,b) \in A^2 : \gamma a \equiv_{\phi_2} b\}$ which proves the implication $(1) \Rightarrow (2)$.

The proof of the implication $(2) \Rightarrow (3)$ is an easy exercise and it remains to prove that (3) implies (1). Let B be an arbitrary subuniverse in $\mathbf{A} \times \mathbf{A}$. By the condition (3), **B** is isomorphic to a congruence of **A**; hence it must contain a subalgebra $\mathbf{A}'$ isomorphic to the diagonal of **A** and hence isomorphic to **A**. Since **A** is finite and has no proper subalgebras, the algebra $\mathbf{A}'$ can be nothing other than the graph of an automorphism of **A**.   •

Theorem 4.4.15 *Let* **A** *be a finite weakly diagonal arithmetical affine complete algebra. Then* $V(\mathbf{A})$ *is arithmetical and* Clo **A** *consists of all functions on* A *that preserve both the congruences and the automorphisms of* **A**. *Hence a polynomial function of* **A** *is a term iff it preserves all automorphisms of* **A**.

PROOF By Theorem 3.4.11 the algebra **A** admits a Pixley function $f(x,y,z)$ which is a polynomial of **A**. Since **A** is weakly diagonal, it has no proper subalgebras and therefore we can represent $f(x,y,z)$ as $t(x,y,z,a)$ where a is an arbitrary element of A and t is a suitable 4-ary term. The rest of the proof that $V(\mathbf{A})$ is arithmetical is the same as the proof of Lemma 3.1.16.

Now let g be an m-ary function on A that preserves all congruences and all automorphisms of **A**. We shall show that g preserves every rectangular subuniverse C of $\mathbf{A} \times \mathbf{A}$; then it certainly also preserves all isomorphisms between quotients of **A**. By Theorem 4.4.12 this will imply that g is a term function.

By Proposition 4.4.14 a rectangular subuniverse C has the form $\{(a,b) \in A \times A : \gamma a \equiv_\rho b\}$ for suitable automorphism γ and congruence ρ. Hence if $\mathbf{a} = (a_1,\ldots,a_m)$ and $\mathbf{b} = (b_1,\ldots,b_m)$ are m-tuples such that $\gamma a_i \equiv_\rho b_i$, $i = 1,\ldots,m$, then

$$\gamma g(\mathbf{a}) = g(\gamma \mathbf{a}) \equiv_\rho g(\mathbf{b})$$

because the function g preserves the automorphism γ and the congruence ρ. Obviously this means that g preserves C.   •

From Theorems 4.4.12 and 4.4.15 we have the following diagram describing the inclusions (larger classes above) for three classes of finite arithmetical algebras **A** having no proper subalgebras:

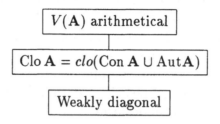

Figure 4.2

In Section 4.4.4 below we shall give Examples (4.4.22 and 4.4.23) which show that each of these inclusions is proper.

In the remainder of this subsection we establish three conditions which imply that a finite arithmetical affine complete algebra having no proper subalgebras is weakly diagonal. Two of them are also necessary. These conditions will help us construct algebras with the given properties. The basic tool for obtaining these results is the following "Cross Lemma" which generalizes Lemma 3.1.13. To explain the name of the lemma, let us agree that a *cross* of the "rectangle" $A \times B$ is any of its subsets which has the form $(\{a\} \times B) \cup (A \times \{b\})$ where $a \in A$ and $b \in B$ are fixed.

Lemma 4.4.16 (Cross Lemma) *Let* **A** *be a finite arithmetical affine complete algebra which has no proper subalgebras and* **B** < **A** × **A**. *Then every rectangle of the rectangular hull of* **B** *has a cross which is contained in* B.

PROOF Elements of the subuniverse B may be represented as the pairs (x, y) with $x, y \in A$. Consider the set of all pairs $((x, y), (x', y'))$ of elements of B with $x = x'$. Specifically let these be enumerated as follows:

$$((a_1, b_1), (a_1, c_1)), \ldots, ((a_m, b_m), (a_m, c_m))$$

(where a_i need not be distinct). Suppose that the rectangular hull of **B** is determined by an isomorphism $\alpha : A/\phi_1 \to A/\phi_2$ where ϕ_1 and ϕ_2 are congruences of **A**. By the formula established in Section 1.2.4 we have that ϕ_2 is precisely the join of all principal congruences $Cg(b_i, c_i)$, $1 \le i \le m$. Let N be any ϕ_2-block and choose the integer n so large that $2^n \ge |N|$.

Now consider mn-tuples $\mathbf{x}$ of elements of A, thinking of $\mathbf{x}$ as consisting of m constituent n-tuples, numbered from left to right. Let K be the subset of all such mn-tuples $\mathbf{x}$ satisfying the following special properties:

- for each $i = 1, \ldots, m$, the elements of the i-th constituent n-tuple of $\mathbf{x}$ are either b_i or c_i;

- the same pattern of b's and c's occurs in each of the m constituent n-tuples occurring in $\mathbf{x}$.

Because of these conditions, the size of K is exactly 2^n and, if $\mathbf{x}$ and $\mathbf{y}$ are any two distinct elements of K, then

$$Cg(\mathbf{x}, \mathbf{y}) = \bigvee_{1 \le i \le mn} Cg(x_i, y_i) = \bigvee_{1 \le i \le m} Cg(b_i, c_i) = \phi_2.$$

Because of the inequality $|K| \ge |N|$, there is a surjective function $f : K \to N$. This function is a partial function in $A^{mn} \to A$ and by what we just proved, it is compatible: indeed, since N is a ϕ_2-block and $Cg(\mathbf{x}, \mathbf{y}) = \phi_2$ for any distinct elements $\mathbf{x}$ and $\mathbf{y}$ in K, we have $(f(\mathbf{x}), f(\mathbf{y})) \in Cg(\mathbf{x}, \mathbf{y})$ for all $\mathbf{x}, \mathbf{y} \in K$. Since **A** is arithmetical, by Theorem 2.2.2 the partial function f has a compatible extension to all of A^{mn} (which we still denote by f). Since **A** is affine complete with no proper subalgebras, we have $f(\mathbf{x}) = t(\mathbf{x}, a)$ for some $(mn+1)$-ary term t and an element $a \in A$. Since **A** has no proper subalgebras, B is subdirect in A^2 and we can find $b \in A$ such that $(b, a) \in B$.

Now consider the elements of B obtained as values of the polynomial

$$t((x_1, y_1), \ldots, (x_{mn}, y_{mn}), (b, a)) =$$
$$(t(x_1, \ldots, x_{mn}, b), t(y_1, \ldots, y_{mn}, a))$$

where the i-th constituent n-tuples of $((x_1, y_1), \ldots, (x_{mn}, y_{mn}))$ range over all possible choices from the pairs (a_i, b_i) and (a_i, c_i). These are the elements of B of the form

$$(t(a_1,\ldots,a_1,\ldots,a_m,\ldots,a_m,b),t(y_1,\ldots,y_{mn},a))$$

where the i-th constituent n-tuples of $(y_1,\ldots,y_{mn})$ consist of all possible choices from $\{b_i,c_i\}$. Let

$$c = t(a_1,\ldots,a_1,\ldots,a_m,\ldots,a_m,b).$$

Then since f maps K onto N, the elements of B so obtained are just $\{c\} \times N$.

Thus we have shown that for any ϕ_2-block N there exists $c \in A$ such that $\{c\} \times N \subseteq B$. In a completely analogous way we can show that for any ϕ_1-block M there exists $d \in A$ such that $M \times \{d\} \subseteq B$. Every rectangle of the rectangular hull of $\mathbf{B}$ has the form

$$a_1/\phi_1 \times a_2/\phi_2 \quad \text{with} \quad \alpha(a_1/\phi_1) = a_2/\phi_2.$$

Taking $a_1/\phi_1 = M$ and $a_2/\phi_2 = N$, we obtain the result. $\qquad\bullet$

Theorem 4.4.17 *Let $\mathbf{A}$ be a finite arithmetical affine complete algebra having no proper subalgebras. Then the following are equivalent:*

(1) $\mathbf{A}$ *is weakly diagonal;*

(2) *each rectangular subuniverse of $\mathbf{A} \times \mathbf{A}$ contains the graph of an automorphism of $\mathbf{A}$;*

(3) *every subalgebra of $\mathbf{A} \times \mathbf{A}$ is isomorphic to a congruence of $\mathbf{A}$.*

PROOF The implication (1) $\Rightarrow$ (2) is true by definition. Assume (2) and consider an arbitrary subalgebra $\mathbf{B} < \mathbf{A} \times \mathbf{A}$. If $\mathbf{H}$ is the rectangular hull of $\mathbf{B}$ then $\mathbf{H}$ must contain the graph G of an automorphism of $\mathbf{A}$. It easily follows from the Cross Lemma that G intersects B. Indeed, if R is any constituent rectangle of $\mathbf{H}$ then $G \cap R$ intersects any cross of R, in particular one contained in B. Since the algebra $\mathbf{G}$ is isomorphic to $\mathbf{A}$ and the latter has no proper subalgebras, the whole G must be contained in B. This proves the implication (2) $\Rightarrow$ (1).

The equivalence of conditions (2) and (3) follows from Proposition 4.4.14. $\qquad\bullet$

Now we shall prove a theorem which partly generalizes Lemma 3.1.14 and provides a sufficient condition for weak diagonality. First we need a simple lemma which also generalizes Lemma 3.1.14

Lemma 4.4.18 *Let* **A** *be a finite arithmetical affine complete algebra which has no proper subalgebras and* θ *be an atomic congruence of* **A**. *If* $(a_1, a_2) \in \theta$ *then the subalgebra* **B** $<$ **A** $\times$ **A** *generated by* (a_1, a_2) *is either the graph of a nontrivial automorphism of* **A** *or contains the diagonal (i.e., the graph of the congruence* 0_A*) of* **A**.

PROOF Let **H** be the rectangular hull of **A**. Since $B \subseteq \theta$ and θ is a rectangular subuniverse in **A** $\times$ **A**, we have $H \subseteq \theta$. Therefore if H is the graph subuniverse which is determined by the isomorphism $\alpha : A/\phi_1 \to A/\phi_2$, then both of the congruences ϕ_1 and ϕ_2 must be below θ. Since θ is an atom and **A** is finite, we have two possibilities:

1. $\phi_1 = \phi_2 = 0_A$ and B is the graph of an automorphism of **A**;

2. $\phi_1 = \phi_2 = \theta$ and α is an automorphism of **A**$/\theta$.

In case 2 the inclusion $H \subseteq \theta$ immediately yields the equality $H = \theta$; hence every cross of every rectangle of H intersects 0_A. Then by the Cross Lemma B intersects 0_A too. Since 0_A is a minimal subuniverse in **A** $\times$ **A**, we conclude $0_A \subseteq B$. ●

Theorem 4.4.19 *Let* **A** *be a finite arithmetical affine complete algebra which has no proper subalgebras and* θ *be the join of all atoms of the lattice* **Con A**. *If there exists a unary term* $u(x)$ *such that* $u(A)$ *is contained in some* θ-*block then* **A** *is weakly diagonal.*

PROOF Among all unary terms $u(x)$ for which $u(A)$ lies in some θ-block, choose one with the property that for any term $v(x)$ and each $\rho \in$ Con **A**, $|u(A/\rho)| \leq |v(A/\rho)|$. This can be done by choosing, for each congruence ρ, a term u_ρ such that $|u_\rho(A/\rho)|$ is minimal, and then taking u to be the composition of all such u_ρ.

Let B be an arbitrary subuniverse in **A**$\times$**A** and pick some element $(a_1, a_2) \in B$. We show that the subalgebra **C** $<$ **B** generated by $(u(a_1), u(a_2))$ is the graph of an automorphism of **A**. Let $u(a_i) = b_i$, $i = 1, 2$. Then $C = \{(t(b_1), t(b_2)) : t \in T\}$ where T is the set of all unary terms, and we have to show that

$$v(b_1) = w(b_1) \quad \Longleftrightarrow \quad v(b_2) = w(b_2)$$

for all $v, w \in T$. Suppose, on the contrary, that for some v and w, $v(a_1) = w(a_1)$ but $v(a_2) \neq w(a_2)$. Now, since θ is the join of atomic congruences of $\mathbf{A}$, there must exist $\rho \in \operatorname{Con} \mathbf{A}$, covered by θ, such that $v(b_1/\rho) = w(b_1/\rho)$ but $v(b_2/\rho) \neq w(b_2/\rho)$. The latter means that the natural homomorphic image of $\mathbf{C}$ in $\mathbf{A}/\rho \times \mathbf{A}/\rho$ is not a graph of an automorphism of $\mathbf{A}/\rho$.

The quotient algebra $\mathbf{A}/\rho$ obviously is finite, arithmetical, and has no proper subalgebras. Also $\mathbf{A}/\rho$ inherits the Pixley polynomial of $\mathbf{A}$ and thus is affine complete by Theorem 3.4.10. This allows us to apply Lemma 4.4.18 to the pair $(b_1/\rho, b_2/\rho)$ contained in the atomic congruence θ/ρ of $\mathbf{A}/\rho$. By the choice of ρ, we have $b_1/\rho \neq b_2/\rho$. Moreover, since $b_1/\rho, b_2/\rho \in u(A/\rho)$, we conclude that $t(b_1/\rho) \neq t(b_2/\rho)$ for all $t \in T$, for otherwise $|tu(A/\rho)| < |u(A/\rho)|$, contradicting the minimality of $|u(A/\rho)|$. Therefore the subalgebra $\mathbf{C}' < \mathbf{A}/\rho \times \mathbf{A}/\rho$ generated by $(b_1/\rho, b_2/\rho)$ does not contain the diagonal of $0_{A/\rho}$. Hence by Lemma 4.4.18, $\mathbf{C}'$ must be a graph of an automorphism of $\mathbf{A}/\rho$. To obtain a contradiction, notice that $\mathbf{C}'$ is the natural homomorphic image of $\mathbf{C}$.   •

4.4.4 Examples

We first discuss the possibilities of construction of locally finite affine complete varieties. Then we present two examples (Examples 4.4.20 and 4.4.21) which show that the conditions of Theorem 4.4.6 (characterizing locally finite affine complete varieties) are essential. Finally we present two examples which apply to Section 4.4.3.

A basic problem is: what are the finite algebras $\mathbf{A}$ that generate affine complete varieties? Theorem 4.4.8 provides necessary and sufficient conditions for such algebras. Obviously the algebra $\mathbf{A}$ has to be hereditarily affine complete which implies that the algebra $\mathbf{A}^+$ is hereditarily congruence primal. By Corollary 4.4.9, the variety $V(\mathbf{A}^+)$ is affine complete as well. This means that if we want to be able to construct the finite algebras generating affine complete varieties, then we certainly must be able to construct the finite hereditarily congruence primal algebras. The latter would generate affine complete varieties and any finite algebra with this property would be a reduct of one of them.

The construction of finite hereditarily affine complete algebras is, however, equivalent to the construction of equivalence lattices (on finite sets) which have the (CFL) property. Indeed, obviously the congruence lattice of any hereditarily congruence primal algebra has the (CFL) property. On the other hand, if A is a finite set and $\mathbf{L}$ is a 0-1 equivalence lattice on A having the (CFL) property, then it is easy to see that the algebra $\langle A; F \rangle$ where $F = clo(L)$ is hereditarily congruence primal. One only has to observe that $\mathrm{Con}\,\mathbf{A} = L$ which follows from the theorem of Quackenbush and Wolk ([Quackenbush, Wolk, 1971]) since F contains a near unanimity function by Theorem 2.3.9. While there is no known general method for constructing all equivalence lattices having the (CFL) property, we have presented one method in Example 2.3.1 of Chapter 2 which produced a class of such lattices which are Boolean and possibly nonarithmetical. We also discussed this class further in Example 4.1.3. Some other related methods can be found in [Kaarli, 1990].

Now we present several examples designed to show that the conditions of Theorem 4.4.6 are essential.

Example 4.4.20 *A finite congruence primal algebra which generates a congruence distributive variety but which is not hereditarily affine complete.*

Let A, B, and $\mathbf{L}$ be as in Example 2.3.14, and $F = clo(L)$. We know that F contains a majority theorem, so the algebra $\mathbf{B} = \langle B; F \rangle$ is term equivalent to an algebra of finite type and the variety $V(\mathbf{B})$ is CD. However, $\mathbf{B}$ is not an hereditarily affine complete algebra because $L = \mathrm{Con}\,\mathbf{B}$ does not have the (CFL) property.

There is an alternative way to show that $\mathbf{B}$ is not hereditarily affine complete. Clearly $\mathbf{B}$ is congruence primal. Moreover one can easily check that π_1 and π_2 are maximal congruences of $\mathbf{B}$; in other words $\mathbf{B}$ is a subdirect product of two simple algebras: $\mathbf{B}/\pi_1$ and $\mathbf{B}/\pi_2$. Now suppose that $\mathbf{B}/\pi_1$ is affine complete. Then actually it must be functionally complete and by Lemma 3.1.13 there must exist $a \in A$ such that $A \times \{a\} \subseteq B$. This contradiction shows that $\mathbf{B}$ has a homomorphic image which is not affine complete.

The next example shows that the CD variety generated by a finite algebra, which has no proper subalgebras and is hereditarily affine complete, need not be affine complete.

Example 4.4.21 *A finite arithmetical hereditarily affine complete algebra with no proper subalgebras and generating a congruence distributive variety which is not affine complete.*

We construct a variety V of type $F = \{f, g, h, m\}$ where the arities of the operation symbols f, g, h, m are 4,1,1,3, respectively. The variety V is generated by an algebra $\mathbf{A}$ with universe $A = \{a, b, c, d\}$ having the equivalence relation θ with blocks $\{a, b\}$ and $\{c, d\}$ as one of its congruences. We interpret the operation symbols f, g, h and m in $\mathbf{A}$ as follows:

- $g(a) = g(b) = b, \quad g(c) = g(d) = d,$

- $h(a) = h(b) = c, \quad h(c) = h(d) = a,$

- for every fixed $w \in A$, the function $f_w(x, y, z) = f(x, y, z, w)$ induces the ternary discriminator modulo θ, the functions f_a and f_b are Pixley functions, and $f(x, y, z, w) \in \{a, d\}$ provided $w \in \{c, d\}$ or $|\{x, y, z\}| = 3$,

- m is the majority function compatible with θ and satisfying $m(x, y, z) \in \{a, d\}$ if $|\{x, y, z\}| = 3$.

It is easy to check that the given conditions determine the functions f and m uniquely. Since the algebra $\mathbf{A}$ has a Pixley polynomial, by Theorem 3.4.11 it is arithmetical and affine complete. Actually the easy calculations show that θ is the only nontrivial congruence of $\mathbf{A}$. Obviously $\mathbf{A}$ generates a CD variety.

In what follows we need a description of the unary term functions of $\mathbf{A}$. By Theorem 3.2.2, the term functions of a finite algebra with a majority term are characterized as those which preserve all subuniverses of the direct square of this algebra. Therefore we first find all subuniverses of $\mathbf{A}^2$. Suppose that a subuniverse S of $\mathbf{A}^2$ contains a diagonal element. Then, since $\mathbf{A}$ has no proper subalgebras, the whole diagonal is contained in S. Hence, S is a subuniverse of $(\mathbf{A}^+)^2$. Clearly $\mathbf{A}^+$ is a congruence primal arithmetical algebra. Hence by Theorem 3.4.2 the only subuniverses of $(\mathbf{A}^+)^2$

are the congruences of **A**. This gives $S \in \{0_A, \theta, 1_A\}$. Using the unary operations g and h we see that every subuniverse of $\mathbf{A}^2$ generated by $(x, y) \in \theta$ intersects the diagonal. Hence, the subuniverses of $\mathbf{A}^2$ which are not congruences must be contained in $B = A^2 \setminus \theta$. By straightforward computations, $f(B^4)$, $g(B)$ and $h(B)$ are contained in $C = \{(a, d), (b, d), (a, c), (d, b), (c, a), (d, a)\}$ meaning that C, $C \cup \{(b, c)\}$, $C \cup \{(c, b)\}$, and B are subuniverses of $\langle A^2; f, g, h \rangle$. All four of these subsets are closed with respect to m, so they are subuniverses of $\mathbf{A}^2$ as well. Since $g(b, c)$ and $g(c, b)$ are contained in C, every subuniverse of $\mathbf{A}^2$ contained in B intersects C. On the other hand, the scheme

$$(a, d) \xrightarrow{g} (b, d) \xrightarrow{h} (c, a) \xrightarrow{g} (d, b) \xrightarrow{h} (a, c) \xrightarrow{fh} (d, a) \xrightarrow{fh} (a, d)$$

shows that C is a minimal subuniverse of $\mathbf{A}^2$. (Here fh is defined by $(fh)(x) = f(h(x), h(x), h(x), h(x))$.) All this yields that

$$0_A, \quad \theta, \quad 1_A, \quad C, \quad C \cup \{(b, c)\}, \quad C \cup \{(c, b)\}, \quad B$$

is the complete list of subuniverses of $\mathbf{A}^2$.

It is an easy exercise to check that there are exactly 15 unary functions on A which preserve these 7 subsets of A^2. Only one of them is bijective (the identity map), 8 functions have 3 as the size of their range and the remaining 6 functions map A onto two element subsets $\{a, c\}$, $\{b, d\}$ or $\{a, d\}$. In view of Proposition 4.4.7, these 6 functions must form a minimal subalgebra of the free algebra $\mathbf{F}_1$. Indeed, clearly a minimal left ideal of the monoid F_1 must consist of functions of the minimal possible range. Since our variety contains a six element algebra $\mathbf{C}$ with no proper subalgebras, $\mathbf{F}_1$ must contain at least a six element minimal subalgebra.

Now we show that $\mathbf{C}$ is not affine complete. Since V is CD, $\mathbf{C}$ has exactly 3 nontrivial congruences: two kernels of projections $\mathbf{C} \to \mathbf{A}$ and their join. Obviously the function α given by the scheme

$$(a, d) \xleftrightarrow{\alpha} (d, a), \quad (b, d) \xleftrightarrow{\alpha} (c, a), \quad (d, b) \xleftrightarrow{\alpha} (a, c)$$

is compatible with all of them. However, as we see below, α is not a polynomial of $\mathbf{C}$ since there is a subuniverse of $\mathbf{C}^2$ which contains the diagonal but is not preserved by α.

Let the quadruples $(x, y, z, w) = ((x, y), (y, w))$ represent the elements of C^2, where $x, y, z, w \in A$ and $(x, y), (z, w) \in C$. Consider the subuniverse D of $\mathbf{C}^2$ generated by (a, c, b, d). Applying all 15 unary term functions to this quadruple, we see that D contains the diagonal. Though not important for our purposes, we mention that $|D| = 15$ and thus $\mathbf{D}$ is isomorphic to the free algebra $\mathbf{F_1}$. Since there is only one bijective term function, for every $(x, y, z, w) \in D$ either $(x, y, z, w) = (a, c, b, d)$ or $|\{x, y, z, w\}| \leq 3$. Hence, $\alpha(a, c, b, d) = (d, b, c, a) \notin D$ yielding $\alpha(D) \not\subseteq D$.

Now we present an example of a finite rigid algebra with no proper subalgebras which is not congruence primal but generates an arithmetical variety. It shows, in particular, that a finite algebra with no proper subalgebras and which generates an arithmetical variety need not have its clone of term functions consist of just the functions which preserve congruences and automorphisms. This means that the general test for a function to be a term function as provided by Theorem 4.4.12 (preservation of all isomorphisms between quotients of $\mathbf{A}$) cannot be weakened.

Example 4.4.22 *A finite rigid algebra which has no proper subalgebras and generates arithmetical variety but is not congruence primal.*

The universe of the algebra we construct is $A = \{a, b, c\}$ and its type consists of two unary operation symbols g and h and a ternary operation symbol t. The symbols g and h are interpreted in $\mathbf{A}$ as follows:

$$g(a) = c, \quad g(b) = g(c) = a,$$
$$h(a) = b, \quad h(b) = h(c) = a.$$

Then because of the equalities

$$g(a) = c, \quad hg(c) = b, \quad h(b) = a$$

$\mathbf{A}$ has no proper subalgebras. Obviously a has to be fixed by any automorphism of $\mathbf{A}$. Since the set of fixed points of any automorphism is a subuniverse, we have that $\mathbf{A}$ is rigid. By checking cases one easily sees that the algebra $\langle A; g, h \rangle$ has only one nontrivial congruence θ, with blocks $\{a\}$ and $\{b, c\}$.

We interpret t in $\mathbf{A}$ as a θ-compatible Pixley function. This requires that

$$t(a, b, c) = t(a, c, b) = t(b, c, a) = t(c, b, a) = a$$

and the values of $t(b, a, c)$ and $t(c, a, b)$ must be taken from $\{b, c\}$. Let, for example, $t(b, a, c) = t(c, a, b) = b$. Hence the variety $V(\mathbf{A})$ is arithmetical.

Now suppose that $\mathrm{Clo}\mathbf{A}$ consists of all functions which preserve congruences and automorphisms of $\mathbf{A}$. Since $\mathbf{A}$ is rigid, this would mean, in fact, that $\mathbf{A}$ is congruence primal. To show that this is not the case observe that clearly the quotient algebra $\mathbf{A}/\theta$ has a nontrivial automorphism. The corresponding graph subuniverse is

$$B = \{(a, b), (a, c), (b, a), (c, a)\}.$$

Since this set does not contain 0_A, all constant functions on A cannot be term functions which means that $\mathbf{A}$ is not congruence primal.

By Theorem 4.4.12 $\mathrm{Clo}\mathbf{A}$ consists of all functions that preserve all isomorphisms between quotients of $\mathbf{A}$. In particular the term functions of $\mathbf{A}$ must preserve B but the constant functions do not.

Finally we give an example which shows that Theorem 4.4.15 is not reversible.

Example 4.4.23 *A finite algebra* $\mathbf{A}$ *with no proper subalgebras generating an arithmetical variety which is not weakly diagonal though* $\mathrm{Clo}\mathbf{A}$ *consists of all functions which preserve all congruences and automorphisms of* $\mathbf{A}$.

The form of the example we are seeking suggests we define $\mathbf{A}$ as follows: we select a set A, an equivalence lattice $\mathbf{L} < \mathbf{Eqv}\,A$, a group Γ acting on A, and consider the algebra $\langle A; F \rangle$ where F is the clone of all functions which preserve all $\rho \in L$ and all actions of elements $\gamma \in \Gamma$. The algebra we get in this way is of infinite type. However, if we show that F contains a majority function then by Theorem 3.2.4 F is generated by one of its finite subsets and hence the algebra $\mathbf{A}$ is term equivalent to an algebra of finite type. It is clear from the above discussion that neither $\mathbf{L}$ nor Γ can be trivial.

Let $A = \{a, b, c, a', b', c'\}$ where $'$ is an involution on A; that is, $a'' = a$, $b'' = b$ and $c'' = c$. We consider two equivalence relations of A: ρ with blocks $\{a, b, c\}$ and $\{a', b', c'\}$ and σ with blocks $\{a, b, c'\}$ and $\{a', b', c\}$. Let $L = \{0_A, \rho \wedge \sigma, \rho, \sigma, 1_A\}$ and Γ be the permutation group of A consisting of the involution $'$ and the identical permutation. Now the clone $F = clo(L \cup \Gamma)$ consists of all L-compatible functions which permute with $'$, in particular, $' \in F$. Hence, if $\mathbf{A} = \langle A; F \rangle$ then $'$ is an automorphism of $\mathbf{A}$ and $L \subseteq \mathrm{Con}\,\mathbf{A}$ implying that F also can be characterized as the clone of all functions which preserve all automorphisms and all congruences of $\mathbf{A}$. Since $'$ is an automorphism of $\mathbf{A}$, every term function $t(x_1,\ldots,x_n)$ of $\mathbf{A}$ is uniquely determined by its values for $x_1 \in \{a,b,c\}$.

We first show that $\mathbf{A}$ has no proper subalgebras. For that purpose it suffices to observe that the unary functions

$$t_1(a) = t_1(b) = c, \qquad t_1(c) = a$$
$$t_2(a) = t_2(b) = c, \qquad t_2(c) = b$$

are members of F. We leave this easy exercise for the reader. That $\mathbf{A}$ has no proper subalgebras, now follows from the following scheme:

$$a \xrightarrow{t_2 t_1} b \xrightarrow{t_1} c \xrightarrow{\;'\;} c' \xrightarrow{t_2} b' \xrightarrow{t_1 t_2} a' \xrightarrow{\;'\;} a\,.$$

Next we show that $\mathrm{Aut}\,\mathbf{A} = \Gamma$. It is enough to observe that there is no automorphism α of $\mathbf{A}$ such that $\alpha a = c$ or $\alpha b = c$. Suppose, for example, that $\alpha a = c$. Then

$$a = t_1(c) = t_1(\alpha a) = \alpha t_1(a) = \alpha t_2(a) = t_2(\alpha a) = t_2(c) = b\,,$$

a contradiction. The case $\alpha b = c$ is similar.

The proof of that $\mathrm{Con}\,\mathbf{A} = L$ is a little bit more complicated. One approach is to apply the method described in Section 2.2.4 (Prescribed automorphisms) to show that L, which is clearly an arithmetical equivalence lattice, supports a principal Pixley function admitting $'$ as an automorphism, and hence that $V(\mathbf{A})$ is arithmetical. This involves a straightforward analysis of cases which we leave as an exercise.

Another approach is to demonstrate directly that all principal congruences of $\mathbf{A}$ belong to L. We do this in what follows. Since $\mathrm{Cg}(x,y) = \mathrm{Cg}(x',y')$, it suffices to consider the congruences $\mathrm{Cg}(x,y)$ with $x \in \{a,b,c\}$. It is easy to see using only unary term functions t_1, t_2 and $'$ that

$$\mathrm{Cg}(a,b) = \rho \wedge \sigma, \quad \mathrm{Cg}(a,c) = \mathrm{Cg}(b,c) = \rho$$

and

$$\mathrm{Cg}(a,c') = \mathrm{Cg}(b,c') = \mathrm{Cg}(c,a') = \mathrm{Cg}(c,b') = \sigma\,.$$

Now because of symmetry, it is enough to show that

$$Cg(a, a') = Cg(a, b') = Cg(b, a') = Cg(b, b') = Cg(c, c') = 1_A.$$

Using the term functions t_1 and t_2 we see that all these congruences contain the pairs (a, a'), (b, b') and (c, c'). Hence we shall be done if we show that the congruence $\phi = Cg((a, a'), (b, b'), (c, c'))$ contains the pairs (a, c) and (b, c). For that purpose we define the binary term function u by the following Cayley table:

u	a	b	c	a'	b'	c'
a	a	a	c	c	c	a
b	a	a	c	c	c	a
c	a	a	c	c	c	a
a'	c'	c'	a'	a'	a'	c'
b'	c'	c'	a'	a'	a'	c'
c'	c'	c'	a'	a'	a'	c'

Obviously this function permutes with $'$ and it is easy to check that it is **L**-compatible as well. Hence $(a, c) = (u(a, a), u(a, a')) \in \phi$ and $(c, b) = (t_2(a), t_2(c)) \in \phi$. Thus all principal congruences belong to L.

Since F consists of all functions which preserve both all automorphisms and all congruences, it follows as before, by Theorem 4.4.12, that the variety $V(\mathbf{A})$ is arithmetical.

It remains to show that $\mathbf{A}$ is not weakly diagonal. In view of Theorem 4.4.17 it is sufficient to exhibit a rectangular subuniverse of $\mathbf{A} \times \mathbf{A}$ which contains neither Δ nor the graph of $'$. We observe that just changing the notation, we may interchange the congruences ρ and σ. Consequently the quotient algebras $\mathbf{A}/\rho$ and $\mathbf{A}/\sigma$ must be isomorphic. In fact any of the two mappings from $\mathbf{A}/\rho$ to $\mathbf{A}/\sigma$ is an isomorphism, because $'$ induces a nontrivial automorphism of each of them. One of the graph universes arising in this way consists of two rectangles:

$$\{a, b, c\} \times \{a, b, c'\} \quad \text{and} \quad \{a', b', c\} \times \{a', b', c\}.$$

Obviously this subuniverse contains neither Δ nor the graph of $'$.

Exercise

> With reference to Example 4.4.23 above, apply the method of
> Section 2.2.4 to show that $\text{Con } \mathbf{A} = L$.

4.4.5   Categorical equivalence

We return to the discussion of categorical equivalence in Section 3.5.

We have seen that every arithmetical affine complete variety of
finite type is generated by a finite algebra having no proper subalgebras. Thus it is of special interest to know when two such algebras
are categorically equivalent. This problem is completely solved in
[C. Bergman, 1998]. As in Section 3.5 Bergman's main tool is his
Theorem 3.5.2 which showed that in determining if two finite algebras with a ternary majority term are categorically equivalent it is
sufficient to consider the structure of the subuniverses of their direct
squares. In particular, algebras $\mathbf{A}$ and $\mathbf{B}$ are categorically equivalent iff the algebras $S_2(\mathbf{A})$ and $S_2(\mathbf{B})$ are isomorphic. Recall that
the universe of $S_2(\mathbf{A})$ is $\text{Sub}(\mathbf{A}^2)$ and the operations are intersection,
relation product, converse, 0_A, and 1_A. For such an algebra $\mathbf{A}$ the
complete set of invariants in this case is most briefly described as
the category $\mathbf{Q}(\mathbf{A})$ consisting of all quotients $\mathbf{A}/\theta$ of $\mathbf{A}$ obtained as
θ runs through $\text{Con } \mathbf{A}$. The category $\mathbf{Q}(\mathbf{A})$ is considered as a full
subcategory of $V(\mathbf{A})$. The morphisms are just the homomorphisms
between the quotients of $\mathbf{A}$. Thus Bergman proves:

Theorem 4.4.24 *Let $\mathbf{A}$ be finite, have no proper subalgebras, and
generate an arithmetical variety. Then $\mathbf{A} \equiv_c \mathbf{B}$ iff $\mathbf{B}$ is also finite,
has no proper subalgebras, generates an arithmetical variety and the
categories $\mathbf{Q}(\mathbf{A})$ and $\mathbf{Q}(\mathbf{B})$ are isomorphic.*

Also, as noted in Section 3.5 ([Davey, Werner, 1983]) there is an
appropriate topological duality for the class of algebras considered
above. Hence these basic elements of the theory of Boolean algebras
extend in a natural way to all arithmetical affine complete varieties
of finite type.

Another interesting observation from [C. Bergman, 1998] is the
following: using McKenzie's description of categorical equivalence,
in [Bergman, Berman, 1996] it is shown that for any algebras $\mathbf{A}$ and
$\mathbf{B}$ (finite or infinite), if $\mathbf{A} \equiv_c \mathbf{B}$ and $\mathbf{A}$ is congruence primal, then so

is **B**. Using this Bergman proves, using only elementary properties of categories, that

$$\mathbf{A} \equiv_c \mathbf{B} \quad \Longrightarrow \quad \mathbf{A}^+ \equiv_c \mathbf{B}^+.$$

But **A** is affine complete iff $\mathbf{A}^+$ is congruence primal. It follows then that if **A** is affine complete and $\mathbf{A} \equiv_c \mathbf{B}$, then **B** is also affine complete, i.e., the property of being affine complete is preserved by categorical equivalence. (Since an algebra is functionally complete iff it is simple and affine complete, and both these properties are preserved by categorical equivalence, we likewise have that the property of being functionally complete is preserved by categorical equivalence.)

From these observations if **A** is a finite algebra having no proper subalgebras and generating an arithmetical variety, then as Bergman observes, the property of **A** being affine complete should somehow be reflected as a property of the category $\mathbf{Q}(\mathbf{A})$. It would be interesting to isolate this property explicitly.

Problem 4.4.25 *For a finite algebra* **A** *having no proper subalgebras and generating an arithmetical variety, find a property of the category* $\mathbf{Q}(\mathbf{A})$ *which reflects the affine completeness of* **A**.

We have also seen in the present chapter that every affine complete variety of finite type is generated by a finite algebra having no proper subalgebras and having a near unanimity term. We also observed (Problem 4.4.10) that the problem of intrinsically characterizing the generating algebras of such varieties is not yet solved. Along with this problem we thus also have the following:

Problem 4.4.26 *Find a complete set of invariants for the relation of categorical equivalence among the generating algebras of affine complete varieties of finite type (i.e., generalize Theorem 4.4.24 to the generators of affine complete varieties of finite type).*

Problems 4.4.10 and 4.4.26 are at present two major outstanding problems in the theory of affine complete varieties. Solutions to them would extend the generalized Boolean theory already embodied in Theorems 4.4.11 and 4.4.24 for the case of arithmetical affine complete varieties.

The natural first step in solving Problem 4.4.26 would be to generalize Bergman's Theorem 3.5.2 to the case of a finite algebra $\mathbf{A}$ having an arbitrary $(n+1)$-ary near unanimity term, and this has recently been done by J. Snow ([Snow, 1998]). As is suggested by Lüders' Theorem 3.5.1, Snow replaces Bergman's algebra $S_2(\mathbf{A})$ by an algebra $S_n(\mathbf{A})$ with universe $\mathrm{Sub}(\mathbf{A}^n)$. The operations of the algebra now, in general, act on subuniverses of $\mathbf{A}^n$ and are of course less familiar than the operations of (binary) relation product and converse and include operations which significantly extend the usual projections. Snow's theorem is the following:

Theorem 4.4.27 *If* $\mathbf{A}$ *is a finite algebra having an* $(n+1)$-ary near unanimity term and $\mathbf{B}$ is any algebra, then $\mathbf{A} \equiv_c \mathbf{B}$ iff $\mathbf{B}$ is finite, has an $(n+1)$-ary near unanimity term and $S_n(\mathbf{A}) \cong S_n(\mathbf{B})$.

It is to be emphasized that because we do not (yet) have a solution to the intrinsic characterization Problem 4.4.10, a solution to Problem 4.4.26, if possible, is likely to be a far less straightforward passage from Theorem 4.4.27 than Bergman's proof of Theorem 4.4.24 from Theorem 3.5.2.

On the other hand we should remark here that the topological duality theory is at the present time in many respects quite satisfactory. In particular in the book [Clark, Davey, 1998] it is shown that every finite algebra with a near unanimity term enjoys a topological duality which naturally extends the Boolean case. Even more, it is also shown there that a finite algebra in a CD variety has a topological duality theory iff the algebra has a near unanimity term. Therefore the Stone topological duality theory for the variety of Boolean algebras extends to every affine complete variety of finite type. Of course any particular details of the duality theory which are special in the context of affine completeness, even in the arithmetical case, are yet to be described.

Chapter 5

Polynomial Completeness in Special Varieties

In this chapter we shall examine affine completeness and its generalizations in terms of polynomial interpolation for single algebras and thus return to the topics introduced in Section 3.4.3. The definitions of local affine and strictly local affine completeness presented there should be recalled.

5.1 Strictly locally affine complete algebras

We begin our discussion by characterizing strictly locally affine complete algebras. The results we obtain will imply, in particular, that all simple rings with nontrivial multiplication and all simple nonabelian groups are locally functionally complete. This will extend the results presented in Section 3.1.4 to the infinite case. The principal results of this section were first obtained by J. Hagemann and Ch. Herrmann ([Hagemann, Herrmann, 1982]). Our development follows the recent new proof by E. Aichinger ([Aichinger, 2000]).

5.1.1 Interpolation and the class $SP_f(\mathbf{A})$

In Section 1.2.1 we formulated Mal'cev conditions for congruence permutability and arithmeticity of a variety. In particular, a variety V is congruence permutable iff there exists a ternary term $m(x, y, z)$ which induces a Mal'cev function on any member of V. Analyzing the proof of the sufficiency part of this statement we saw that the

221

same proof applies in order to show that an algebra $\mathbf{A}$ is congruence permutable whenever it admits a compatible ternary Mal'cev function. In fact, even more, we saw that all we need is to have for every pair $a, b \in A$ a compatible ternary function m_{ab} such that the equalities $m_{ab}(a, b, b) = a$ and $m_{ab}(a, a, b) = b$ hold. Now we show that congruence permutability of the class $SP_f(\mathbf{A})$ can be characterized in terms of interpolation of Mal'cev function, and in terms of rectangular subuniverses as well.

For technical reasons, it is convenient to introduce a special notation for what could be called the intersection of all Mal'cev functions on A. Namely, we denote by $\mathsf{Mal}\, A$ the partial ternary function on A whose domain consists of all ordered triples of the forms (a, a, b) and (a, b, b), where $a, b \in A$, and which is defined there like every Mal'cev function.

The following theorem is a parallel to Theorems 1.2.1 and 1.2.13.

Theorem 5.1.1 *For an arbitrary algebra $\mathbf{A}$ the following conditions are equivalent:*

(1) *The class $SP_f(\mathbf{A})$ is congruence permutable.*

(2) *For every $\mathbf{B}, \mathbf{C} \in SP_f(\mathbf{A})$, all subuniverses of the direct product $\mathbf{B} \times \mathbf{C}$ are rectangular.*

(3) *$\mathsf{Mal}\, A$ can be interpolated by a term function at every finite subset of its domain.*

PROOF If $\mathbf{B}, \mathbf{C} \in SP_f(\mathbf{A})$ then also $\mathbf{B} \times \mathbf{C} \in SP_f(\mathbf{A})$. By (1), the algebra $\mathbf{B} \times \mathbf{C}$ is congruence permutable; so its subuniverses must be rectangular because the kernels of the projections $\mathbf{B} \times \mathbf{C} \to \mathbf{B}$ and $\mathbf{B} \times \mathbf{C} \to \mathbf{C}$ permute. This proves the implication (1) $\Rightarrow$ (2).

Let X consist of triples (a_i, a_i, b_i) and (c_i, d_i, d_i), $1 \leq i \leq m$. (We may repeat some triple in order to have the same number m of triples of both types.) Let $\mathbf{a} = (a_1, \ldots, a_m) \in A^m$ and similarly $\mathbf{b}, \mathbf{c}, \mathbf{d} \in A^m$. Assuming (2), the subuniverse S generated by $(\mathbf{a}, \mathbf{c})$, $(\mathbf{a}, \mathbf{d})$, and $(\mathbf{b}, \mathbf{d})$ in $\mathbf{A}^m \times \mathbf{A}^m$ must be rectangular; hence $(\mathbf{b}, \mathbf{c}) \in S$. This implies the existence of a term function $t : X \to A$ such that

$$t((\mathbf{a}, \mathbf{c}), (\mathbf{a}, \mathbf{d}), (\mathbf{b}, \mathbf{d})) = (\mathbf{b}, \mathbf{c}),$$

that is

$$t(a_i, a_i, b_i) = b_i \quad \text{and} \quad t(c_i, d_i, d_i) = c_i$$

for $i = 1, \ldots, m$. Hence t interpolates $\mathsf{Mal}\, A$ at X and we have proved the implication $(2) \Rightarrow (3)$.

It remains to prove that (3) implies (1). Thus, we assume (3) and prove that every $\mathbf{B} < \mathbf{A}^m$ is congruence permutable. It is sufficient to show that for every two m-tuples $\mathbf{a}, \mathbf{b} \in B$, there exists a ternary term t such that $t(\mathbf{a}, \mathbf{a}, \mathbf{b}) = \mathbf{b}$ and $t(\mathbf{a}, \mathbf{b}, \mathbf{b}) = \mathbf{a}$. Define the set $X \subseteq A^m$ as the collection of all triples (a_i, a_i, b_i) and (a_i, b_i, b_i) with $1 \leq i \leq m$. By (3) there exists a ternary term t inducing the Mal'cev function on X. Obviously this term satisfies the required condition. •

Next we show that the arithmeticity of the class $SP_f(\mathbf{A})$ can be characterized similarly to Theorem 5.1.1, in terms of interpolation of $\mathsf{Pix}\, A$, the partial ternary function on A whose domain is the set of all ordered triples of the form (a, a, b), (a, b, a), and (b, a, a) and which agrees with every Pixley function on A.

The sufficiency part of this result is based on the observation that in the proof of Theorem 1.2.2, all we need to prove the arithmeticity of $\mathbf{A}$ is the existence of ternary compatible functions p_{abc}, for every $a, b, c \in A$ such that p_{abc} induces $\mathsf{Pix}\, A$ on

$$X = \{(a, b, a), (c, b, c), (a, b, b), (b, b, c)\}.$$

In particular, an algebra $\mathbf{A}$ is arithmetical if $\mathsf{Pix}\, A$ can be interpolated by a polynomial function at every finite subset of its domain.

Theorem 5.1.2 *For an arbitrary algebra* $\mathbf{A}$ *the following conditions are equivalent:*

(1) *The class* $SP_f(\mathbf{A})$ *is arithmetical.*

(2) $\mathsf{Pix}\, A$ *can be interpolated at every finite subset* X *of its domain by a term function.*

PROOF The proof of the implication $(2) \Rightarrow (1)$ precisely follows the proof of the implication $(3) \Rightarrow (1)$ of the preceding theorem. Thus assume (1) and construct an interpolating term for $\mathsf{Pix}\, A$ at X. Let $\mathbf{B}$ be the subalgebra of $\mathbf{A}^X$ consisting of all term functions $X \to A$. We consider the following three equivalence relations on B:

$$\rho = \{(u, v) \in B^2 : (\forall (x, y, y) \in X)\ u(x, y, y) = v(x, y, y)\},$$
$$\sigma = \{(u, v) \in B^2 : (\forall (x, x, y) \in X)\ u(x, x, y) = v(x, x, y)\},$$
$$\tau = \{(u, v) \in B^2 : (\forall (x, y, x) \in X)\ u(x, y, x) = v(x, y, x)\}.$$

All three are congruences of **B** because in fact they are kernels of projection maps. For example, ρ is the kernel of the projection of **B** into $\mathbf{A}^{X_1}$ where $X_1 = \{(x, y, z) \in X : y = z\}$.

Let $p_i = p_i^3$, $i = 1, 2, 3$, be the ternary projections considered as members of B. Then

$$p_3(x, y, y) = y = p_2(x, y, y),$$
$$p_2(y, y, x) = y = p_1(y, y, x),$$
$$p_3(x, y, x) = x = p_1(x, y, x),$$

implying

$$p_3 \equiv_\rho p_2 \equiv_\sigma p_1 \quad \text{and} \quad p_3 \equiv_\tau p_1.$$

Hence we have $(p_3, p_1) \in (\rho \circ \sigma) \wedge \tau$ and by the arithmeticity of **B**, $(p_3, p_1) \in (\sigma \wedge \tau) \circ (\rho \wedge \tau)$. This implies the existence of a ternary term function t such that

$$(p_3, t) \in \sigma \wedge \tau \quad \text{and} \quad (t, p_1) \in \rho \wedge \tau.$$

Then, given $(x, y, z) \in X$, we have

$$t(x, y, z) = p_3(x, y, z) = z \quad \text{if} \quad x = y,$$
$$t(x, y, z) = p_3(x, y, z) = z \quad \text{if} \quad x = z,$$
$$t(x, y, z) = p_1(x, y, z) = x \quad \text{if} \quad y = z,$$

and we see that $t(x, y, z)$ interpolates $\mathsf{Pix}\, A$ at X.   •

In Theorem 3.2.2 we proved that the existence of a near unanimity term can be characterized in terms of interpolation. However, analyzing the proof of how the existence of a near unanimity term implies the interpolation property for algebra **A**, we see again that much less is needed. It suffices to have, for any $a, b \in A$, a term u_{ab} which agrees with a near unanimity function at all $(n + 1)$-tuples $(a, \ldots, a, b, a, \ldots, a)$, for any position of b. We state this result, specialized to the case $n = 2$, explicitly since we need it in the sequel. In what follows the symbol $\mathsf{Maj}\, A$ denotes the ternary partial function on A whose domain coincides with that of $\mathsf{Pix}\, A$ and which agrees with every majority function on A.

Lemma 5.1.3 *Let* **A** *be an algebra such that for every* $a, b \in A$ *there exists a ternary term which agrees with a majority function at*

$\{(a,a,b),(a,b,a),(b,a,a)\}$. *Then the following conditions are equivalent for a finite partial function f on A:*

(1) *There exists a term function which agrees with f on its domain.*

(2) *f has interpolating term functions for every two element subset of its domain.*

(3) *f preserves all subuniverses of* $\mathbf{A}^2$.

Theorem 5.1.4 *An algebra* $\mathbf{A}$ *is strictly locally affine complete iff the class* $SP_f(\mathbf{A}^+)$ *is arithmetical.*

PROOF Assume first that $\mathbf{A}$ is strictly locally affine complete. Then Pix A can be interpolated at every finite subset of its domain by a polynomial function. Since polynomial functions of $\mathbf{A}$ are term functions of $\mathbf{A}^+$, Theorem 5.1.2 yields the arithmeticity of $SP_f(\mathbf{A}^+)$.

Now assume that the class $SP_f(\mathbf{A}^+)$ is arithmetical. Then by Theorem 5.1.2 Pix A can be interpolated by a function $p \in \mathrm{Pol}\mathbf{A}$ on the set

$$\{(a,a,b),(a,b,b),(a,b,a),(a,a,a),(b,a,a),(b,b,a)\}$$

where $a,b \in A$. It is then easy to check that the polynomial function $p(x,p(x,y,z),z)$ interpolates Maj A on the set

$$\{(a,a,b),(a,b,a),(b,a,a)\}.$$

Hence Lemma 5.1.3 applies to $\mathbf{A}^+$.

Let $f : X \to A$ be any compatible function where X is a finite subset of A^m. We want to prove that there is a polynomial function of $\mathbf{A}$ which agrees with f on X. By Lemma 5.1.3 it is sufficient to show that f preserves all subuniverses of $\mathbf{A}^+ \times \mathbf{A}^+$. Since $SP_f(\mathbf{A}^+)$ is arithmetical, it follows from Theorem 5.1.1 that all subuniverses of $\mathbf{A}^+ \times \mathbf{A}^+$ are rectangular. Since each of those subuniverses is diagonal, all of them are, in fact, congruences of $\mathbf{A}$. Since the function f preserves all congruences of $\mathbf{A}$ by assumption, the proof is complete.●

5.1.2   Neutrality

The characterization of strictly locally affine complete algebras obtained in Theorem 5.1.4 is not very efficient: there is no very reasonable way to check that the class $SP_f(\mathbf{A}^+)$ is arithmetical. Since our primary aim in this section is to describe strictly locally affine complete groups and rings, and the latter are known to be CP, we are interested in more efficient conditions that would yield arithmeticity of a class $SP_f(\mathbf{A}^+)$ provided it is CP. It will turn out that this can be done in terms of neutrality. (See Section 1.2.6.)

We start with two technical lemmas where p_i and p_{ij} denote the projection maps of A^3 onto its direct factors. For example, $p_1(a_1, a_2, a_3) = a_1$, $p_{23}(a_1, a_2, a_3) = (a_2, a_3)$, etc.

Lemma 5.1.5 *Let* $\mathbf{A}$ *be an algebra such that* $SP_f(\mathbf{A})$ *is congruence permutable. Let* $\mathbf{C}$ *and* $\mathbf{D}$ *be subalgebras of* $\mathbf{A}^3$ *all of whose 2-fold coordinate projections coincide and* $\mathbf{D} < \mathbf{C}$. *Then* $\sigma = p_1(\ker(p_{23}|_C))$ *and* $\tau = p_1(\ker(p_{23}|_D))$ *are congruences of* $\mathbf{B} = p_1(\mathbf{C}) = p_1(\mathbf{D})$, *and* $[\sigma, \sigma] \leq \tau$ *in* $\mathbf{Con}\,\mathbf{B}$.

Proof Since $\mathbf{C}$ and $\mathbf{D}$ have the same 2-fold coordinate projections, we have $p_1(C) = p_1(D)$. Therefore $p_1|_C$ is a surjective homomorphism $\mathbf{C} \to \mathbf{B}$ and maps all congruences of $\mathbf{C}$ to congruences of $\mathbf{B}$. Hence $\sigma \in \mathrm{Con}\,\mathbf{B}$. Similarly $\tau \in \mathrm{Con}\,\mathbf{B}$.

We select $a, b \in B$ and $\mathbf{c}, \mathbf{d} \in B^{k-1}$ such that $a \equiv_\sigma b$ and $\mathbf{c} \equiv_\sigma \mathbf{d}$, and assume that $t(a, \mathbf{c}) \equiv_\tau t(a, \mathbf{d})$ where t is some k-ary term. We have to show that $t(b, \mathbf{c}) \equiv_\tau t(b, \mathbf{d})$.

By definition of σ and in view of $p_{12}(C) = p_{12}(D)$, there exist $x = (x_1, x_2, x_3)$ and $y = (y_1, y_2, y_3)$ in D such that

$$x_1 = a, \; y_1 = b \quad \text{and} \quad x_2 = y_2. \tag{5.1}$$

Similarly, in view of $p_{13}(C) = p_{13}(D)$, there exist $\mathbf{g} = (\mathbf{g}_1, \mathbf{g}_2, \mathbf{g}_3)$ and $\mathbf{h} = (\mathbf{h}_1, \mathbf{h}_2, \mathbf{h}_3)$ in D^{k-1} such that

$$\mathbf{g}_1 = \mathbf{c}, \; \mathbf{h}_1 = \mathbf{d} \quad \text{and} \quad \mathbf{g}_3 = \mathbf{h}_3. \tag{5.2}$$

Since $SP_f(\mathbf{A})$ is CP, by Theorem 5.1.1 $\mathbf{Mal}\,A$ can be interpolated at any finite subset X of its domain by some term function m. The following calculations involve only finitely many elements of

A. Therefore we can select a ternary term m which induces $\mathsf{Mal}\, A$ wherever necessary.

Let

$$
\begin{aligned}
d_1 &= m(t(y,\mathbf{g}), m(t(x,\mathbf{h}), t(x,\mathbf{g}), t(y,\mathbf{g})), t(y,\mathbf{h})), \\
d_2 &= t(y,\mathbf{g}).
\end{aligned}
$$

Since $x, y \in D$ and $\mathbf{g}, \mathbf{h} \in D^{k-1}$, we have

$$(x,\mathbf{g}), (x,\mathbf{h}), (y,\mathbf{g}), (y,\mathbf{h}) \in D^k \,;$$

hence $d_1, d_2 \in D$. We are going to show that $p_{23}(d_1) = p_{23}(d_2)$ implying $p_1(d_1) \equiv_\tau p_1(d_2)$.

Keeping in mind the equalities (5.1) and (5.2), we have

$$
\begin{aligned}
p_2(d_1) &= m(t(y_2,\mathbf{g}_2), m(t(x_2,\mathbf{h}_2), t(x_2,\mathbf{g}_2), t(y_2,\mathbf{g}_2)), t(y_2,\mathbf{h}_2)) \\
&= m(t(y_2,\mathbf{g}_2), m(t(y_2,\mathbf{h}_2), t(y_2,\mathbf{g}_2), t(y_2,\mathbf{g}_2)), t(y_2,\mathbf{h}_2)) \\
&= m(t(y_2,\mathbf{g}_2), t(y_2,\mathbf{h}_2), t(y_2,\mathbf{h}_2)) \\
&= t(y_2,\mathbf{g}_2) = p_2(d_2)
\end{aligned}
$$

and

$$
\begin{aligned}
p_3(d_1) &= m(t(y_3,\mathbf{g}_3), m(t(x_3,\mathbf{h}_3), t(x_3,\mathbf{g}_3), t(y_3,\mathbf{g}_3)), t(y_3,\mathbf{h}_3)) \\
&= m(t(y_3,\mathbf{g}_3), m(t(x_3,\mathbf{h}_3), t(x_3,\mathbf{h}_3), t(y_3,\mathbf{h}_3)), t(y_3,\mathbf{h}_3)) \\
&= m(t(y_3,\mathbf{g}_3), t(y_3,\mathbf{h}_3), t(y_3,\mathbf{h}_3)) \\
&= t(y_3,\mathbf{g}_3) = p_3(d_2)\,;
\end{aligned}
$$

hence $p_{23}(d_1) = p_{23}(d_2)$.

Now

$$
\begin{aligned}
p_1(d_1) &= m(t(y_1,\mathbf{g}_1), m(t(x_1,\mathbf{h}_1), t(x_1,\mathbf{g}_1), t(y_1,\mathbf{g}_1)), t(y_1,\mathbf{h}_1)) \\
&= m(t(b,\mathbf{c}), m(t(a,\mathbf{d}), t(a,\mathbf{c}), t(b,\mathbf{c})), t(b,\mathbf{d}))
\end{aligned}
$$

and

$$p_1(d_2) = t(y_1,\mathbf{g}_1) = t(b,\mathbf{c}).$$

Hence, in view of $p_1(d_1) \equiv_\tau p_1(d_2)$ and $t(a,\mathbf{c}) \equiv_\tau t(a,\mathbf{d})$, we get

$$
\begin{aligned}
t(b,\mathbf{c}) &\equiv_\tau m(t(b,\mathbf{c}), m(t(a,\mathbf{d}), t(a,\mathbf{c}), t(b,\mathbf{c})), t(b,\mathbf{d})) \\
&\equiv_\tau m(t(b,\mathbf{c}), m(t(a,\mathbf{d}), t(a,\mathbf{d}), t(b,\mathbf{c})), t(b,\mathbf{d})) \\
&= m(t(b,\mathbf{c}), t(b,\mathbf{c}), t(b,\mathbf{d})) \\
&= t(b,\mathbf{d}).
\end{aligned}
$$

This proves the lemma. ●

Lemma 5.1.6 *Let* **A**, **C**, *and* **D** *be as in Lemma 5.1.5. If all sub-algebras of* **A** *are neutral then* **C** = **D**.

PROOF Let **B**, σ, and τ have the same meaning as in the preceding lemma. If $x = (x_1, x_2, x_3)$ is an arbitrary element of C then there exists $y = (y_1, y_2, y_3) \in D$ such that

$$x_2 = y_2 \quad \text{and} \quad x_3 = y_3 \,. \tag{5.3}$$

By definition of σ, we have $x_1 \equiv_\sigma y_1$. But by Lemma 5.1.5 $[\sigma, \sigma] \leq \tau$, so the neutrality of **B** implies $\sigma = \tau$. Hence $x_1 \equiv_\tau y_1$. The latter yields the existence of $z = (z_1, z_2, z_3)$ and $w = (w_1, w_2, w_3)$ in D such that

$$z_1 = x_1, \quad w_1 = y_1, \quad z_2 = w_2 \quad \text{and} \quad z_3 = w_3 \,. \tag{5.4}$$

Now, as in Lemma 5.1.5, we choose a ternary term m which induces Mal A wherever necessary, that is, on a finite set of elements of domain of Mal A which occur in the following calculations. It remains to show that $m(z, w, y) = x$ or, equivalently, $m(z_i, w_i, y_i) = x_i$ for $i = 1, 2, 3$. Using the equalities (5.3) and (5.4) we get

$$\begin{aligned}
m(z_1, w_1, y_1) &= m(x_1, w_1, w_1) = x_1 \,, \\
m(z_2, w_2, y_2) &= m(z_2, z_2, x_2) = x_2 \,, \\
m(z_3, w_3, y_3) &= m(z_3, z_3, x_3) = x_3 \,.
\end{aligned}$$

The proof of lemma is complete.   ●

Theorem 5.1.7 *For an arbitrary algebra* **A**, *the class* $SP_f(\mathbf{A})$ *is arithmetical iff it is congruence permutable and all subalgebras of* **A** *are neutral.*

PROOF First assume that $SP_f(\mathbf{A})$ is arithmetical and prove that every subalgebra **B** < **A** is neutral. By Lemma 1.2.26 we have to show that the equality $[\rho, \rho] = \rho$ holds in Con **B**. Let $\rho \in$ Con **B**, $a, b \in B$, and $(a, b) \in \rho$. We are going to prove $(a, b) \in [\rho, \rho]$.

Applying Theorem 5.1.2 we find a ternary term function t which interpolates Pix A at the set $\{(a, a, b), (b, a, b), (a, a, a), (b, a, a)\}$. We notice that

$$(a, b) \in \rho, \quad ((a, b), (a, a)) \in \rho$$

and
$$(b,b) = (t(b,a,a), t(b,a,b)) \in [\rho, \rho].$$

Hence $(a,b) = (t(a,a,a), t(b,a,a)) \in [\rho, \rho]$ by the definition of the commutator.

Now assume that $SP_f(\mathbf{A})$ is CP and all subalgebras of $\mathbf{A}$ are neutral. We shall prove that $\mathsf{Pix}\, A$ can be interpolated by a term function at every finite subset of its domain. We first observe that $\mathsf{Pix}\, A$ can be interpolated by a term function at every two element subset X of its domain. Indeed, for triples of form (a,a,b) and (c,d,c) this can be done by the projection function p_3^3, and, symmetrically, for triples of the form (a,b,b) and (c,d,c) by p_1^3. The remaining case $X = \{(a,a,b),(c,d,d)\}$ is settled by a term which interpolates $\mathsf{Mal}\, A$ at X and which exists by Theorem 5.1.1.

Now the proof will be complete if we show that $\mathsf{Maj}\, A$ can be interpolated by a term function at every subset of A having the form $X = \{(a,a,b),(a,b,a),(b,a,a)\}$, because then Lemma 5.1.3 applies. We consider the following two subalgebras of $\mathbf{A}^X$. The algebra $\mathbf{D}$ consists of all term functions $X \to A$ and the algebra $\mathbf{C}$ consists of all such functions $X \to A$ which can be interpolated by a term function at every two element subset of X. Obviously $D \subseteq C$ and we need to establish the reverse inclusion. Since obviously the restriction of $\mathsf{Mal}\, A$ to X belongs to C, this will prove the theorem. However, clearly C and D have the same 2-fold projections in A^3; hence the Lemma 5.1.6 implies the equality $C = D$.   •

Corollary 5.1.8 *For any algebra* $\mathbf{A}$ *the class* $SP_f(\mathbf{A}^+)$ *is arithmetical iff it is congruence permutable and the algebra* $\mathbf{A}$ *is neutral.*

PROOF We observe that A is the only subuniverse of $\mathbf{A}^+$ and the algebras $\mathbf{A}$ and $\mathbf{A}^+$ have the same congruences and the same commutator function.   •

Corollary 5.1.9 *An algebra of a congruence permutable variety is strictly locally affine complete iff it is neutral.*

In view of examples of Section 1.2.6, we now have the following results for groups, rings, and modules. From them we conclude, in particular, that every simple ring with nonzero multiplication and every simple nonabelian group is locally functionally complete, i.e., has the interpolation property (Section 3.4.3). These conclusions generalize the corresponding conclusions for (finite) functionally complete groups and rings found in Section 3.1.4.

Corollary 5.1.10 *A group* **A** *is strictly locally affine complete iff* $[\mathbf{R}, \mathbf{R}] = \mathbf{R}$ *for every normal subgroup* **R** *of* **A**.

Corollary 5.1.11 *A ring* **A** *is strictly locally affine complete iff the equality* $\mathbf{R}\mathbf{R} = \mathbf{R}$ *holds for every ideal* **R** *of* **A**.

Corollary 5.1.12 *Only the zero modules are strictly locally affine complete.*

Finally, we have the following theorem, which directly generalizes Theorem 3.1.21. It is often called the McKenzie-Gumm theorem and is a fundamental fact about simple algebras in congruence permutable varieties.

Theorem 5.1.13 *Every simple algebra in a congruence permutable variety either has the interpolation property or is affine over an abelian group which is either an elementary p-group or is a torsion free divisible group.*

The theorem first appears in [Gumm, 1979] where it is a consequence of Gumm's geometric analysis of affine algebras, an analysis important in the development of commutator theory. A key result in that analysis is that any algebra **A** in a CP variety is affine over some abelian group iff in $\mathbf{Con}(\mathbf{A} \times \mathbf{A})$ there is a congruence θ which is a complement of the two projection kernels, i.e., such that these three congruences together with $0_{A \times A}$ and $1_{A \times A}$ form the smallest nonmodular lattice usually denoted by $\mathbf{M_3}$. If **A** is simple then the underlying abelian group is either a torsion free divisible group or is an elementary abelian p-group. In [Pixley, 1982] it is shown that **A** in a CP variety has the interpolation property iff **A** is simple and **A** × **A** has no skew congruences. Combining these results proves Theorem 5.1.13.

Still another proof of Theorem 5.1.13 is obtained in the paper [Rosenberg, Szabó, 1984] mentioned in Section 3.4.3. In that paper the proof of Rosenberg's theorem (Theorem 1.1.5) is modified to obtain for any (infinite) set A a collection of finitary relations R on A

with the property that for each $r \in R$, the algebra $\langle A; clo(\{r\}) \rangle$ is not locally primal and for every algebra $\langle A; F \rangle$ which is not locally primal, $F \subseteq clo(\{r\})$ for some $r \in R$. The collection R falls into eight types which, though not claimed to be optimal, naturally extend the six types of Rosenberg's original theorem. In particular one of the types is the same as type (4) (prime affine) of Rosenberg's theorem except that now the abelian group is either an elementary p-group, as before, or is torsion free and divisible. As in Szendrei's proof of Theorem 3.1.21 congruence permutability eliminates certain possibilities, leading to the following result.

Theorem 5.1.14 *An algebra* **A** *in a congruence permutable variety is locally primal iff the following conditions are satisfied:*

(i) **A** *has no proper subalgebras,*

(ii) **A** *is simple,*

(iii) **A** *has no automorphisms whose cycles are either all of the same prime length or all infinite, and*

(iv) **A** *is affine over no abelian elementary p-group or torsion free and divisible group.*

If we consider polynomials instead of term functions, i.e., the interpolation property instead of local primality, this eliminates conditions (i) and (iii) so that we obtain Theorem 5.1.13.

Exercises

1. Prove that the ring of integers is not affine complete.

2. An associative ring **A** is called *von Neumann regular* if for every $a \in A$ there exists $b \in A$ such that $a = aba$. Prove that all von Neumann regular rings are strictly locally affine complete.

3. Prove that none of the symmetric groups $\mathbf{S}_n$ is affine complete but in case $n \geq 5$ all unary compatible functions on $\mathbf{S}_n$ are polynomial.

4. Prove that the wreath product of two finite simple nonabelian groups is affine complete.

5.2   Modules

5.2.1   General observations

In this section we consider polynomial completeness problems in the case of modules. In particular we consider only left unital modules over an associative ring with a unity element. If **A** is an **R**-module then every m-ary polynomial function on **A** has the form

$$f(x_1, \ldots, x_m) = r_1 x_1 + \cdots + r_m x_m + a$$

where $r_1, \ldots, r_m \in R$ and $a \in A$. Since congruences of modules are determined by their submodules, the compatibility criterion from Section 1.1.3 takes the following form. A function f is compatible iff

$$f(\mathbf{a}) - f(\mathbf{b}) \in \mathrm{Sg}(b_1 - a_1, \ldots, b_m - a_m) \qquad (5.5)$$

for all $\mathbf{a}, \mathbf{b} \in A^m$. Note that

$$\mathrm{Sg}(a_1, \ldots, a_m) = Ra_1 + \cdots + Ra_m = \{r_1 a_1 + \cdots + r_m a_m : r_i \in R\}.$$

Because of the existence of a simple canonical form for polynomials and relatively simple compatibility criterion, it may seem that the polynomial completeness problems for modules are quite easy to solve. Actually this is not the case. Very much depends on the structure of the ring **R**. Here we present complete solutions of the affine completeness and the local affine completeness problem in two important cases: for semisimple modules and abelian groups. Note that by Corollary 5.1.12 no nonzero module can be strictly locally affine complete. We also show that semisimple rings and Artinian simple rings may be defined by certain completeness properties of their modules. The results of this section are due to H. Werner, W. Nöbauer, K. Kaarli, and A. Saks. For obvious reasons, the first result, Theorem 5.2.9, is the description of affine complete vector spaces ([Werner, 1971]). W. Nöbauer initiated the study of affine completeness of abelian groups ([Nöbauer, 1976], [Nöbauer, 1978]). He mainly studied affine completeness of direct sums; in particular, the important Proposition 5.2.19 and the description of affine complete finitely generated abelian groups were obtained by him. The very basic Lemma 5.2.3 is also due to W. Nöbauer. The classes of affine complete abelian groups and locally affine complete abelian groups were completely described in [Kaarli, 1982] and in [Kaarli, 1985]. Note that the compatible function extension property was discovered when trying to solve the affine completeness problem of torsion free abelian groups of rank 1 (see the proof of Theorem 5.2.22). The results presented in Sections 5.2.2 and 5.2.3 are from [Saks, 1985a] and [Saks, 1985b].

We start with a few general observations which remain valid for all rings. Obviously every compatible function on a module can be represented as a sum of a zero preserving function and a constant function. Since the constant functions are polynomials, it follows that when studying affine completeness or local affine completeness of modules, we may restrict to the case of zero preserving functions. The other important observation is that all zero preserving compatible functions on modules preserve all submodules. This is an easy consequence from the criterion (5.5). Therefore, if f is an m-ary zero preserving compatible function on an **R**-module **A** and $a_1, \ldots, a_m \in A$, then

$$f(a_1, \ldots, a_m) = r_1 a_1 + \cdots + r_m a_m$$

where the coefficients r_i depend on the m-tuple $(a_1, \ldots, a_m)$.

Another useful observation is that the investigation of (local) affine completeness of modules may be reduced to the case of faithful modules. Let **A** be an arbitrary **R**-module and

$$I = \text{Ann}\mathbf{A} = \{r \in R : rA = 0\}.$$

The subset I is an ideal of **R**. Then **A** can be regarded, in a natural way, as a module over $\mathbf{R}_1 = \mathbf{R}/I$, and the latter module is faithful. It is easy to see that the **R**-module **A** is (locally) affine complete iff so is the $\mathbf{R}_1$-module **A**. The reason is that these two modules have the same term functions, hence the same submodules, the same compatible functions, and the same polynomial functions.

Now we shall explicitly show that the variety of modules over a given ring **R** cannot be locally affine complete. In fact we have the following specific result.

Theorem 5.2.1 *A subdirectly irreducible **R**-module is never locally affine complete.*

PROOF Let **A** be a SI **R**-module and **B** be its monolith, that is, a unique minimal submodule. Take an arbitrary nonzero element $b \in B$ and define the binary function f on A as follows:

$$f(x,y) = \begin{cases} b & \text{if } x \neq 0 \text{ and } y \neq 0, \\ 0 & \text{if } x = 0 \text{ or } y = 0. \end{cases}$$

The compatibility of f is obvious. Suppose f is interpolated by the polynomial $rx + sy + a$ at the set $\{(0,0),(0,b),(b,0),(b,b)\}$. Since f preserves the zero, $a = 0$. Then

$$b = f(b,b) = rb + sb = f(b,0) + f(0,b) = 0,$$

a contradiction. ●

The following example shows that it was necessary to use a binary function in the last proof.

Example 5.2.2 *A module which is not locally affine complete though all its unary compatible functions are polynomials.*

Let **A** be a two element abelian group, that is, a module over the ring of integers **Z**. This module is simple; so all functions on it are compatible. By Theorem 5.2.1 **A** is not locally affine complete. On the other hand, there are exactly 4 unary functions on A and 4 unary polynomial functions on **A** as well. If $A = \{0,1\}$ then the latter are x, $x+1$, 0 and 1. ●

Now we show that when studying (local) affine completeness of modules it suffices to consider just binary functions.

Lemma 5.2.3 *An **R**-module **A** is (locally) affine complete iff all binary compatible functions on **A** are (local) polynomials.*

PROOF Clearly, if all binary compatible functions on a given algebra are (local) polynomials then so are all unary compatible functions. We proceed by induction on the arity of function. We give the proof for affine completeness and then indicate how minor changes allow us to use the same argument in the case of local affine completeness.

Suppose all compatible functions on **A** in less than m variables $(m \geq 3)$ are polynomials and let f be an m-ary compatible function on **A**. Without loss of generality, $f(0,\ldots,0) = 0$. Consider the $(m-1)$-ary function $f(x_1, \ldots, x_{m-1}, a)$ where $a \in A$ is a fixed element. Obviously this function is compatible, so it must be polynomial. Thus there exist $r_j = r_j(a) \in R$ and $c = c(a) \in A$ such that

$$f(x_1, \ldots, x_{m-1}, a) = r_1 x_1 + \cdots + r_{m-1} x_{m-1} + c \qquad (5.6)$$

for every $x_1, \ldots, x_{m-1} \in A$. In particular, $f(0, \ldots, 0, a) = c(a)$ for every $a \in A$. Since the unary function $f(0, \ldots, 0, y)$ is compatible, there exist $s \in R$ and $b \in A$ such that $c(a) = sa + b$ for every $a \in A$.

Now consider the function $f(0, x_2, \ldots, x_m)$ which is again a polynomial by our induction hypothesis. Let

$$f(0, x_2, \ldots, x_m) = t_2 x_2 + \cdots + t_m x_m + d$$

where $t_2, \ldots, t_m \in R$ and $d \in A$. Comparing this with (5.6) and taking $x_2 = \cdots = x_{m-1} = 0$, we have $t_m x_m + d = s x_m + b$. Consequently,

$$t_2 x_2 + \cdots + t_{m-1} x_{m-1} = r_2 x_2 + \cdots + r_{m-1} x_{m-1}$$

for arbitrary $x_2, \ldots, x_{m-1} \in A$. Taking here

$$x_2 = \cdots = x_{j-1} = x_{j+1} = \cdots = x_{m-1} = 0,$$

we have $t_j x_j = r_j x_j$ for every $x_j \in A$ and $j = 2, \ldots, m-1$. This means that $r_j(a)$ actually can be chosen the same for all $a \in A$. Similarly, considering the function $f(x_1, 0, x_3, \ldots, x_m)$, we see that the coefficient r_1 also can be chosen the same for all $a \in A$. This settles the affine completeness case.

Now suppose that all $(m-1)$-ary compatible functions on **A** are local polynomials and we want to prove that a given m-ary compatible function on **A** is a local polynomial, too. Obviously it suffices to show that f can be interpolated by a polynomial function on any set X^m where X is a finite subset of A. We may assume, without loss of generality, that $0 \in X$. It is easy to observe that the above arguments again work to produce a required polynomial. $\bullet$

5.2.2 Completeness of weak direct powers

Let A_i, $i \in I$, be a family of **R**-modules. We recall that the *direct sum* of this family is the submodule of the **R**-module $\prod(A_i : i \in I)$, denoted by $\Sigma^{\oplus}(\mathbf{A}_i : i \in I)$, whose universe is the set of all families $(a_i)_{i \in I}$ where $a_i \in A_i$ for all $i \in I$ and only a finite number of the components a_i are nonzero. A *weak direct power* $\mathbf{A}^{(I)}$ of an $\mathbf{R}$-module $\mathbf{A}$ is defined as the direct sum $\Sigma^{\oplus}(\mathbf{A}_i : i \in I)$ where $\mathbf{A}_i = \mathbf{A}$ for all $i \in I$. If the index set I is finite, say $I = \{1, \dots, n\}$, then we write $\mathbf{A}_1 \oplus \cdots \oplus \mathbf{A}_n$ instead of $\Sigma^{\oplus}(\mathbf{A}_i : i \in I)$. Obviously

$$\mathbf{A}_1 \oplus \cdots \oplus \mathbf{A}_n = \mathbf{A}_1 \times \cdots \times \mathbf{A}_n \,,$$

but it is more customary to speak in this situation about a direct sum rather than a direct product. Note that the direct sum of rings might be defined similarly. However, since we restricted ourselves to the rings with unity element, only the direct sums of finitely many rings would be rings in our sense. Nevertheless, following the tradition of ring theory, in what follows we shall refer to a direct product of finitely many rings as to their direct sum.

If $\mathbf{A}$ is an $\mathbf{R}$-module and $\mathbf{R}' = \mathbf{End}\,_{\mathbf{R}}\mathbf{A}$ then $\mathbf{A}$ becomes, in a natural way, an $\mathbf{R}'$-module and the endomorphisms of this module are called the *biendomorphisms* of the $\mathbf{R}$-module $\mathbf{A}$. The ring of all biendomorphisms of the given $\mathbf{R}$-module will be denoted by $\mathbf{R}''$. There is a natural homomorphism $\mathbf{R} \to \mathbf{R}''$. Clearly, if this homomorphism is injective, the $\mathbf{R}$-module $\mathbf{A}$ is faithful and $\mathbf{R}$ may be identified with a subring of $\mathbf{R}''$. The $\mathbf{R}$-module $\mathbf{A}$ is said to be *balanced* if the natural homomorphism $\mathbf{R} \to \mathbf{R}''$ is surjective. We say that $\mathbf{R}$ *is dense* in $\mathbf{R}''$, if for every $f \in R''$, and every finite subset $X \subseteq A$, there exists $r \in R$ such that $f(x) = rx$ for every $x \in X$. Note that the density of $\mathbf{R}$ in $\mathbf{R}''$ is equivalent to the requirement that every $f \in R''$ must be a local polynomial function of the $\mathbf{R}$-module $\mathbf{A}$.

Next we shall consider compatible functions on weak direct powers of an $\mathbf{R}$-module $\mathbf{A}$. It will turn out that every such function is determined by a unique compatible function on $\mathbf{A}$. An arbitrary m-ary function g on A extends componentwise to A^I. Let us denote this extension by $\tilde{g}$. If $\mathbf{A}$ is an $\mathbf{R}$-module then the function $\tilde{g}$ preserves $\mathbf{A}^{(I)}$.

Lemma 5.2.4 *Let $\mathbf{A}$ be an $\mathbf{R}$-module, I be a set with $|I| \geq 2$, and f be an m-ary zero preserving compatible function on $\mathbf{A}^{(I)}$. Then there exists an m-ary function g on A such that $f = \tilde{g}$ and g has the following properties:*

(i) *g is compatible on* **A**;

(ii) *g is additive, that is, a group homomorphism from* **A**^{m} *to* **A**;

(iii) *if m* = 1 *then g is a biendomorphism of the* **R**-*module* **A**.

In view of Lemma 5.2.3 we shall not need the lemma for functions of arity greater than 2. Since the result for unary functions follows directly from that for binary functions, we shall give the proof only for case $m = 2$. The proof for arbitrary m is similar but requires only more space.

PROOF Let **B** = **A**$^{(I)}$. Since f is compatible on **B**, we have $f = (f_i)_{i \in I}$ where every f_i is a binary compatible function on **A**. We shall show that all the f_i are equal; thus any of them may be taken in the role of g. Then obviously g has the property (i). Let $j, k \in I$, $j \neq k$, and take arbitrary $x, y \in A$. Clearly there exist $b, c \in B$ such that $b_j = b_k = x$ and $c_j = c_k = y$. Since f is compatible and zero preserving there exist $r, s \in R$ such that $f(b, c) = rb + sc$. On the other hand, $f(b, c) = (f_i(b_i, c_i))_{i \in I}$. Hence

$$f_j(x, y) = rx + sy = f_k(x, y).$$

Thus g has property (i).

Take arbitrary $x, y, u, v \in A$, and distinct elements $j, k \in I$. Then choose $b, c, d, e \in B$ such that

$$b_j = x + u, \quad c_j = y + v, \quad d_j = x, \quad e_j = 0,$$
$$b_k = x, \quad c_k = y, \quad d_k = 0, \quad e_k = 0.$$

Since $f = \tilde{g}$, we have

$$f(b, c) - f(d, e) = (g(b_i, c_i) - g(d_i, e_i))_{i \in I}.$$

On the other hand, since f is compatible, there exist $r, s \in R$ such that

$$f(b, c) - f(d, e) = r(b - d) + s(c - e) = (r(b_i - d_i) + s(c_i - e_i))_{i \in I}.$$

Taking in this equality first $i = j$ and then $i = k$, we get

$$g(x + u, y + v) - g(u, v) = rx + sy = g(x, y)$$

implying $g(x + u, y + v) = g(x, y) + g(u, v)$. Thus g has the property (ii).

If $m = 1$ then g is a unary function on A. By (ii) g is an endomorphism of the additive group of $\mathbf{A}$. So, in order to prove (iii), we have to show that g permutes with every endomorphism ϕ of the $\mathbf{R}$-module $\mathbf{A}$. Take an arbitrary element $a \in A$ and distinct elements $j, k \in I$. There exists $b \in B$ such that $b_j = a$ and $b_k = \phi(a)$. Since f is compatible and zero preserving, there exists $r \in R$ such that $f(b) = rb$. Then, in particular, $g(b_j) = g(a) = ra$ and $g(b_k) = g(\phi(a)) = r\phi(a)$. Hence,

$$g(\phi(a)) = r\phi(a) = \phi(ra) = \phi(g(a)),$$

as required for property (iii).   •

Corollary 5.2.5 *Let an $\mathbf{R}$-module $\mathbf{A}$ be the direct sum of $\mathbf{R}$-modules $\mathbf{A}_i$, $i \in I$, and assume that for every $j \in I$ there exists $k \in I$, $k \neq j$, such that $\mathbf{A}_j$ and $\mathbf{A}_k$ are isomorphic. Then $\mathbf{A}$ is (locally) affine complete iff all unary zero preserving compatible functions on $\mathbf{A}$ are (local) polynomials.*

PROOF In view of Lemma 5.2.3 all we need is to prove that every binary zero preserving compatible function f on $\mathbf{A}$ is a (local) polynomial whenever all unary zero preserving compatible functions on $\mathbf{A}$ are (local) polynomials. Since f is compatible, it has the form $(f_i)_{i \in I}$ where every f_i is a binary zero preserving compatible function on $\mathbf{A}_i$. Lemma 5.2.4 together with the assumption on the family of modules $\mathbf{A}_i$ easily yields that every f_i is additive, in particular, $f_i(x, y) = f_i(x, 0) + f_i(0, y)$ for all $x, y \in A_i$. Clearly then also $f(x, y) = f(x, 0) + f(0, y)$ for all $x, y \in B$. Since f is compatible and zero preserving, so are the unary functions $f(x, 0)$ and $f(0, y)$. Now the sum $f(x, y)$ of these functions is a local polynomial, too.   •

Corollary 5.2.6 *Let $\mathbf{A}$ be an $\mathbf{R}$-module and $\mathbf{B} = \mathbf{A}^{(I)}$, $|I| \geq 2$. Then every unary zero preserving compatible function f on $\mathbf{B}$ is a biendomorphism of $\mathbf{B}$.*

PROOF We need to prove that $f(\phi(b)) = \phi(f(b))$ for every $b \in B$ and every endomorphism ϕ of $\mathbf{B}$. Since every $b \in B$ is a sum of finitely many elements belonging to different direct summands $\mathbf{A}_i$, the general case easily reduces to one with $b \in A_j$, $\phi(A_j) \subseteq A_k$,

$j, k \in I$. Now the desired result becomes obvious if we take into account that $f = \tilde{g}$ where g is a biendomorphism of **A** and the restriction of ϕ to A_j can be considered as an endomorphism of **A**.•

Now we produce certain sufficient conditions for a weak direct power to be (locally) affine complete.

Theorem 5.2.7 *Let* **A** *be an* **R**-*module,* **R**″ *be the ring of biendomorphism of this module, and* I *be a set of size at least 2.*

(a) *If* **R** *is dense in* **R**″ *then the module* $\mathbf{A}^{(I)}$ *is locally affine complete.*

(b) *If* **A** *is balanced then* $\mathbf{A}^{(I)}$ *is affine complete.*

PROOF (a) By the preceding corollary we have to show that every unary zero preserving compatible function f on $\mathbf{A}^{(I)}$ is a local polynomial. By Lemma 5.2.3 we have $f = \tilde{g}$ where $g \in R''$ is a local polynomial function of the **R**-module **A** because of the density condition. Clearly then f is a local polynomial function of $\mathbf{A}^{(I)}$.

(b) As in the preceding case, we represent the given compatible function f on $\mathbf{A}^{(I)}$ in the form $\tilde{g}$ and show that g is a polynomial function of the **R**-module **A**. The latter is a direct consequence of the fact ([Faith, 1973]) that **A** is a balanced **R**-module.   •

Theorem 5.2.8 *All free modules with more than one free generator are affine complete.*

PROOF It is well known that every free **R**-module is isomorphic to some weak direct power of the **R**-module **R** and the size of the set of free generators of $\mathbf{R}^{(I)}$ is just the size of I. Thus our claim follows from Theorem 5.2.7 and the well known fact that **R** is a balanced **R**-module.   •

Note that the free **R**-module in one generator, that is, the ring **R** considered as a module over itself, need not be locally affine complete, though sometimes it is even affine complete. For example, let **R** be a commutative ring. Then both the ring **R** and the **R**-module **R** have exactly the same congruences but different polynomial functions. Hence such a module cannot be affine complete. In particular no cyclic module over a commutative ring is affine complete. In Section 3.4.2 we already noted the geometric significance of the following consequence, first proved by Werner.

Theorem 5.2.9 *A vector space is affine complete iff its dimension is not 1.*

PROOF This follows from Theorems 5.2.1 and 5.2.8, and the obvious fact that the one element modules are affine complete.   •

5.2.3   Semisimple modules and semisimple rings

An **R**-module **A** is called *semisimple* if it satisfies one of the following equivalent conditions (see [Kasch, 1982]):

1. **A** is a direct sum of simple **R**-modules;

2. **A** coincides with the sum of all its minimal submodules;

3. every submodule of **A** is a direct summand of **A**.

If **A** is a simple **R**-module then all weak direct powers $\mathbf{A}^{(I)}$ are clearly semisimple. Such semisimple **R**-modules are said to be *homogeneous*. It is well known that every semisimple **R**-module **B** uniquely decomposes into a direct sum of homogeneous semisimple **R**-modules. The latter are called *homogeneous components* of **B**. For easy reference, we state some basic properties of semisimple modules as a separate lemma. The proof of (i) is an easy exercise; the proof of (ii) can be found in [Kasch, 1982].

Lemma 5.2.10 *Let **A** be a semisimple **R**-module and $\mathbf{A}_i$, $i \in I$, be homogeneous components of **A**. Let $\mathbf{R}''$ and $\mathbf{R}''_i$ be the biendomorphism ring of **A** and $\mathbf{A}_i$, respectively, for every $i \in I$. Then:*

(i) *$\mathbf{R}''$ and $\prod(\mathbf{R}''_i : i \in I)$ are canonically isomorphic;*

(ii) *$\mathbf{R}$ is dense in $\mathbf{R}''$.*

The following two theorems characterize (locally) affine complete semisimple modules.

Theorem 5.2.11 *A semisimple **R**-module is locally affine complete iff it has no simple homogeneous components.*

PROOF Assume first that a semisimple **R**-module **A** has a simple homogeneous component **B** and let **C** be the sum of the other homogeneous components of **A**. Then $\mathbf{A} = \mathbf{B} \oplus \mathbf{C}$ and **A** has no skew submodules with respect to this decomposition. Hence every pair (g, h) where g and h are compatible functions on **B** and **C**, respectively, is a compatible function on **A**. Since **B** is not locally affine complete by Theorem 5.2.1, it is easy to produce a compatible function on **A** which is not a local polynomial.

Now let **A** be without simple homogeneous components. We want to prove that **A** is locally affine complete. By Corollary 5.2.5 it suffices to show that every unary zero preserving compatible function f on **A** is a local polynomial. Let $\mathbf{A}_i$, $i \in I$, be the homogeneous components of **A**. Then $f = (f_i)_{i \in I}$ where every f_i is a zero preserving compatible function on $\mathbf{A}_i$. By Corollary 5.2.6 all f_i are biendomorphisms of $\mathbf{A}_i$, $i \in I$, and then the first claim of Lemma 5.2.10 implies that f is a biendomorphism of **A**. Now the second claim of Lemma 5.2.10 gives that f is a local polynomial function of **A**. •

Theorem 5.2.12 *A semisimple* **R***-module is affine complete iff it is balanced and has no simple homogeneous components.*

PROOF Suppose first that a semisimple **R**-module **A** has no simple homogeneous components and is balanced. As in the proof of the preceding theorem, we need to check only unary zero preserving compatible functions and the latter turn out to be biendomorphisms of the **R**-module **A**. Since this module is balanced, all its biendomorphisms have the form rx where $r \in R$. This proves that **A** is affine complete.

Now let **A** be affine complete. Then **A** is locally affine complete and by Theorem 5.2.11 it has no simple homogeneous components. It remains to show that **A** is balanced. By Lemma 5.2.10 the ring **R** is dense in $\mathbf{R}''$, the ring of biendomorphisms of the **R**-module **A**. Hence every $f \in R''$ is a zero preserving local polynomial function of **A**, hence a compatible function on that module. Since **A** is affine complete, f is a polynomial function rx where $r \in R$. This proves the theorem. •

Recall that a ring **R** is (left) Artinian if every nonempty set of left ideals of **R** contains a minimal element. The Jacobson radical $J(\mathbf{R})$ of **R** is the intersection of all maximal left ideals of **R** and **R** is *Jacobson semisimple* if $J(\mathbf{R}) = 0$. A ring **R** is said to be *semisimple* if it satisfies one of the following equivalent conditions:

1. the **R**-module **R** is semisimple;

2. every **R**-module is semisimple;

3. **R** is Jacobson semisimple and Artinian;

4. **R** is isomorphic to a direct sum of finitely many simple Artinian rings.

Theorem 5.2.13 *Let* **A** *be a module over a semisimple ring* **R**. *Then the following conditions are equivalent:*

(1) **A** *is affine complete;*

(2) **A** *is locally affine complete;*

(3) **A** *has no simple homogeneous components.*

PROOF Clearly (1) implies (2) and by Theorem 5.2.11 (2) implies (3).

Finally, assume that **A** has no simple homogeneous components. We want to prove that **A** is affine complete. By Theorem 5.2.12 it suffices to observe that **A** is balanced. However, all modules over a semisimple ring are known to be balanced ([Faith, 1973]). Hence (3) implies (1). ●

Now we want to prove that the class of semisimple rings, as well as the class of simple Artinian rings, can be characterized by requiring certain modules to be affine complete. We start with a lemma.

Lemma 5.2.14 *If* **A** *is a subdirectly irreducible* **R**-*module,* **B** *is its monolith, and* **A** $\neq$ **B**, *then* **A** $\oplus$ **B** *is not affine complete.*

PROOF Let us fix an arbitrary nonzero element $b \in B$ and define

$$f(x + y, u + v) = \begin{cases} b & \text{if } x, u \in A \setminus B, \\ 0 & \text{otherwise,} \end{cases}$$

for all $x, u \in A$, $y, v \in B$. To observe that f is compatible we have to check that

$$f(x_1 + y_1, u_1 + v_1) - f(x_2 + y_2, u_2 + v_2)$$

is always an element of

$$\text{Sg}((x_1 + y_1) - (x_2 + y_2), (u_1 + v_1) - (u_2 + v_2)).$$

Clearly, the only nontrivial case is $x_1, u_1 \notin B$, and $x_2 \in B$ or $u_2 \in B$. To settle this case, it is sufficient to show that $b \in \text{Sg}(x+y)$ for every $x \in A \backslash B$, $y \in B$. Since $\mathbf{A}$ is SI and $b \in B$, it is sufficient to show that $\text{Sg}(x+y)$ contains a nonzero element of $\mathbf{A}$; that is, there exists $r \in R$ such that $ry = 0$ but $rx \neq 0$. Suppose this is not the case. Then $y \neq 0$ and the mapping $ry \to rx$ is a surjective homomorphism of R-modules $\mathbf{B} = \mathbf{R}y \to \mathbf{R}x$. This, however, contradicts the choice of x. Indeed, since $x \notin B$ and $\mathbf{A}$ is SI, the R-module $\mathbf{R}x$ is not simple, and it cannot be a homomorphic image of the simple R-module $\mathbf{B}$.

It remains to show that f is not a polynomial function. Assume, on the contrary, that there exists $r, s \in R$ such that f is the polynomial function $rx + sy$. Then, taking $a \in A \setminus B$ and identifying it with the element $a + 0 \in \mathbf{A} \oplus \mathbf{B}$, we have

$$b = f(a, a) = ra + sa = rf(a, 0) + sf(0, a) = 0 + 0,$$

a contradiction. ●

Recall that a submodule $\mathbf{B}$ of an R-module $\mathbf{A}$ is called *essential* if it has a nonzero intersection with every nonzero submodule of $\mathbf{A}$. Clearly every module is an essential submodule of itself. A semisimple module has no other essential submodules, because they all are direct summands. Another important example of an essential submodule is the monolith of a SI R-module.

Theorem 5.2.15 *A ring $\mathbf{R}$ is semisimple iff every R-module $\mathbf{A} \oplus \mathbf{B}$ where $\mathbf{B}$ is an essential submodule of $\mathbf{A}$ is affine complete.*

PROOF Let $\mathbf{R}$ be a semisimple ring, $\mathbf{A}$ be an R-module and $\mathbf{B}$ be an essential submodule of $\mathbf{A}$. Then $\mathbf{A}$ is a semisimple module and by the above remark, $\mathbf{B} = \mathbf{A}$. Now $\mathbf{A} \oplus \mathbf{B}$ is affine complete by Theorem 5.2.13.

To prove the sufficiency, we first apply the condition of the theorem to the direct sum $\mathbf{A} \oplus \mathbf{B}$ where $\mathbf{B}$ is the monolith of a SI R-module $\mathbf{A}$, and take in account Lemma 5.2.14. The result is that all SI R-modules are simple. This shows that $\mathbf{R}$ is Jacobson semisimple. Indeed, the R-module $\mathbf{R}$ is a subdirect product of SI R-modules.

Since the latter are simple, the zero ideal of $\mathbf{R}$ is the intersection of maximal left ideals of $\mathbf{R}$.

Let $\mathbf{A}_i$, $i \in I$, be a maximal family of simple, pairwise nonisomorphic $\mathbf{R}$-modules and put $\mathbf{A} = \sum^{\oplus}(\mathbf{A}_i : i \in I)$. The $\mathbf{R}$-module $\mathbf{A}$ is faithful since $\mathbf{R}$ is Jacobson semisimple, and balanced by Theorem 5.2.12. Hence $\mathbf{R}$ can be identified with the biendomorphism ring $\mathbf{R}''$ of $\mathbf{A}$. On the other hand, by Lemma 5.2.10, $\mathbf{R} = \mathbf{R}'' = \prod(\mathbf{R}''_i : i \in I)$ where $\mathbf{R}''_i$ is the biendomorphism ring of the $\mathbf{R}$-module $\mathbf{A}_i$, $i \in I$.

By Schur's Lemma, the endomorphism ring $\mathbf{R}'_i$ of the $\mathbf{R}$-module $\mathbf{A}_i$ is a division ring, for every $i \in I$. Thus every $\mathbf{R}''_i$ is the ring of all linear transformations of the vector space $\mathbf{A}_i$ over $\mathbf{R}'_i$. It is well known that such a ring contains minimal left ideals, and each of the latter is isomorphic to the $\mathbf{R}$-module $\mathbf{A}_i$. Let S be the sum of all minimal left ideals of $\mathbf{R}$. It is easy to see that S is an ideal of $\mathbf{R}$. On the other hand, S is a semisimple $\mathbf{R}$-module and we shall be done if we show that $S = R$.

If $S \neq R$ then we can find a maximal left ideal L of $\mathbf{R}$ containing S. Then the simple $\mathbf{R}$-module $\mathbf{R}/L$ is annihilated by S but S does not annihilate any of the simple $\mathbf{R}$-modules $\mathbf{A}_i$, $i \in I$. Hence $\mathbf{R}/L$ is not isomorphic to any of $\mathbf{A}_i$, a contradiction. This contradiction proves the theorem. $\bullet$

Theorem 5.2.16 *A ring $\mathbf{R}$ is simple and Artinian iff the direct sum of every two nonzero $\mathbf{R}$-modules is affine complete.*

PROOF If $\mathbf{R}$ is simple and Artinian then it is semisimple and has only one simple module $\mathbf{A}$, up to isomorphism. Hence any direct sum of two nonzero $\mathbf{R}$-modules is isomorphic to $\mathbf{A}^{(I)}$ where $|I| \geq 2$. The latter is affine complete by Theorem 5.2.13.

If the direct sum of every two nonzero $\mathbf{R}$-modules is affine complete, then $\mathbf{R}$ is semisimple by Theorem 5.2.15. Moreover, it follows from Theorem 5.2.13 that $\mathbf{R}$ has only one, up to isomorphism, simple module. Clearly then $\mathbf{R}$ must be simple. $\bullet$

5.2.4   Abelian groups

For the basics of the abelian group theory we refer the reader to [Fuchs, 1970, 1973]. For our purposes it is useful to know that the variety of abelian groups $\mathcal{A}$ has countably many subvarieties: every natural number n defines the subvariety $\mathcal{A}_n$ by the equation $nx \approx 0$

and those are all proper subvarieties of the variety $\mathcal{A}$. If A is an abelian group and $V(\mathbf{A}) = \mathcal{A}$ then the group $\mathbf{A}$ is called *unbounded*. All other abelian groups are said to be *bounded*. In particular, if $V(\mathbf{A}) = \mathcal{A}_n$ then n is said to be the *exponent* of $\mathbf{A}$ and denoted by $\exp \mathbf{A}$. Also, it is convenient to write $\exp \mathbf{A} = \infty$ if $\mathbf{A}$ is an unbounded group.

Abelian groups can be considered as modules over the ring of integers $\mathbf{Z}$; therefore the basic facts from section 5.2.1 apply. In particular, an abelian group is affine complete iff all binary compatible functions on it are polynomial. We start with an observation which shows that in fact in many cases it is just enough to consider only unary functions.

Lemma 5.2.17 *Let* $\mathbf{A} = \mathbf{B} \oplus \mathbf{C}$ *where* $\mathbf{B}$ *and* $\mathbf{C}$ *are abelian groups of the same exponent, and let* $\mathbf{D}$ *be any cyclic subgroup of* $\mathbf{A}$. *Then* $\exp \mathbf{A} = \exp (\mathbf{A}/\mathbf{D})$.

PROOF Suppose first that both $\mathbf{B}$ and $\mathbf{C}$ are of finite exponent n. The structure theory of abelian groups implies that both can be represented as direct sums of cyclic groups among which one of order n occurs at least once. Hence $\mathbf{B} \oplus \mathbf{C}$ has a direct summand $\mathbf{S}$ isomorphic to the direct square of the cyclic group of order n. Now if $\exp (\mathbf{A}/\mathbf{D}) < n$ then also $\exp ((\mathbf{S} + \mathbf{D})/\mathbf{D}) = \exp (\mathbf{S}/(\mathbf{S} \cap \mathbf{D})) < n$. However, the latter is impossible because $\mathbf{S} \cap \mathbf{D}$ is a cyclic subgroup of $\mathbf{S}$.

Now assume that both $\mathbf{B}$ and $\mathbf{C}$ are unbounded and suppose that $a \in A$ generates a cyclic subgroup $\mathbf{D}$ such that the quotient group $\mathbf{A}/\mathbf{D}$ is of finite exponent n. Let $a = b + c$ where $b \in B$, $c \in C$.

By the direct decomposition, $nA \subseteq D$ implies $nB \subseteq \mathrm{Sg}(b)$ and $nC \subseteq \mathrm{Sg}(c)$. Now, if b has a finite order r then $rnB = 0$, contradicting the assumption that $\mathbf{B}$ is unbounded. Hence b, and similarly c, must be of infinite order. Thus $nb \neq 0$. Since $nB \subseteq nA \subseteq \mathrm{Sg}(a)$, there exists a nonzero integer s such that $nb = sb + sc$. Then, however, $sc = 0$, a contradiction. $\bullet$

Lemma 5.2.18 *Let* $\mathbf{A}$ *be an abelian group with the property that* $\exp \mathbf{A} = \exp (\mathbf{A}/\mathbf{D})$ *for every cyclic subgroup* $\mathbf{D} < \mathbf{A}$. *Then* $\mathbf{A}$ *is affine complete iff all unary compatible functions on* $\mathbf{A}$ *are polynomials.*

PROOF The necessity being obvious, we only prove the sufficiency. Let all unary compatible functions on **A** be polynomials. By Lemma 5.2.3 we only have to prove that every binary compatible function $f(x, y)$ on **A** is a polynomial. Clearly we may assume without loss of generality that $f(0,0) = 0$. By our assumption, all the induced functions $f(a, y)$ and $f(x, a)$, where $a \in A$ is fixed, are polynomial functions of **A**. Therefore there exist integers k_z, l_z and elements $b_z, c_z \in A$, for every $z \in A$ such that

$$f(x, y) = k_y x + b_y \qquad (5.7)$$
$$f(x, y) = l_x y + c_x \qquad (5.8)$$

for every $x, y \in A$. Since $f(0,0) = 0$, we obviously have $b_0 = c_0 = 0$. Hence, $c_x = f(x, 0) = k_0 x$ and $b_y = f(0, y) = l_0 y$ for every $x, y \in A$ which together with (5.7) and (5.8) will yield

$$(k_y - k_0)x = (l_x - l_0)y \qquad (5.9)$$

for every $x, y \in A$.

Clearly, we are done if the both sides of the equality (5.9) are constantly zero, for then $f(x, y)$ is the polynomial function $k_0 x + l_0 y$. Suppose, on the contrary, that the sides of (5.9) are not constantly zero. Thus there exists $d \in A$ such that $k_d x \neq k_0 x$ for some $x \in A$. Clearly then $k_d \neq k_0$. Let **D** be the subgroup of **A** generated by d. Then by (5.9) the quotient group **A/D** satisfies the equation $(k_d - k_0)x \approx 0$ and by the hypothesis of our lemma, the same must be true for the group **A** as well. This means that $k_d x = k_0 x$ for every $x \in A$, a contradiction. One can similarly prove that $l_x y = l_0 y$ for all $x, y \in A$.   ●

We continue with a proposition which actually yields all positive results about affine completeness of abelian groups.

Proposition 5.2.19 *Let* **B** *and* **C** *be abelian groups of the same exponent and assume that one of them contains an element of order* exp **C**. *Then the direct sum* **B** ⊕ **C** *is affine complete.*

This proposition covers two different situations: first if both **B** and **C** are unbounded and one of them contains an element of infinite order, and second, both **B** and **C** are of the same finite exponent. It is well known that every abelian group of finite exponent n contains an element of order n.

PROOF Let $\mathbf{A} = \mathbf{B} \oplus \mathbf{C}$ and assume that $\mathbf{C}$ contains an element of order $e = \exp \mathbf{C}$. Note that here the symbol e denotes either a positive integer or ∞. By Lemma 5.2.17 $\mathbf{A}$ satisfies the hypothesis of Lemma 5.2.18; so it is sufficient to prove only that every unary zero preserving compatible function f on $\mathbf{A}$ is polynomial. Because of compatibility of f, there are functions $g : B \to B$ and $h : C \to C$, compatible on $\mathbf{B}$ and $\mathbf{C}$, respectively, such that

$$f(x + y) = g(x) + h(y) \tag{5.10}$$

for every $x \in B$, $y \in C$. Since f is zero preserving, so are the functions g and h. Hence all these functions preserve the subgroups. In particular, for every $x \in B, y \in C$ there exist $r_{x+y}, s_x, t_y \in \mathbf{Z}$ such that

$$f(x + y) = r_{x+y}(x + y), \quad g(x) = s_x x, \quad h(y) = t_y y.$$

Thus, the formula (5.10) implies

$$r_{x+y}x = s_x x, \quad r_{x+y}y = t_y y \tag{5.11}$$

for all $x \in B$, $y \in C$. In particular, if we take for y any element of order e then the second of equalities (5.11) yields the equality $r_{x+y}z = t_y z$ for every $x \in B$ and every element z of any abelian group of exponent e. Thus the first of the equalities (5.11) gives

$$g(x) = s_x x = r_{x+y}x = t_y x$$

for every $x \in B$ and every $y \in C$ of order e. Consequently there exists an integer t such that g is the polynomial function tx and $h(y) = ty$ for all elements $y \in C$ whose order is e. Note that t is unique if $e = \infty$ and unique modulo e otherwise. At this point the proof essentially splits into two cases.

1. e is finite. Changing the roles of $\mathbf{B}$ and $\mathbf{C}$ we find an integer s such that h is the polynomial function sy and $g(x) = sx$ for all $x \in B$ of order e. Then clearly $s = t$ and we are done.

2. $e = \infty$. The argument we used in the former case does not work now because $\mathbf{B}$ need not contain an element of infinite order. Instead we exploit the compatibility of h to show that $h(c) = tc$ also

for the elements c of finite order. Consider $h'(y) = h(y) - ty$ and let $d \in C$ be of infinite order. Then the order of $c + d$ is also infinite. Consequently $h'(c + d) = 0$. By the compatibility of h',

$$h'(c) = h'(c) - h'(c + d) \in \mathrm{Sg}(d) \quad \text{and} \quad h'(c) \in \mathrm{Sg}(c).$$

Since 0 is the only element of finite order in $\mathrm{Sg}(d)$, we get $h'(c) = 0$, hence $h(c) = tc$.   •

In what follows we call a finite subset $X = \{x_1, \ldots, x_n\}$ of an abelian group **A** *independent* if the equality $k_1 x_1 + \cdots + k_n x_n = 0$ where $k_1, \ldots, k_n$ are integers holds only if $k_1 = \cdots = k_n = 0$. An infinite subset of **A** is said to be independent if so are all its finite subsets. The *rank* of an abelian group **A** is the minimal upper bound of the cardinalities of independent subsets of **A**. For example, if **T** is the torsion part (the largest torsion subgroup) of **A** then **A** and **A/T** have the same rank. In particular, the torsion abelian groups are precisely the abelian groups of rank 0.

Theorem 5.2.20 *Every abelian group of rank greater than 1 is affine complete.*

PROOF Let **A** be an abelian group with rank at least 2 and f be a binary zero preserving compatible function on **A**. We consider the set S of all finitely generated subgroups $\mathbf{B} < \mathbf{A}$ of rank at least 2. Every $\mathbf{B} \in S$ is a direct sum of cyclic groups and because of the rank condition, there must be at least 2 infinite cyclic direct summands. Hence, by Proposition 5.2.19 all subgroups $\mathbf{B} \in S$ are affine complete. Since f is zero preserving and the variety of abelian groups has congruence extension property, the restriction of f to any subgroup of **A** is a compatible function on that subgroup. Hence, for any $\mathbf{B} \in S$, there exist integers k and l such that $f(x, y) = kx + ly$ for all $x, y \in B$. Note that these integers are uniquely determined by **B** because **B** is unbounded. It remains to notice that every two subgroups $\mathbf{B}, \mathbf{C} \in S$ are contained in $\mathbf{D} = \mathbf{B} + \mathbf{C} \in S$ and every two elements of **A** are contained in some subgroup $\mathbf{B} \in S$.   •

Theorem 5.2.21 *Every abelian group of rank 1 with unbounded torsion part is affine complete.*

PROOF The idea of the proof of Theorem 5.2.20 applies here, too. The only difference is that now S is the set of all subgroups $\mathbf{B} < \mathbf{A}$

which are generated by the join of the torsion part $\mathbf{T}$ of $\mathbf{A}$ and a finite nonempty subset of $A \setminus T$. Since $\mathbf{A}$ is of rank 1, for every $\mathbf{B} \in \mathcal{S}$ the quotient group $\mathbf{B}/\mathbf{T}$ is infinite cyclic. It is well known that the torsion part of such groups splits off; that is, $\mathbf{B}$ is isomorphic to the direct sum of $\mathbf{T}$ and the infinite cyclic group. Since $\mathbf{T}$ is unbounded, by Proposition 5.2.19 $\mathbf{B}$ is affine complete. The rest repeats the final arguments of the proof of the preceding theorem. $\bullet$

The next question is what can be said about abelian groups of rank 1 whose torsion part is bounded. The answer is given by the following theorem.

Theorem 5.2.22 *An abelian group of rank 1 with bounded torsion part is not locally affine complete.*

PROOF Let $\mathbf{A}$ be a torsion free group of rank 1 and let $\mathbf{T}$ be its torsion part. Then $\mathbf{A}/\mathbf{T}$ is a torsion free abelian group of rank 1. It is well known that such groups are countable and have a distributive lattice of subgroups. In fact every such group can be embedded into the additive group of rationals. Thus, the group $\mathbf{A}/\mathbf{T}$ is a countable arithmetical algebra and by Theorem 3.4.12, if it were locally affine complete, it also would be strictly locally affine complete. However abelian groups, as modules in general, cannot be strictly locally affine complete. Consequently, there exists a compatible function g on $\mathbf{A}/\mathbf{T}$ which is not a local polynomial function. Since it is easier to work with unary functions, we notice that g can be taken to be unary. For example, take $a \in A \setminus T$, define $g(ka + T) = k^2 a + T$, $k \in \{0, 1, 2\}$, and extend this, using the compatible function extension property to all of $\mathbf{A}/\mathbf{T}$.

Let $\exp \mathbf{T} = e$. Then $e(x + t) = ex$ for every $x \in A$ and $t \in T$. Hence, $e(x + T)$ is a well defined element of A for every $x \in A$. This fact allows us to define a function $f : A \to A$ by the formula $f(x) = e(g(x + T))$. This function induces the function eg on the quotient group $\mathbf{A}/\mathbf{T}$. So, if f were a local polynomial function on $\mathbf{A}$ then eg would be a local polynomial function on $\mathbf{A}/\mathbf{T}$. In particular, there would exist an integer s such that $sk(a + T) = ek^2(a + T)$, $k \in \{0, 1, 2\}$. Since $a + T$ is an element of infinite order, the latter would imply $s = e$ and $2s = 4e$, a contradiction.

Hence it remains to prove that f is a compatible function on $\mathbf{A}$. Let $b, c \in A$ be arbitrary and $g(b + T) = b_1 + T$, $g(c + T) = c_1 + T$.

Then $f(b) = eb_1$ and $f(c) = ec_1$. Since g is compatible, there exists an integer s such that $g(b+T) - g(c+T) = s(b-c+T)$. Consequently there also exists $t \in T$ such that $b_1 - c_1 = s(b-c) + t$. Then

$$f(b) - f(c) = e(s(b-c) + t) = es(b-c)$$

proving that f is compatible.   •

It remains to classify abelian torsion groups from the point of view of (local) affine completeness. As we have seen, every nontorsion group is either affine complete or not locally affine complete. The situation is different in the case of torsion groups. There do exist locally affine complete torsion abelian groups which are not affine complete. We first describe affine complete torsion abelian groups and then use these results to describe locally affine complete torsion abelian groups.

We already know that the direct sum of two bounded abelian groups of the same exponent is always affine complete. Hence it remains to consider unbounded torsion groups and bounded groups which cannot be decomposed into the direct sum of two groups of the same exponent. It will turn out that none of these abelian groups is affine complete. Recall that a *p-group* is a group $\mathbf{A}$ such that the order of every element of $\mathbf{A}$ is a power of the prime number p and a group is said to be *primary* if it is a p-group for some p. It is well known that every torsion abelian group decomposes into a direct sum of p-groups, for distinct prime numbers p. The latter are called *primary components* of that group.

Lemma 5.2.23 *Unbounded abelian p-groups are not affine complete.*

PROOF Let $\mathbf{A}$ be an unbounded p-group. Then $\mathbf{A}$ can be considered as a module over the ring of p-adic integers. If ξ is any p-adic integer and $a, b \in A$ then there exists an integer k such that $\xi a = ka$ and $\xi b = kb$. This implies that the function $f(x) = \xi x$ is a compatible function on the group $\mathbf{A}$. However, if ξ is not an integer then there is no such integer k that $\xi x = kx$ for all $x \in A$. Therefore the function f is not a polynomial.   •

Lemma 5.2.24 *Let $\mathbf{A} = \sum^{\oplus}(\mathbf{A}_i : i \in I)$ where $\mathbf{A}_i$ are p_i-groups, for distinct primes p_i. If f_i are m-ary compatible functions on $\mathbf{A}_i$, $i \in I$, then $f = (f_i)_{i \in I}$ is an m-ary compatible function on $\mathbf{A}$.*

PROOF This is a straightforward consequence of the fact that **A** has no skew subgroups with respect to the given decomposition. That is, every subgroup $\mathbf{B} < \mathbf{A}$ has the form $\sum^{\oplus}(\mathbf{B}_i : i \in I)$, where $\mathbf{B}_i < \mathbf{A}_i$, $i \in I$. $\bullet$

Lemma 5.2.25 *Let* $\mathbf{A} = \mathbf{B} \oplus \mathbf{C}$ *where* $\mathbf{C}$ *is a cyclic group of order* p^n *and* $\mathbf{B}$ *is an abelian group of exponent* p^m, $m < n$. *Then* $\mathbf{A}$ *is not locally affine complete.*

PROOF For technical reasons it is convenient to identify $\mathbf{C}$ with the additive group of integers modulo p^n. Now define the binary function f on $\mathbf{A}$ by the rule:

$$f(x + y, u + v) = p^m yv$$

(multiplication modulo p^n) where $x, u \in B$, $y, v \in C$.

That f is compatible, becomes clear by the following calculations:

$$f(x_1 + y_1, u_1 + v_1) - f(x_2 + y_2, u_2 + v_2)$$
$$= p^m y_1 v_1 - p^m y_2 v_2 = p^m y_1 v_1 - p^m y_1 v_2 + p^m y_1 v_2 - p^m y_2 v_2$$
$$= p^m y_1((u_1 + v_1) - (u_2 + v_2)) + p^m v_2((x_1 + y_1) - (x_2 + y_2)).$$

It remains to prove that f is not a local polynomial function. Let us identify $1 \in C$ by the element $0 + 1 \in A$. It is now an easy exercise to show that no polynomial interpolates f at all of the points of $\{(0,0), (0,1), (1,0), (1,1)\}$. $\bullet$

Note that in the proof of Lemma 5.2.25 a binary function was involved in order to incorporate the case $p = 2$, $n - m = 1$. For all other cases the function $g(x + y) = p^m y^2$ would have been as good as f.

Theorem 5.2.26 *For a torsion abelian group* $\mathbf{A}$, *the following conditions are equivalent:*

(1) $\mathbf{A}$ *is affine complete;*

(2) $\mathbf{A}$ *is a direct sum of two groups, both having the same finite exponent;*

(3) $\mathbf{A}$ *has finitely many nonzero primary components, each of the latter is bounded and has a subgroup isomorphic to the direct square of the cyclic group whose order is the exponent of that primary component.*

PROOF That (2) implies (1) was mentioned above. The equivalence of (2) and (3) is clear by abelian group theory. Hence all we need to show is that (1) implies (3).

Let $\mathbf{A}$ be affine complete and let $\mathbf{A} = \sum^{\oplus}(\mathbf{A}_i : i \in I)$ where the $\mathbf{A}_i$, $i \in I$, are nonzero primary components of $\mathbf{A}$. Then it easily follows from Lemma 5.2.24 that all $\mathbf{A}_i$ must be affine complete. Thus by Lemmas 5.2.23 and 5.2.25 they all must meet the requirement on primary components in condition (3).

It remains to show that I is finite. Pick $j \in I$ and define f_j to be the identity function on A_j. For all other $k \in I$, let f_k be the zero function on A_k. Since all the f_i are compatible, by Lemma 5.2.24 the function $f = (f_i)_{i \in I}$ must be a compatible function on $\mathbf{A}$. However, this function is a polynomial only in case I is finite.   ●

The following theorem summarizes the basic results on the affine completeness of abelian groups which were obtained above.

Theorem 5.2.27 *An abelian group is affine complete iff it belongs to one of the following classes of abelian groups:*

  (i) *groups of rank at least 2;*

 (ii) *groups of rank 1 having unbounded torsion part;*

(iii) *direct sums of two bounded groups of the same exponent.*

The discussion above shows that if there exist locally affine complete abelian groups which are not affine complete, then they must be unbounded torsion groups. In order to characterize them, we need the following notion. A subgroup $\mathbf{B}$ of an abelian p-group $\mathbf{A}$ is called **basic** if it is a direct sum of cyclic groups and the quotient group $\mathbf{A}/\mathbf{B}$ is divisible. A given abelian p-group may have many basic subgroups but they are all isomorphic. Also the quotient group $\mathbf{A}/\mathbf{B}$ where $\mathbf{B}$ is a basic subgroup of $\mathbf{A}$ is unique, up to isomorphism. We shall denote by $\mathbf{Z}_{p^{\infty}}$ the quasicyclic group which can be defined as a direct limit of groups $\mathbf{Z}_{p^n}$, n a positive integer. Recall that every divisible abelian p-group is isomorphic to a weak direct power of the quasicyclic p-group.

Theorem 5.2.28 *An abelian torsion group $\mathbf{A}$ is locally affine complete iff each of its primary components $\mathbf{P}$ satisfies one of the following conditions:*

(i) **P** *is affine complete;*

(ii) *a basic subgroup* **B** *of* **P** *is unbounded;*

(iii) *the quotient group* **P**/**B** *is isomorphic to a direct sum of at least two copies of the quasicyclic group.*

PROOF Suppose first that each of the primary components of **A** satisfies one of the three conditions. We show that then every finite subset X of A belongs to an affine complete subgroup **C** of **A**. Then if f is a zero preserving compatible function on **A**, its restriction to C is compatible for the subgroup **C** (by the congruence extension property). Consequently f coincides with some polynomial function p on C. Then clearly p interpolates f at X.

For a natural number n let $\mathbf{A}[n] = \{x \in A : nx = 0\}$. Clearly every $\mathbf{A}[n]$ is a bounded subgroup of A and every finite subset of A is contained in one of these subgroups. We show that all these subgroups are affine complete. Let $n = p_1^{k_1} \cdots p_m^{k_m}$ where p_i are distinct primes, and let $\mathbf{P}_i$ be the p_i-component of **A**, $i = 1, \ldots, m$. Then it is easy to see that

$$\mathbf{A}[n] = \mathbf{P}_1[p_1^{k_1}] \oplus \cdots \oplus \mathbf{P}_m[p_m^{k_m}].$$

Also it is not difficult to observe that under the requirements of the theorem all the subgroups $\mathbf{P}_i[p_i^{k_i}]$ are affine complete; that is, each of them contains a subgroup isomorphic to the direct square of the cyclic group of order $p_i^{k_i}$. This proves the sufficiency part of the theorem.

Now assume that **A** is locally affine complete. By Lemma 5.2.24 then all the primary components of **A** must be locally affine complete. Let **P** be one of them, and let it be a p-group. If **P** is bounded then by Lemma 5.2.25 and Theorem 5.2.26 it must be affine complete. If **P** is not bounded and does not satisfy any of the conditions of the theorem, there is only one possibility: **P** must have a bounded basic subgroup **B** and the quotient group **P**/**B** must be isomorphic to $\mathbf{Z}_{p^\infty}$. To complete the proof we need to prove that such a group is not locally affine complete. Here we leave the details for the reader because the idea of the proof is the same as in Theorem 5.2.22. The same approach works because the group $\mathbf{Z}_{p^\infty}$ is a countable arithmetical algebra. ●

Since the theory of modules over any commutative principal ideal domain is very similar to abelian group theory, most of the results of this section remain valid for modules over such rings. There is only one exception: when proving that abelian groups of rank 1 are not affine complete, we relied on the countability of the ring of integers. This argument will not work if the ring **R** is uncountable. This leads to the following problem.

Problem 5.2.29 *Does there exist an affine complete torsion free module of rank 1 over a commutative principal ideal domain?*

Exercises

1. Describe affine complete modules over the ring of p-adic integers.

2. Let **R** be any ring lying between the ring of integers and the field of rationals and **A** be an **R**-module. Prove that **A** is affine complete iff so is its additive group.

3. Prove that for any **R**-module **A** (the ring **R** is arbitrary) there exists a set I such that the **R**-module A^I is affine complete.

5.3   Lattices

5.3.1   General observations

Not very much is known about affine completeness of general lattices and, in fact, all positive results have been obtained for distributive lattices and we shall describe these. However, several constructions of compatible functions which we shall make use of do not require distributivity and therefore we shall present them separately in this section and then apply them in the next section to characterize all affine complete and locally affine complete distributive lattices. Also we observe here that there are no strictly locally affine complete lattices. This is true since every nontrivial lattice contains a two element chain and the finite partial function which interchanges these two elements is obviously compatible on its domain. But this function clearly cannot be realized by a lattice polynomial.

We start with a general lemma which will be useful beyond lattices as well.

Lemma 5.3.1 *Let* **A** *be an algebra and* e *be a unary idempotent compatible function on* **A** *such that* $C = e(A)$ *is a subuniverse of some reduct of* **A**. *Then, if* f *is an m-ary compatible function of that subreduct* **C** *then the function*

$$g(x_1, \ldots, x_m) = f(e(x_1), \ldots, e(x_m))$$

is a compatible function on **A** *and extends* f.

PROOF Obviously the restriction of g to C^m coincides with f. Now, let $\rho \in \mathrm{Con}\,\mathbf{A}$ and $a_i, b_i \in A$, $a_i \equiv_\rho b_i$, $i = 1, \ldots, m$. Since e is compatible for **A**, $e(a_i) \equiv_\rho e(b_i)$, $i = 1, \ldots, m$. Since the restriction of a congruence to a subreduct is a congruence of the latter, and f is compatible for **C**, we have

$$f(e(a_1), \ldots, e(a_m)) \equiv_\rho f(e(b_1), \ldots, e(b_m)).$$

This proves the lemma. ●

The next lemma will be used for construction of compatible functions on lattices.

Lemma 5.3.2 *The congruences of any Boolean algebra are the same as the congruences of the underlying Boolean lattice.*

PROOF Let **L** be a Boolean lattice. We have only to prove that the complementation operation $'$ is compatible with all congruences of **L**. Obviously it is sufficient to prove the compatibility with only principal congruences. Let $a, b \in L$ and $a \equiv_\rho b$ where $\rho = \mathrm{Cg}(c, d)$, and without loss of generality, $c \leq d$. Applying the formula of Section 1.2.2 for principal congruences in distributive lattices we have in this case

$$(a, b) \in \mathrm{Cg}(c, d) \iff (a \vee d = b \vee d)\&(a \wedge c = b \wedge c). \quad (5.12)$$

It easily follows from this formula that $(c', d') \in \mathrm{Cg}(c, d)$ and therefore $\mathrm{Cg}(c', d') \leq \mathrm{Cg}(c, d)$. Application of De Morgan laws to (5.12) completes the proof. ●

Now we give a useful necessary condition for local affine completeness of lattices. Let **L** be a lattice and $[a, b]$ be an interval in **L**. We call this interval *trivial* if $a = b$ and *Boolean* if it is a Boolean lattice.

Proposition 5.3.3 *If a lattice has a nontrivial Boolean interval then this lattice has a compatible function which does not preserve the order relation. Consequently a locally affine complete lattice cannot have nontrivial Boolean intervals.*

PROOF Assume that a lattice **L** has a nontrivial Boolean interval $[a, b]$. Let $p(x) = (x \wedge b) \vee a$ be the polynomial function of **L** which projects L onto $[a, b]$ and denote by $'$ the complementation operation of the interval $[a, b]$. By Lemma 5.3.2 the operation $'$ is a compatible function on the lattice $[a, b]$. Now put $f(x) = p(x)'$. Since p is an idempotent compatible function on **L**, by Lemma 5.3.1 the function f is also compatible on **L**. However, obviously f does not preserve the order. •

Corollary 5.3.4 *A locally affine complete lattice has no covering pairs of elements; in particular it has neither atoms nor coatoms. All locally affine complete lattices are infinite.*

Next we find two necessary conditions for affine completeness of lattices. These conditions will be given in terms of almost principal ideals and filters. An ideal I of a lattice **L** is said to be *almost principal* if its intersection with every principal ideal of **L** is a principal ideal of **L**. An *almost principal filter* is defined dually. Clearly all principal ideals (filters) are almost principal ideals (filters) and in bounded lattices the two are the same. In what follows, if a is an element of a lattice **L** then $\downarrow a$ denotes the principal ideal of **L** generated by a. Also, if A is a subset of **L** then define

$$\downarrow A = \bigcup_{a \in A} \downarrow a .$$

The notations $\uparrow a$ and $\uparrow A$ are defined dually. Note that $\downarrow A$ ($\uparrow A$) need not be an ideal (a filter), in general.

We are going to show that every almost principal ideal (filter) of any lattice gives rise to a compatible function on this lattice. If I is an almost principal ideal of **L** then for every $a \in L$ there exists a uniquely defined element $b \in I$ such that $\downarrow a \cap I = \downarrow b$. Thus, I defines a function $f_I : L \to I$ such that

$$\downarrow f_I(a) = \downarrow a \cap I$$

for every $a \in L$. Dually, every almost principal filter D of **L** defines the function $f^D : L \to D$ such that

$$\uparrow f^D(a) = \uparrow a \cap D$$

for every $a \in L$.

Lemma 5.3.5 *Let I (D) be an almost principal ideal (filter) of the lattice* **L**. *Then the function f_I (f^D) is a compatible function on* **L**.

PROOF We claim that

$$f_I(a \wedge b) = f_I(a) \wedge a \wedge b \tag{5.13}$$

and in particular

$$f_I(a) = f_I(a) \wedge a$$

for all $a, b \in L$. Clearly f_I is order preserving, so $f_I(a \wedge b) \le f_I(a)$. Also $f_I(a \wedge b) \le a \wedge b$ directly follows from the definition of f_I; hence $f_I(a \wedge b) \le f_I(a) \wedge a \wedge b$. On the other hand, $f_I(a) \in I$ implies $f_I(a) \wedge a \wedge b \in I$ and obviously $f_I(a) \wedge a \wedge b \le a \wedge b$. Thus, $f_I(a) \wedge a \wedge b \in \downarrow(a \wedge b) \cap I = \downarrow f_I(a \wedge b)$, that is, $f_I(a) \wedge a \wedge b \le f_I(a \wedge b)$.

Now let $\theta \in$ Con **L** and $a \equiv_\theta b$. Then $f_I(a) \wedge a \equiv_\theta f_I(a) \wedge a \wedge b$ and by (5.13) we have $f_I(a) \equiv_\theta f_I(a \wedge b)$. Similarly $f_I(b) \equiv_\theta f_I(a \wedge b)$ and by transitivity, $f_I(a) \equiv_\theta f_I(b)$. Thus the function f_I is compatible. The compatibility of f^D is proved by a dual argument. ●

5.3.2 Distributive lattices

In this section we provide answers to three basic questions about polynomial completeness of distributive lattices. First we describe local polynomial functions as those preserving all congruence relations and the order relation. Then we establish a condition which forces all compatible functions to preserve the order relation thus obtaining a description of locally affine complete distributive lattices. Finally we introduce further conditions which force all local polynomial functions to be polynomials and in this way obtain a complete characterization of affine complete distributive lattices.

The results of the present subsection are due to G. Grätzer, D. Dorninger, D. Eigenthaler, and M. Ploščica. G. Grätzer discovered the role of Boolean intervals, proved that all order preserving

compatible functions on bounded distributive lattices are polynomials, and described the affine complete members of the class of
bounded distributive lattices ([Grätzer, 1964]). The description of
local polynomial functions on arbitrary distributive lattices appears
in [Dorninger, Eigenthaler, 1982]. The affine completeness problem
for distributive lattices was completely solved in [Ploščica, 1994].

Obviously all local polynomial functions of an arbitrary lattice
are compatible and order preserving. Now we shall show that the
converse holds in the case of distributive lattices.

Theorem 5.3.6 *A function on a distributive lattice is a local polynomial iff it is compatible and order preserving.*

PROOF Let **L** be a distributive lattice and f be an m-ary compatible function on **L** which preserves the order relation of **L**. We
shall show that, given a finite subset X in L^m, there exists an m-ary
polynomial p of **L** such that the restrictions of f and p to X coincide.
Since **L** admits a majority term, by Theorem 3.2.2 it is sufficient to
consider only two element sets X.

We first consider the case $X = \{\mathbf{a}, \mathbf{b}\}$ where the m-tuples $\mathbf{a}$ and $\mathbf{b}$
are comparable. Let $\mathbf{a} \leq \mathbf{b}$ (i.e., all $a_i \leq b_i$); hence also $f(\mathbf{a}) \leq f(\mathbf{b})$.
We let $\underline{m} = \{1, \ldots, m\}$. For every $\alpha \subseteq \underline{m}$ put $k_\alpha = f(c_1, \ldots, c_m)$
where

$$c_k = \begin{cases} b_k & \text{if } k \in \alpha, \\ a_k & \text{if } k \notin \alpha, \end{cases}$$

and $\mathbf{x}^\alpha = \bigwedge_{k \in \alpha} x_k$. Note that $\mathbf{x}^\emptyset$ is not defined by this rule but it is
reasonable to think of $\mathbf{x}^\emptyset$ as something greater than every element of
L, that is, $a \wedge \mathbf{x}^\emptyset = a$ for every $a \in L$. Now we define the polynomial

$$p(\mathbf{x}) = \bigvee_{\alpha \subseteq \underline{m}} (k_\alpha \wedge \mathbf{x}^\alpha)$$

and prove that

$$f(\mathbf{a}) = p(\mathbf{a}) \quad \text{and} \quad f(\mathbf{b}) = p(\mathbf{b}). \tag{5.14}$$

Since **L** is a subdirect product of two element lattices and the constants k_α in p are the values of a compatible function f at combina-
tions of $\mathbf{a}$ and $\mathbf{b}$, it suffices to check the equalities (5.14) only in case
$\mathbf{L} = \mathbf{D_2}$. We have to consider three subcases.

1. $f(\mathbf{a}) = 1$. Here $f(\mathbf{a}) = k_{\emptyset} = p(\mathbf{a})$ by the definition. Since $f(\mathbf{a}) \leq f(\mathbf{b})$, we have $f(\mathbf{b}) = 1$ as well. Consequently the equalities (5.14) hold.

2. $f(\mathbf{b}) = 0$. Here $f(\mathbf{b}) = k_{\underline{m}}$ and f preserves the order so all of the coefficients k_{α} are zero; so p is the constant function 0. Since $f(\mathbf{b}) = 0$ implies $f(\mathbf{a}) = 0$, the equalities (5.14) hold again.

3. $f(\mathbf{a}) = 0$ and $f(\mathbf{b}) = 1$. We select a subset $\alpha \subseteq \underline{m}$ such that $k_{\alpha} = 1$ but $k_{\beta} = 0$ for all proper subsets β of α. Then $\mathbf{b}^{\alpha} = 1$; that is, $b_i = 1$ for all $i \in \alpha$. Indeed, if we would have $b_i = 0$ for some $i \in \alpha$ then we would get $k_{\beta} = 1$ where $\beta = \alpha \setminus \{i\}$, a contradiction. Hence $k_{\alpha} \wedge \mathbf{b}^{\alpha} = 1$ and consequently also $p(\mathbf{b}) = 1$. It remains to show that $p(\mathbf{a}) = 0$. Supposing the contrary would imply the existence of a subset $\gamma \subseteq \underline{m}$ such that $k_{\gamma} = \mathbf{a}^{\gamma} = 1$. If $\mathbf{a}^{\gamma} = 1$ then $a_i = 1$ for every $i \in \gamma$ and also $b_i = 1$ for every $i \in \gamma$. This means that the value of k_{γ} does not change if we replace b_i's by a_i's for $i \in \gamma$. Hence $1 = k_{\gamma} = k_{\emptyset} = f(\mathbf{a}) = 0$, a contradiction.

To conclude the proof, we consider the general case, with $\mathbf{a}$ and $\mathbf{b}$ possibly incomparable. Let $\mathbf{c} = \mathbf{a} \wedge \mathbf{b}$, that is, $\mathbf{c} = (c_1, \ldots, c_m)$ where $c_i = a_i \wedge b_i$, $i = 1, \ldots, m$. By the above proof, there exist polynomials p_1 and p_2 such that p_1 meets f at $\{\mathbf{c}, \mathbf{a}\}$ and p_2 at $\{\mathbf{c}, \mathbf{b}\}$. Now define

$$p(\mathbf{x}) = p_1(\mathbf{x} \wedge \mathbf{a}) \vee p_2(\mathbf{x} \wedge \mathbf{b}).$$

It is easy to check that p agrees with f at $\{\mathbf{a}, \mathbf{b}\}$.   ●

A lattice **L** is called **(locally) order affine complete** if all compatible order preserving functions on **L** are (local) polynomials. Hence Theorem 5.3.6 can be restated as follows.

Theorem 5.3.7 *Every distributive lattice is locally order affine complete.*

Now we shall show that in the case of distributive lattices the condition established in Proposition 5.3.3 is also sufficient for local affine completeness. We start with a lemma which shows how a compatible function on a distributive lattice yields Boolean intervals.

Lemma 5.3.8 *Let f be a unary compatible function on a distributive lattice **L**. Then, given arbitrary $a, b \in L$ with $a \leq b$ the interval $[f(b), f(a) \vee f(b)]$ is Boolean.*

PROOF We show that the complement in $[f(b), f(a) \vee f(b)]$ of an arbitrary element c of this interval can be computed by the formula

$$c' = (f(c) \wedge f(a)) \vee f(b),$$

that is

$$c \vee (f(c) \wedge f(a)) \vee f(b) = f(a) \vee f(b) \qquad (5.15)$$

and

$$c \wedge ((f(c) \wedge f(a)) \vee f(b)) = f(b) \qquad (5.16)$$

for every $c \in [f(b), f(a) \vee f(b)]$.

Since both sides of the formulas (5.15) and (5.16) are values of a compatible function at (a, b, c) and $\mathbf{L}$ is a subdirect product of two element lattices, it suffices to check these formulas only in case $\mathbf{L} = \mathbf{D_2}$. The case $f(a) \leq f(b)$ is trivial. Thus it remains to consider the case $f(a) = 1$, $f(b) = 0$ and straightforward calculations show that the formulas (5.15) and (5.16) hold for $c \in \{0, 1\}$. •

It follows from Lemma 5.3.8 that in a distributive lattice without nontrivial Boolean intervals all compatible functions are order preserving. This fact is not valid in case of modular lattices as an example constructed by E. T. Schmidt ([Schmidt, 1987]) shows.

Theorem 5.3.9 *A distributive lattice is locally affine complete iff it does not contain nontrivial Boolean intervals.*

PROOF Due to Proposition 5.3.3 we have to prove only the sufficiency. Let $\mathbf{L}$ be a distributive lattice which has no proper Boolean intervals and let f be an m-ary compatible function on $\mathbf{L}$. By Theorem 5.3.6 it suffices to prove that f is order preserving. Suppose the contrary, i.e., that there exist $\mathbf{a}, \mathbf{b} \in L^m$ such that $\mathbf{a} \leq \mathbf{b}$ but $f(\mathbf{a}) \not\leq f(\mathbf{b})$. It is easy to see that then $\mathbf{L}$ also admits a unary compatible function which does not preserve the order. Indeed, putting $\mathbf{c}^i = (a_1, \ldots, a_i, b_{i+1}, \ldots, b_m)$, we have

$$\mathbf{a} = \mathbf{c}^m \leq \mathbf{c}^{m-1} \leq \mathbf{c}^{m-2} \leq \cdots \leq \mathbf{c}^1 \leq \mathbf{c}^0 = \mathbf{b};$$

so there must exist i such that $f(\mathbf{c}^i) \not\leq f(\mathbf{c}^{i-1})$. Hence, for the unary compatible function

$$g(x) = f(a_1, \ldots, a_{i-1}, x, b_{i+1}, \ldots, b_m)$$

we have $g(a_i) \nleq g(b_i)$. Then by Lemma 5.3.8 the nontrivial interval $[g(b_i), g(a_i) \vee g(b_i)]$ is Boolean. This contradiction proves the theorem. $\qquad\bullet$

Our next step is to determine which distributive lattices are affine complete. We first prove a lemma which shows that it is sufficient to consider only unary functions. A crucial point in the proof of this lemma is that there are only four types of unary polynomial functions on distributive lattices and we may distinguish between these types using the boundedness of the ranges of the functions. The types are the following:

1. $p(x) = a \wedge (b \vee x) \quad (b \leq a)$,

2. $p(x) = a \wedge x$,

3. $p(x) = b \vee x$,

4. $p(x) = x$.

It is easy to see that p is of type 1 iff $p(L)$ has both a greatest and a smallest element. If $p(L)$ has a greatest (smallest) element but does not have a smallest (greatest) element then p is of type 2 (3). Note that a smallest or a greatest element of $p(L)$, if it exists, is uniquely determined. For example, if $p(x) = a \wedge (b \vee x)$ then a and b are a greatest and a smallest element of $p(L)$, respectively.

Lemma 5.3.10 *A distributive lattice* **L** *is affine complete iff every unary compatible function on* **L** *is a polynomial.*

PROOF Of course we have to prove only the sufficiency. Assume that all $(m-1)$-ary compatible functions on **L** are polynomials and consider an m-ary compatible function f on **L**. For every $\mathbf{a} \in L^{m-1}$ introduce the unary function $f_{\mathbf{a}}(x) = f(\mathbf{a}, x)$. By our hypothesis, all $f_{\mathbf{a}}$ are polynomial functions of **L**. We are going to show that the type of the polynomial $f_{\mathbf{a}}$ depends only on f and not on the choice of $\mathbf{a}$. In view of the above remark it is sufficient to show that, given two tuples $\mathbf{a}, \mathbf{b} \in L^{m-1}$, the set $f_{\mathbf{a}}(L)$ has a smallest (greatest) element iff so does $f_{\mathbf{b}}(L)$. Clearly the two cases are dual to each other; so we restrict to the case with smallest elements. Assume the contrary, that is, the set $f_{\mathbf{a}}(L)$ has a smallest element, say s, and

$f_{\mathbf{b}}(L)$ has no smallest element. Obviously this is only possible if $\mathbf{L}$ itself has no smallest element and $f_{\mathbf{b}}$ is of type 2 or 4. Hence there exists $t \in L$ such that $f_{\mathbf{b}}(x) = x$ whenever $x \leq t$. Now we can find elements $u, v \in L$ such that $u < v$ and v is smaller than s, t, and all of the components of $\mathbf{a} \wedge \mathbf{b}$. Let $\phi : \mathbf{L} \to \mathbf{D_2}$ be the lattice homomorphism such that $\phi(u) = 0$, $\phi(v) = 1$ and put $\theta = \ker \phi$. Then $\phi(a_i) = 1 = \phi(b_i)$ for $i = 1, \ldots, m - 1$; so $\mathbf{a} \equiv_\theta \mathbf{b}$, and by compatibility

$$f(\mathbf{a}, u) \equiv_\theta f(\mathbf{b}, u).$$

On the other hand,

$$f(\mathbf{a}, u) = f_{\mathbf{a}}(u) \geq s > v \quad \text{and} \quad f(\mathbf{b}, u) = f_{\mathbf{b}}(u) = u,$$

implying

$$f(\mathbf{a}, u) \not\equiv_\theta f(\mathbf{b}, u).$$

This completes the proof that the type of $f_{\mathbf{a}}$ depends only on f.

Assume now, for example, that all unary polynomials $f_{\mathbf{a}}$ are of type 1; that is, all the sets $f_{\mathbf{a}}(L)$ have both a smallest and a greatest element. Hence there exist two functions $g, h : L^{m-1} \to L$ such that

$$h(x_1, \ldots, x_{m-1}) \leq g(x_1, \ldots, x_{m-1})$$

and

$$f(x_1, \ldots, x_m) = g(x_1, \ldots, x_{m-1}) \wedge (h(x_1, \ldots, x_{m-1}) \vee x_m) \quad (5.17)$$

for all $x_1, \ldots, x_m \in L$. The latter formula shows that whenever the functions g and h are polynomials then so is f. Thus, by our induction hypothesis, it suffices to show that the functions g and h are compatible. Let $\mathbf{a}, \mathbf{b} \in L^{m-1}$ and put $c = h(\mathbf{a}) \wedge h(\mathbf{b})$. Then it is easy to check that $h(\mathbf{a}) = f(\mathbf{a}, c)$ and $h(\mathbf{b}) = f(\mathbf{b}, c)$. Thus the compatibility of f implies that of h. Similarly, using the following dual form of the formula (5.17):

$$f(x_1, \ldots, x_m) = h(x_1, \ldots, x_{m-1}) \vee (g(x_1, \ldots, x_{m-1}) \wedge x_m),$$

one can prove that the function g is compatible. The cases of types 2 and 3 can be handled similarly but more simply and the case of type 4 is immediate.   •

Before stating and proving the main result we need one more technical lemma.

Lemma 5.3.11 *Let f be a unary order preserving compatible function on a distributive lattice* **L**. *Then*

(i) *f is idempotent;*

(ii) *the set $f(L)$, that is, the set of all fixed points of f, is convex;*

(iii) *the set $\downarrow f(L)$ is an almost principal ideal of* **L**; *in fact we have the equality $\downarrow f(L) \cap \downarrow a = \downarrow(a \wedge f(a))$ for every $a \in L$;*

(iv) *the set $\uparrow f(L)$ is an almost principal filter of* **L**; *in fact we have the equality $\uparrow f(L) \cap \uparrow a = \uparrow(a \vee f(a))$ for every $a \in L$.*

PROOF Let **L** be a subdirect product of two element lattices L_i, $i \in I$. The elements of L are represented as systems $a = (a_i)_{i \in I}$ with $a_i \in L_i$. Because of the compatibility, the function f induces the coordinate functions $f_i : L_i \to L_i$, $i \in I$, and the system $(f_i)_{i \in I}$ completely determines f. Obviously f is idempotent iff so are all f_i. However, clearly all order preserving unary functions on $\mathbf{D_2}$ are idempotent. This proves (i).

The proof of (ii) similarly reduces to the case of two-element lattice. Indeed, let $a, b, c \in L$, $a \leq b \leq c$, and let a and b be fixed points of f. Then a_i and c_i are fixed points of f_i and $a_i \leq b_i \leq c_i$. However, the set of all fixed points of f_i is obviously convex (actually all subsets in $\mathbf{D_2}$ are convex). Thus b_i is a fixed point of f_i and since this is valid for all $i \in I$, we have $f(b) = b$.

Properties (iii) and (iv) are dual; so it is enough to prove (iii). First we show that

$$\downarrow f(L) \cap \downarrow a = \downarrow(a \wedge f(a))$$

for every $a \in L$. Clearly $a \wedge f(a) \in \downarrow f(L) \cap \downarrow a$ and hence

$$\downarrow(a \wedge f(a)) \subseteq \downarrow f(L) \cap \downarrow a.$$

To prove the reverse inclusion, we let b be an arbitrary element of $\downarrow f(L) \cap \downarrow a$. We have to show that $b \leq a \wedge f(a)$ or, equivalently, that $b_i \leq a_i \wedge f_i(a_i)$ for every $i \in I$. The latter is trivial if $b_i = 0$ or $a_i = f_i(a_i) = 1$. If $b_i = 1$ then $b \in \downarrow a$ implies $a_i = 1$. Suppose that $f_i(a_i) = 0$. Then, since f preserves the order relation, $f_i(0) = 0$ as well. Consequently $f_i(L_i) = \{0\}$ and $b \in \downarrow f(L)$ implies $b_i = 0$, a contradiction. This proves that $f_i(a_i) = 1$; thus $b_i \leq a_i \wedge f_i(a_i)$ holds.

It remains to prove that $\downarrow f(L)$ is closed with respect to joins. Let $a, b \in \downarrow f(L)$ and $c = (a \vee b) \wedge f(a \vee b)$. Then by the first part of the proof $a, b \leq c$ and obviously $c \leq a \vee b$. This implies $a \vee b = c \in \downarrow f(L)$ and we are done.   ●

Theorem 5.3.12 *A distributive lattice* **L** *is affine complete iff each of the following conditions is satisfied:*

  (i) **L** *does not contain nontrivial Boolean intervals;*

 (ii) *every proper almost principal ideal of* **L** *is a principal ideal of* **L***;*

(iii) *every proper almost principal filter of* **L** *is a principal filter of* **L***.*

PROOF The condition (i) is necessary by Theorem 5.3.9. We prove the necessity of the condition (ii); the necessity of (iii) can be proved by a dual argument. Let I be a proper almost principal ideal of **L**. By Lemma 5.3.5 the function f_I is compatible, so it must be a polynomial. Since the range of the function f_I is the ideal I and $I \neq L$, this polynomial is of type 2. Let $p(x) = a \wedge x$ for some $a \in L$. Then $I = f_I(L) = p(L) = \downarrow a$; i.e., condition (ii) is fulfilled.

Assume now that **L** satisfies the conditions (i)–(iii) and let f be a compatible function on **L**. We wish to show that f is a polynomial function of **L**. Due to Lemma 5.3.10, it suffices to consider only unary functions; thus $f : L \to L$. By Theorem 5.3.9 we know that f is a local polynomial function so it preserves the order relation and Lemma 5.3.11 applies. Hence the sets $\downarrow f(L)$ and $\uparrow f(L)$ are an almost principal ideal and an almost principal filter of **L**, respectively. We distinguish 4 cases.

1. $\downarrow f(L) = L = \uparrow f(L)$. Then for every $b \in L$ there exist $a, c \in L$ such that $f(a) \leq b \leq f(c)$. Since by Lemma 5.3.11 f is idempotent and the set $f(L)$ is convex, we have $f(b) = b$. Hence f is the identity function.

2. $\downarrow f(L) = \downarrow a$ for some $a \in L$, $\uparrow f(L) = L$. Now the convexity of $f(L)$ implies $f(L) = \downarrow a$. Let x be an arbitrary element of L. We show that

$$f(x) \leq x \quad \text{and} \quad f(x) \geq a \wedge x. \tag{5.18}$$

Together with the obvious fact that $f(x) \le a$, this implies

$$f(x) \in \downarrow x \cap \downarrow a \cap \uparrow(a \wedge x) = \{a \wedge x\};$$

thus $f(x) = a \wedge x$.

To prove the inequalities (5.18) we again use the representation of $\mathbf{L}$ in the form of a subdirect product of two element lattices $\mathbf{L}_i$, $i \in I$. If $f(x) \not\le x$ then there is $i \in I$ such that $x_i = 0$ and $f_i(x_i) = 1$. Since f is order preserving, this implies $f_i(L_i) = \{1\}$, contradicting the assumption $\uparrow f(L) = L$. This proves $f(x) \le x$. The inequality $f(x) \ge a \wedge x$ fails only if there exists $i \in I$ such that $f_i(x_i) = 0$ and $x_i = a_i = 1$. Then, since f preserves the order relation, we have $f_i(L_i) = \{0\}$ which contradicts the assumption $\downarrow f(L) = \downarrow a$.

3. $\downarrow f(L) = L$, $\uparrow f(L) = \uparrow b$ for some $b \in L$. In this case, dual to the preceding one, we have $f(x) = b \vee x$ for every $x \in L$.

4. $\downarrow f(L) = \downarrow a$, $\uparrow f(L) = \uparrow b$ for some $a, b \in L$ with $b \le a$. From the convexity of $f(L)$, $f(L) = [b, a]$ and thus $b \le f(x) \le a$ for every $x \in L$. Similarly to the second case we can prove the inequalities $f(x) \le x \vee b$ and $f(x) \ge x \wedge a$. All this implies $f(x) = a \wedge (b \vee x)$ for every $x \in L$.   •

If $\mathbf{L}$ is a bounded lattice then all of its almost principal ideals (filters) are principal ideals (filters). Hence we have the following corollary.

Corollary 5.3.13 *A bounded distributive lattice is affine complete iff it has no nontrivial Boolean intervals.*

We conclude with two examples.

Example 5.3.14 *Unbounded affine complete distributive lattices.*

Obviously every chain satisfies the last two conditions of Theorem 5.3.12. Therefore every chain without covering pairs of elements is affine complete. In particular the real line considered as a lattice is affine complete.

Example 5.3.15 *A distributive lattice without nontrivial Boolean intervals which is not affine complete.*

Let $\mathbf{L}$ be a direct square of the real line with the natural order relation. Clearly this lattice does not contain nontrivial Boolean intervals but it has a proper almost principal ideal I which is not a principal ideal. For example, we may take $I = \{(x, y) : x \le 0\}$.

5.3.3 Order affine completeness of lattices

The notion of order affine completeness was introduced in section 5.3.2 where it turned out that all distributive lattices are locally order affine complete. In fact here also we mainly consider local versions of order affine completeness.

When studying polynomial functions of lattices it is useful to consider tolerances. A **tolerance relation** on a set A is a reflexive and symmetric binary relation on A. A **tolerance of an algebra A** is a tolerance relation on the universe of **A** which is a subuniverse of $\mathbf{A}^2$. In other words, tolerances are symmetric diagonal subuniverses of $\mathbf{A}^2$. The following lemma lists elementary properties of diagonal subuniverses and tolerances of lattices. Its straightforward proof is left for the reader.

Lemma 5.3.16 *Let* **L** *be a lattice and* D *be a diagonal subuniverse of* $\mathbf{L}^2$. *Then:*

(i) *if* $a \leq b \leq c$ *and* $(a,c) \in D$ *then* $(a,b),(b,c) \in D$;

(ii) *if* $(a,b) \in D$ *then* $(a, a \wedge b),(a, a \vee b),(a \wedge b, b),(a \vee b, b) \in D$;

(iii) $D \circ D^{-1}$ *is a tolerance of* **L**.

If **L** is a lattice and S is a subuniverse of $\mathbf{L}^2$, we let

$$S_{\leq} = \{(a,b) \in S : a \leq b\}.$$

The set $S_{\geq}$ is defined dually. The following criteria are very useful when working with tolerances of lattices.

Lemma 5.3.17 *If* S *and* T *are tolerances of a lattice, they are equal iff* $S_{\geq} = T_{\geq}$. *A tolerance* T *of a lattice is transitive iff* $T_{\geq}$ *is transitive.*

PROOF The necessity of both statements is obvious. For the sufficiency of the first we assume that $S_{\geq} = T_{\geq}$. Because of the symmetry then also $S_{\leq} = T_{\leq}$. Let $(a,b) \in S$. Then by Lemma 5.3.16, $(a, a \wedge b) \in S_{\geq}$, $(a \wedge b, b) \in S_{\leq}$. Thus by our assumption, $(a, a \wedge b) \in T_{\geq}$, $(a \wedge b, b) \in T_{\leq}$, yielding $(a,b) = (a, a \wedge b) \vee (a \wedge b, b) \in T$. This proves $S \subseteq T$. The similar arguments show that $T \subseteq S$. Hence $S = T$.

Now assume that $T_\geq$ is transitive. Then clearly $T_\leq$ is transitive also. If $(a,b),(b,c) \in T$ then $(a, a \wedge b),(a \wedge b, a \wedge b \wedge c) \in T_\geq$, hence $(a, a \wedge b \wedge c) \in T$. Similarly, $(a \wedge b \wedge c, c) \in T$, thus

$$(a,c) = (a, a \wedge b \wedge c) \vee (a \wedge b \wedge c, c) \in T,$$

so that T is transitive. ●

In what follows, if S is a subuniverse of $\mathbf{L}^2$, $\mathbf{a} = (a_1, \ldots, a_m)$ and $\mathbf{b} = (b_1, \ldots, b_m)$ are in L^m, we shall write $(\mathbf{a}, \mathbf{b}) \in S$ iff $(a_i, b_i) \in S$, $i = 1, \ldots, m$. From our point of view, the importance of tolerances of lattices is based on the next lemma which, for the finite case, first appeared in [Kindermann, 1979].

Lemma 5.3.18 *An order preserving function on a lattice* $\mathbf{L}$ *is a local polynomial function of* $\mathbf{L}$ *iff it preserves all tolerances of* $\mathbf{L}$.

PROOF The necessity of the condition is obvious since every local polynomial function of any algebra preserves all diagonal (in other words, reflexive) subuniverses of the direct square. For the sufficiency, suppose that f is an m-ary function on L which preserves all tolerances and the order relation of $\mathbf{L}$. In view of Theorem 3.2.2 we have to prove that f preserves every diagonal subuniverse D of $\mathbf{L}^2$. Let $\mathbf{a}, \mathbf{b} \in L^m$, $(\mathbf{a}, \mathbf{b}) \in D$ and consider first the case $\mathbf{a} \leq \mathbf{b}$, that is, $(\mathbf{a}, \mathbf{b}) \in D_\leq$. The case $\mathbf{a} \geq \mathbf{b}$ is similar. Clearly $D_\leq$ is also a diagonal subuniverse of $\mathbf{L}^2$ and it is easy to see that $T = D_\leq \circ D_\geq$ is a tolerance of $\mathbf{L}$ containing $D_\leq$. Hence, if $f(\mathbf{a}) = c$ and $f(\mathbf{b}) = d$ then $c \leq d$ and $(c, d) \in T$. The latter means that there exists $s \in L$ such that $(c, s),(d, s) \in D_\leq$; in particular $(c, s) \in D$ and $d \leq s$. But then $(c, d) \in D$ by Lemma 5.3.16.

Now consider the general case. By Lemma 5.3.16 we have both $(\mathbf{a}, \mathbf{a} \wedge \mathbf{b}),(\mathbf{a} \wedge \mathbf{b}, \mathbf{b}) \in D$. Then $f(\mathbf{a} \wedge \mathbf{b}) \leq f(\mathbf{a})$, $f(\mathbf{a} \wedge \mathbf{b}) \leq f(\mathbf{b})$ since f is order preserving and $(f(\mathbf{a}), f(\mathbf{a} \wedge \mathbf{b})),(f(\mathbf{a} \wedge \mathbf{b}), f(\mathbf{b})) \in D$ by the first part of the proof. Hence also

$$(f(\mathbf{a}), f(\mathbf{b})) = (f(\mathbf{a}), f(\mathbf{a} \wedge \mathbf{b})) \vee (f(\mathbf{a} \wedge \mathbf{b}), f(\mathbf{b})) \in D;$$

so we are done. ●

An algebra is called *tolerance trivial* if all of its tolerances are congruences. Now an obvious corollary from Lemma 5.3.18 is that all tolerance trivial lattices are locally order affine complete. In fact the requirement of tolerance triviality is too strong. What we need is that, given an order preserving compatible function f on a lattice **L**, the function f also preserves all tolerances of **L**. Using the language introduced in Section 1.1, the same condition can be formulated as follows: the coclone R generated by $\{\leq\} \cup \operatorname{Con} \mathbf{L}$ must contain all tolerances of **L**. Moreover, it is enough to require that R contains all finitely generated tolerances of **L**. In particular, in view of the obvious fact that the binary part of every coclone is closed with respect to arbitrary meets and relation product, we conclude that **L** is locally order affine complete provided all its finitely generated tolerances can be represented as meets of relation products of congruences of **L**. A disadvantage of the latter condition is that the relation product of tolerances need not be a tolerance, in general. It is more convenient to work with another operation which we shall call the *symmetrized (relation) product* and denote it by $*$. By definition,

$$S * T = \{(a,c) \in L^2 : (a \vee c, a \wedge c) \in S \circ T\}$$

where S and T are arbitrary tolerances of **L**. Thus, $(a,c) \in S * T$ iff there exists $b \in L$ such that $(a \vee c, b) \in S$ and $(b, a \wedge c) \in T$. It is useful to note that due to the reflexivity of S and T, the element b can be chosen in the interval $[a \wedge c, a \vee c]$. Another helpful observation is that $(S * T)_{\geq} = (S \circ T)_{\geq}$.

Lemma 5.3.19 *Let S and T be tolerances of a lattice **L** and f be an order preserving function on **L**. If f preserves S and T then it also preserves $S * T$.*

PROOF Suppose that f is m-ary and take $\mathbf{a}, \mathbf{c} \in L^m$ such that $(\mathbf{a}, \mathbf{c}) \in S * T$. Thus there is $\mathbf{b} \in L^m$ so that $(\mathbf{a} \vee \mathbf{c}, \mathbf{b}) \in S$ and $(\mathbf{b}, \mathbf{a} \wedge \mathbf{c}) \in T$. Since f preserves S and T, $(f(\mathbf{a} \vee \mathbf{c}), f(\mathbf{b})) \in S$ and $(f(\mathbf{b}), f(\mathbf{a} \wedge \mathbf{c})) \in T$. On the other hand, f preserves the order; therefore $f(\mathbf{a} \vee \mathbf{c}) \geq f(\mathbf{a}) \vee f(\mathbf{c})$ and $f(\mathbf{a} \wedge \mathbf{c}) \leq f(\mathbf{a}) \wedge f(\mathbf{c})$. Now the reflexivity of S and T implies both $(f(\mathbf{a}) \vee f(\mathbf{c}), b) \in S$ and $(b, f(\mathbf{a}) \wedge f(\mathbf{c})) \in T$ where $b = (f(\mathbf{b}) \wedge (f(\mathbf{a}) \vee f(\mathbf{c}))) \vee (f(\mathbf{a}) \wedge f(\mathbf{c}))$. Thus $(f(\mathbf{a}), f(\mathbf{c})) \in S * T$.   •

Corollary 5.3.20 *If S and T are tolerances of a lattice **L** then so is $S * T$.*

PROOF Obviously $S * T$ is a tolerance relation on L. That $S * T$ is closed under the lattice operations directly follows from Lemma 5.3.19. •

Now we establish some elementary computation rules for the symmetrized product which will be useful in the sequel.

Lemma 5.3.21 *The operation $*$ is associative on the set of tolerances of a lattice $\mathbf{L}$.*

PROOF Let S, T and U be tolerances of $\mathbf{L}$. In view of Lemma 5.3.17 it suffices to verify the equalities

$$(S * (T * U))_{\geq} = (S \circ T \circ T)_{\geq} = ((S * T) * U)_{\geq} .$$

Since the proofs of them are similar, we prove only the first of them. First take a pair $(a, d) \in (S \circ T \circ U)_{\geq}$. Since $S, T,$ and U are diagonal, there exist $b, c \in L$ such that $a \geq b \geq c \geq d, (a, b) \in S, (b, c) \in T,$ and $(c, d) \in U$. But then $(b, d) \in (T * U)_{\geq}$ and $(a, d) \in (S * (T * U))_{\geq}$. This proves the inclusion $(S \circ T \circ T)_{\geq} \subseteq (S * (T * U))_{\geq}$. For the opposite inclusion, take $(a, d) \in (S * (T * U))_{\geq}$. Then $(a, d) \in (S \circ (T * U))_{\geq}$ and we may choose $b \in L$ such that $a \geq b \geq d, (a, b) \in S, (b, d) \in T * U$. But then $(b, d) \in (T * U)_{\geq} = (T \circ U)_{\geq}$, implying $(a, d) \in (S \circ T \circ U)_{\geq}$. This proves the lemma. •

Lemma 5.3.22 *For arbitrary tolerances S and T_i, $i \in I$, of a lattice $\mathbf{L}$ of finite height,*

$$(\bigcap\{T_i : i \in I\}) * S = \bigcap\{T_i * S : i \in I\} .$$

PROOF Choose $(a, c) \in \bigcap\{T_i * S : i \in I\}$; therefore there exist $b_i \in [a \wedge c, a \vee c]$ such that $(a \wedge c, b_i) \in T_i, (b_i, a \vee c) \in S, i \in I$. Since $\mathbf{L}$ is of finite height, the meet $b = \bigwedge\{b_i : i \in I\}$ exists and equals to the meet of finitely many of the b_i. Then by Lemma 5.3.16 $(b, a \vee c) \in S$ and $(a \wedge c, b) \in T_i$ for every $i \in I$; hence $(a, c) \in (\bigcap\{T_i : i \in I\}) * S$. We proved the inclusion

$$(\bigcap\{T_i : i \in I\}) * S \supseteq \bigcap\{T_i * S : i \in I\} .$$

Since the opposite inclusion is obvious, we are done. •

We shall call a tolerance of a lattice $\mathbf{L}$ *congruence generated* if it can be obtained from congruences of $\mathbf{L}$ by forming arbitrary meets and symmetrized products. It follows from Lemma 5.3.22 that every congruence generated tolerance of **L** is a meet of tolerances T_i, $i \in I$, each of which is a symmetrized product of finitely many congruences of **L**. Now the following result is obvious.

Theorem 5.3.23 *Assume that every finitely generated tolerance of a lattice **L** is congruence generated. Then **L** is locally order affine complete.*

Our next goal is to show that the converse of Theorem 5.3.23 is true in case of lattices of finite height. Since we do not know the answer in the general case, we state the following problem.

Problem 5.3.24 *Does there exist a locally order affine complete lattice having a tolerance which is not congruence generated?*

In what follows we exploit the fact that in case of lattices of finite height there is a one-to-one correspondence between tolerances and certain unary functions. The same fact is fundamental in tame congruence theory. Hence, following [Hobby, McKenzie, 1988] we shall call a function f on a lattice **L** *decreasing* if $f(a) \leq a$ for every $a \in L$. We define *increasing* functions dually.

Proposition 5.3.25 *Let f be a decreasing $\vee$-endomorphism of **L**. Then the set*

$$T^f = \{(a,b) \in L^2 : f(a \vee b) \leq a \wedge b\}$$

*is a tolerance of **L**. If the height of **L** is finite then the mapping $f \to T^f$ is bijective and its inverse is given by the rule: for every tolerance T of **L**, the function $f = f_T$ is defined by the formula*

$$f(x) = \bigwedge \{y \in L : (x,y) \in T\}.$$

There is a similar correspondence between tolerances and increasing $\wedge$-endomorphisms.

PROOF Let f be a decreasing $\vee$-endomorphism of **L**. Then clearly the set T^f is a reflexive and symmetric binary relation on L. We have to show that T^f is a subuniverse of $\mathbf{L}^2$.

Assume that $(a_1, b_1), (a_2, b_2) \in T^f$. Then

$$
\begin{aligned}
f((a_1 \vee a_2) \vee (b_1 \vee b_2)) &= f((a_1 \vee a_2) \vee (b_1 \vee b_2)) \\
&= f(a_1 \vee b_1) \vee f(a_2 \vee b_2) \\
&\leq (a_1 \wedge b_1) \vee (a_2 \wedge b_2) \\
&\leq (a_1 \vee a_2) \wedge (b_1 \vee b_2)
\end{aligned}
$$

proves that T^f is closed with respect to joins. Here we used the fact that every $\vee$-endomorphism is order preserving. It remains to show that $(a_1 \wedge a_2, b_1 \wedge b_2) \in T^f$, that is,

$$
f((a_1 \wedge a_2) \vee (b_1 \wedge b_2)) \leq (a_1 \wedge a_2) \wedge (b_1 \wedge b_2),
$$

which is equivalent to

$$
f(a_1 \wedge a_2), f(b_1 \wedge b_2) \leq a_1 \wedge a_2 \wedge b_1 \wedge b_2 \tag{5.19}
$$

because f is a $\vee$-endomorphism. Since f is decreasing, we have $f(a_1 \wedge a_2) \leq a_1 \wedge a_2$. On the other hand, using the facts that f is order preserving and $(a_1, b_1), (a_2, b_2) \in T^f$, we get

$$
f(a_1 \wedge a_2) \leq f(a_1 \vee b_1) \leq a_1 \wedge b_1 \leq b_1 ,
$$

$$
f(a_1 \wedge a_2) \leq f(a_2 \vee b_2) \leq a_2 \wedge b_2 \leq b_2 .
$$

This proves the inequality (5.19) for $f(a_1 \wedge a_2)$. The proof of the other inequality is similar.

Now assume that $\mathbf{L}$ is of finite height and take an arbitrary tolerance T of $\mathbf{L}$. Obviously the function $f = f_T$ is decreasing. We prove that f is a $\vee$-endomorphism. Let $a, b \in L$, $a \leq b$. Then $(f(b), b) \in T$ implies $(a \wedge f(b), a) \in T$; hence $f(a) \leq a \wedge f(b) \leq f(b)$. This shows that f is order preserving, in particular $f(a \vee b) \geq f(a) \vee f(b)$ for arbitrary $a, b \in L$. For the opposite inequality observe that $(f(a), a) \in T$ and $(f(b), b) \in T$ imply $(f(a) \vee f(b), a \vee b) \in T$ from which we infer $f(a \vee b) \leq f(a) \vee f(b)$.

It remains to prove the equalities $f_{T^f} = f$ and $T^{f_T} = T$ for arbitrary tolerance T and decreasing $\vee$-endomorphism f. Let $f_{T^f}(a) = b$. Then $(a, b) \in T^f$ which implies $f(a) \leq b$. On the other hand, $(a, f(a)) \in T^f$ implies $b \leq f(a)$. This proves the equality $f_{T^f} = f$. The equality $T = T^{f_T}$ follows from the observation that in case $a \geq b$ then both $(a, b) \in T$ and $(a, b) \in T^{f_T}$ are equivalent to $f_T(a) \leq b$. $\bullet$

If S and T are tolerances of a lattice $\mathbf{L}$ of finite height then the composition $f_T f_S$ of two decreasing $\lor$-endomorphisms must be a decreasing $\lor$-endomorphism, too. Hence in view of Proposition 5.3.25 there must exist a tolerance U such that $f_T f_S = f_U$. The next lemma gives an explicit description of this tolerance.

Lemma 5.3.26 *Let $\mathbf{L}$ be a lattice of finite height and S, T be tolerances of $\mathbf{L}$. Then $f_{S*T} = f_T f_S$.*

PROOF Take any element $a \in L$ and let $b = f_S(a)$, $c = f_T(b)$. Then $(a \lor c, a \land c) = (a, c) \in S \circ T$ implying $(a, c) \in S * T$; hence $f_{S*T}(a) \leq c = f_T f_S(a)$. For the opposite inequality, let $f_{S*T}(a) = d$. Then $(a, d) = (a \lor d, a \land d) \in S \circ T$; that is, there exists $b_1 \in L$ such that $(a, b_1) \in S$ and $(b_1, d) \in T$. Hence $f_S(a) \leq b_1$ and using the fact that f_T is order preserving, we have $f_T f_S(a) \leq f_T(b_1) \leq d$.   •

Now we are almost ready to prove a theorem which gives a characterization of locally order affine complete lattices of finite height. The only thing we still miss is a certain supply of compatible order preserving functions. The following lemma fills this gap.

Lemma 5.3.27 *Let $\mathbf{L}$ be a lattice of finite height, $\mathbf{a} \in L^m$, and $b \in L$. For every $\mathbf{x} \in L^m$ denote by $T_{\mathbf{x}}$ the smallest congruence generated tolerance T of $\mathbf{L}$ such that $(\mathbf{a}, \mathbf{a} \land \mathbf{x}) \in T$. Then the function $g = g_{\mathbf{a},b} : L^m \to L$ defined by $g(\mathbf{x}) = f_{T_{\mathbf{x}}}(b)$ is an order preserving compatible function on $\mathbf{L}$ such that $g(\mathbf{a}) = b$.*

PROOF The equality $g(\mathbf{a}) = b$ is obvious because $T_{\mathbf{a}} = \Delta_L$. Let $\mathbf{x}, \mathbf{y} \in L^m$, $\mathbf{x} \leq \mathbf{y}$. Then $\mathbf{x} \land \mathbf{a} \leq \mathbf{y} \land \mathbf{a} \leq \mathbf{a}$ which implies $T_{\mathbf{y}} \subseteq T_{\mathbf{x}}$. Hence $g(\mathbf{x}) = f_{T_{\mathbf{x}}}(b) \leq f_{T_{\mathbf{y}}}(b) = g(\mathbf{y})$.

It remains to prove that g is compatible. Let $(\mathbf{x}, \mathbf{y}) \in \theta \in \mathrm{Con}\ \mathbf{L}$. Then $(\mathbf{a}, \mathbf{a} \land \mathbf{x}) \in T_{\mathbf{y}} * \theta$. Since $T_{\mathbf{y}} * \theta$ is a congruence generated tolerance, $T_{\mathbf{x}} \subseteq T_{\mathbf{y}} * \theta$; hence also $T_{\mathbf{x}} * \theta \subseteq (T_{\mathbf{y}} * \theta) * \theta = T_{\mathbf{y}} * (\theta * \theta)$. The transitivity of θ yields $\theta * \theta = \theta$; thus $T_{\mathbf{x}} * \theta \subseteq T_{\mathbf{y}} * \theta$. Since $\mathbf{x}$ and $\mathbf{y}$ were arbitrary m-tuples, we have the equality $T_{\mathbf{x}} * \theta = T_{\mathbf{y}} * \theta$. Now, using Lemma 5.3.26 we get

$$f_\theta(g(\mathbf{x})) = f_\theta(f_{T_{\mathbf{x}}}(b)) = f_{T_{\mathbf{x}}*\theta}(b) =$$
$$f_{T_{\mathbf{y}}*\theta}(b) = f_\theta(f_{T_{\mathbf{y}}}(b)) = f_\theta(g(\mathbf{y})).$$

Since θ is a congruence, the latter yields $g(\mathbf{x}) \equiv_\theta g(\mathbf{y})$.   •

The finite case of the next theorem appears in [Wille, 1977b].

Theorem 5.3.28 *A lattice* **L** *of finite height is locally order affine complete iff every finitely generated tolerance of* **L** *is congruence generated.*

PROOF In view of Theorem 5.3.23 only the necessity part needs a proof. Thus assume that **L** is locally order affine complete and consider the tolerance T of **L** generated by $(a_i, b_i) \in L^2$, $i = 1, \ldots, m$. We may assume, without loss of generality, that $b_i = f_T(a_i)$ for $i = 1, \ldots, m$. Suppose that T is not congruence generated and let S be the smallest congruence generated tolerance of **L** such that $T \subseteq S$. Then $f_S \leq f_T$ and these functions must differ on the set $\{a_1, \ldots, a_m\}$. Assume without loss of generality that $f_S(a_1) < f_T(a_1)$. Now consider the compatible order preserving function $g = g_{\mathbf{a}, a_1}$. Clearly $g(\mathbf{a}) = a_1$ and it is easy to see that $T_{\mathbf{b}} = S$. Therefore we have $g(\mathbf{b}) = f_S(a_1) < b_1$. Since $b_1 = f_T(a_1)$, we have $(f_S(a_1), a_1) \notin T$. We see that $(\mathbf{a}, \mathbf{b}) \in T$ but $(g(\mathbf{a}), g(\mathbf{b})) \notin T$; thus g does not preserve the tolerance T and cannot be a local polynomial function. This contradiction shows that T must be congruence generated. •

The last proof raises the following problem.

Problem 5.3.29 *Does there exist a natural number k such that an arbitrary lattice of finite height is locally order affine complete iff all k-ary order preserving compatible functions on this lattice are local polynomials?*

In the rest of this section we consider another local analog of order affine completeness introduced in [Kaarli, Täht, 1993]. We have mentioned that no lattice can be strictly locally affine complete. Still it makes sense to consider strictly locally order affine complete lattices which by analogy should mean the following: for any m and for every finite subset X of L^m, every compatible order preserving function from X to L is a restriction of some polynomial function of **L**. However, such a requirement seems to be too strong because in case X is an antichain, all functions from X to L are order preserving. We get a much more interesting class if we restrict ourselves to finite subsemilattices X of $\langle L^m; \wedge \rangle$ or $\langle L^m; \vee \rangle$.

Definition A lattice **L** is called **strictly locally order affine complete** if for every m and every finite subuniverse X of $\langle L^m; \wedge \rangle$, every compatible order preserving function from X to L is a restriction of a polynomial function of **L**.

Note that while all order affine complete lattices are locally order affine complete, in general they need not be strictly locally order affine complete. The reason is that a compatible order preserving partial function need not have a compatible order preserving extension on the whole lattice. In particular Theorem 5.3.34 will imply that finite nonBoolean distributive lattices are not strictly locally order affine complete but by Theorem 5.3.7 they are order affine complete. The following theorem gives an important characterization of strictly locally order affine complete lattices.

Theorem 5.3.30 *For an arbitrary lattice* **L** *the following are equivalent:*

(1) **L** *is tolerance trivial;*

(2) **L** *is strictly locally order affine complete;*

(3) *for every finite subsemilattice X of $\langle L^2; \wedge \rangle$, every order preserving compatible function $f : X \to L$ is a restriction of a polynomial function of* **L**.

PROOF Let **L** be a tolerance trivial lattice and $f : X \to L$ be a compatible order preserving function where X is a finite subuniverse of $\langle L^m; \wedge \rangle$. Clearly f preserves all tolerances of **L**. Now exactly the same argument as that used in the proof of Lemma 5.3.18 applies to show that f preserves all diagonal subuniverses of $\mathbf{L}^2$. Note that at this point we need the assumption that X is a subsemilattice. Hence by Theorem 3.2.2 f is a restriction of a polynomial function of **L**. This proves the implication (1) $\Rightarrow$ (2).

The implication (2) $\Rightarrow$ (3) is trivial.

Now assume (3) and prove (1). We have to show that an arbitrary tolerance T of **L** is transitive. By Lemma 5.3.17 it suffices to show that $T_{\geq}$ is transitive. Take $(a,b), (b,c) \in T_{\geq}$. Then clearly the set $X = \{(a,b), (b,c)\}$ is a subsemilattice of $\langle L^2; \wedge \rangle$. Consider the function $f : X \to L$ defined by $f(a,b) = a$, $f(b,c) = c$. Obviously f is a compatible order preserving function; hence by (3) it must be the restriction of a polynomial function of **L**. This means that f must preserve T; thus $(a,c) = (f(a,b), f(a,c)) \in T$. ●

Combining Theorems 5.3.23 and 5.3.30 we see that the adjective "strict" is justified, indeed. There is also a simple direct way to understand that every strictly locally order affine complete lattice is locally order affine complete. Namely, it follows from the fact that all finitely generated semilattices are finite.

Another consequence of Theorem 5.3.30 is that the class of all strictly locally order affine complete lattices is closed with respect to homomorphic images. The corresponding question about locally order affine complete lattices is still open.

Problem 5.3.31 *Prove or disprove that every homomorphic image of a locally order affine complete lattice is locally order affine complete.*

It is natural to ask whether we could replace L^2 by just L in condition (3) of Theorem 5.3.30. The answer is "no" as the example of the three element chain shows.

Now we show that strictly locally order affine complete lattices are rather close to relatively complemented lattices. Actually many important lattices (projective geometries, matroid lattices) turn out to be strictly locally order affine complete.

Theorem 5.3.32 *The class of strictly locally order affine complete lattices lies between the class of relatively complemented lattices and the class of congruence permutable lattices.*

PROOF We first prove that every relatively complemented lattice **L** is strictly locally order affine complete. In view of Theorem 5.3.30 it suffices to show that **L** is tolerance trivial. By Lemma 5.3.17 we have to prove that, given any tolerance T of **L**, the subuniverse $T_\geq$ is transitive. Let $(a,b),(b,c) \in T_\geq$. Since **L** is relatively complemented, there exists $d \in L$ such that $b \wedge d = c$ and $b \vee d = a$. Then we have $(a,d) = (b,c) \vee (d,d) \in T$ and $(a,c) = (a,b) \wedge (a,d) \in T$.

Now we prove that every strictly locally order affine complete lattice **L** is CP. Let $\rho, \sigma \in \mathrm{Con}\,\mathbf{L}$. We have to prove that $\rho \circ \sigma$ is a congruence of **L** and we start with transitivity. Let $(a,b),(b,c) \in \rho \circ \sigma$. By Theorem 5.3.30, the tolerance $\rho * \sigma$ is a congruence of **L**. Assume first that $a \geq b \geq c$. Then clearly $(a,b),(b,c) \in \rho * \sigma$; thus the transitivity of $\rho * \sigma$ yields $(a,c) \in \rho * \sigma$ and because of $a \geq c$ also $(a,c) \in \rho \circ \sigma$. In the general case put $b' = a \wedge b$, $c' = a \wedge b \wedge c$. Then

$(a, b'), (b', c') \in \rho \circ \sigma$; hence $(a, c') \in \rho \circ \sigma$. Using the fact that $\rho \circ \sigma$ is a diagonal subuniverse of $\mathbf{L}^2$, we easily derive from this $(a, c) \in \rho \circ \sigma$.

For the symmetry, take $(a, b) \in \rho \circ \sigma$. Then, using transitivity of $\rho \circ \sigma$, we have

$$(b, a) \in \rho \circ \sigma \circ \rho \circ \sigma \subseteq \rho \circ \sigma.$$

This completes the proof.   •

We get even better results in the case of modular lattices of finite height. First we need a lemma from [Herrmann, 1973] which associates to every such lattice a special decreasing ∨-endomorphism.

Lemma 5.3.33 *Let $\mathbf{L}$ be a modular lattice of finite height. For every $0 \neq x \in L$ let x^+ be the meet of all elements of $\mathbf{L}$ covered by x. If, moreover, $0^+ = 0$ then the function $x \rightarrow x^+$ is a decreasing ∨-endomorphism of $\mathbf{L}$.*

PROOF Denote $f(x) = x^+$. Clearly f is decreasing and it follows from the modularity that f is order preserving. Indeed, let $a, b, c \in L$, $a < b$, and let c be covered by b. Then either $a \leq c$ or a covers $a \wedge c$. In both cases $f(a) \leq a \wedge c \leq c$, hence $f(a) \leq f(b)$. Note that all intervals $[f(a), a]$ are relatively complemented lattices because obviously they are dually atomic.

We are going to show that f is a ∨-endomorphism. Suppose $\mathbf{L}$ is a lattice of smallest height in which f fails to be a ∨-endomorphism. Choose $a, b \in L$ such that $f(a) \vee f(b) \neq f(a \vee b)$. By the choice of $\mathbf{L}$ then $a \vee b = 1$ and clearly neither a nor b can be 1. Hence there are coatoms c and d of $\mathbf{L}$ such that $a \leq c$, $b \leq d$. Since $a \vee b = 1$, the modularity of $\mathbf{L}$ yields $a \vee (b \wedge c) = c$ and $b \vee (a \wedge d) = d$. Now, since the claim of the theorem is true in intervals $[0, c]$ and $[0, d]$, we have $f(c) = f(a) \vee f(b \wedge c)$ and $f(d) = f(b) \vee f(a \wedge d)$ which imply $f(c) \vee f(d) = f(a) \vee f(b)$. This means that the theorem will be proved if we verify the equality $f(c) \vee f(d) = f(1)$.

By modularity $c \wedge d$ is covered by both of c and d. Therefore $m = f(c) \vee f(d) \leq c \wedge d$. The lattice $[m, c]$ is relatively complemented because it is an interval of a relatively complemented lattice $[f(c), c]$. Hence this lattice is also atomic, meaning that c is a join of covers of m. Similar arguments prove that d is a join of covers of m but then also 1 is a join of covers of m. Hence the lattice $[m, 1]$ is relatively complemented; in particular, m is a meet of coatoms of $\mathbf{L}$. Then, however, $f(1) \leq m$. The opposite inequality is a consequence of the fact that f is order preserving. •

The following theorem is from [Kaarli, Täht, 1993]. For the simple finite modular case see [Wille, 1977a].

Theorem 5.3.34 *A modular lattice of finite height is strictly locally order affine complete iff it is relatively complemented.*

PROOF The statement is trivial if $|L| = 1$. Thus assume that $|L| > 1$. In view of Theorem 5.3.32 we have to prove only necessity. Let **L** be a modular strictly locally order affine complete lattice and let $T = T^f$ be a tolerance of **L** determined by the function $f(x) = x^+$. By Theorem 5.3.30 **L** is tolerance trivial. Since L has more than one element, $f(1) \neq 1$. Thus $T \neq \Delta_L$ and the only possibility is $T = L^2$. Then, however, $1^+ = 0$, meaning that **L** is dually atomic or, equivalently, relatively complemented. •

We conclude the section by showing that the classes of locally order affine complete lattices and strictly locally order affine complete lattices are closed with respect to finite direct products. This will give us more examples of such lattices. We first prove a lemma which shows that lattices do not have skew tolerances.

Lemma 5.3.35 *For arbitrary lattices $\mathbf{L}_1, \ldots, \mathbf{L}_n$, every tolerance of their direct product has the form $T_1 \times \cdots \times T_n$ where T_i is a tolerance of $\mathbf{L}_i$, $i = 1, \ldots, n$.*

PROOF It suffices to handle the case $n = 2$. Let T_i be a tolerance of $\mathbf{L}_i$ induced by T, $i = 1, 2$. Hence, for example, $(a_1, b_1) \in T_1$ iff there exist $a_2, b_2 \in L_2$ such that $((a_1, a_2), (b_1, b_2)) \in T$. Then clearly $T \subseteq T_1 \times T_2$.

Now take $((a_1, c_2), (b_1, d_2)) \in T_1 \times T_2$ and let $((a_1, a_2), (b_1, b_2))$ and $((c_1, c_2), (d_1, d_2))$ be in T. Also let e be the meet of all a_i, b_i, c_i, d_i, $i = 1, 2$. Then

$$((a_1, e), (b_1, e)) = ((a_1, a_2), (b_1, b_2)) \wedge ((a_1 \vee b_1, e), (a_1 \vee b_1, e)) \in T,$$

$$((e, c_2), (e, d_2)) = ((c_1, c_2), (d_1, d_2)) \wedge ((e, c_2 \vee d_2), (e, c_2 \vee d_2)) \in T.$$

Hence also

$$((a_1, c_2), (b_1, d_2)) = ((a_1, e), (b_1, e)) \vee ((e, c_2), (e, d_2)) \in T,$$

which completes the proof. •

Theorem 5.3.36 *Let* **L** *be the direct product of lattices* $\mathbf{L}_1, \ldots, \mathbf{L}_n$. *The lattice* **L** *is (strictly) locally order affine complete iff all lattices* $\mathbf{L}_i$ *are (strictly) locally order affine complete.*

PROOF Let **L** be locally order affine complete. We want to prove that every $\mathbf{L}_i$ is locally order affine complete. Take an arbitrary compatible order preserving function f on $\mathbf{L}_i$ and consider a function $g = (g_1, \ldots, g_n)$ on L where $g_i = f$ and g_j is some constant function on L_j for $j \neq i$. Since **L** has no skew congruences, the function g is order preserving and compatible; thus it must be a local polynomial. Clearly then its i-th component f is a local polynomial, also.

For the converse, suppose all $\mathbf{L}_i$ are locally order affine complete and let f be a compatible order preserving function on **L**. Then it has the form $f = (f_1, \ldots, f_n)$ where every f_i is a compatible order preserving function on $\mathbf{L}_i$, $i = 1, \ldots, n$. Since every $\mathbf{L}_i$ is locally order affine complete, all f_i are local polynomials; thus each of them preserves all tolerances of $\mathbf{L}_i$. Now Lemma 5.3.35 implies that f preserves all tolerances of **L**. Hence by Lemma 5.3.18 the function f is a local polynomial function of **L**.

The statement about strict local order affine completeness is a direct consequence of Theorem 5.3.30 and Lemma 5.3.35.   •

We have seen that there is a nice description of strictly locally order affine lattices, at least in an important special case of modular lattices of finite height. Unfortunately, in case of locally order affine complete lattices the situation is much worse. In fact, all of the known examples of locally order affine complete modular lattices of finite height have the form $\mathbf{D} \times \mathbf{M}$ where **D** is distributive and **M** is a complemented modular lattice. We feel that this class should have a transparent structural characterization.

Problem 5.3.37 *Characterize modular locally order affine complete lattices of finite height. Does there exist one which is not a direct product of a distributive lattice and a complemented lattice?*[1]

This problem can be made even more specialized. Suppose a lattice **L** of finite height is represented as a subdirect product of

---

[1]In April, 2000, during the publication process, Kaarli found examples of finite lattices which give a positive answer to the question asked in Problem 5.3.37 and a negative answer to Problem 5.3.38.

SI lattices $\mathbf{L}_i$, $i = 1, \ldots, n$, and let $\mathbf{L}_{ij}$ be the 2-fold coordinate projections of $\mathbf{L}$. In view of [Wille, 1977b], $\mathbf{L}$ is locally order affine complete if so are all of the $\mathbf{L}_{ij}$. On the other hand, by Theorem 3.2.1 the projections $\mathbf{L}_{ij}$ uniquely determine $\mathbf{L}$. Consequently, it is important to focus first on subdirect products of two SI lattices. Moreover, it is well known that every SI modular lattice of finite height is simple. Thus the problem reduces to subdirect products of two simple modular lattices of finite height. Among such subdirect products so far only the three element chain and direct products of simple complemented lattices are known to be locally order affine complete. So we have the following problem.

Problem 5.3.38 *Let $\mathbf{L}$ be a subdirect product of two simple modular lattices of finite height and assume that $\mathbf{L}$ is locally order affine complete. Is it true that $\mathbf{L}$ is a direct product of these lattices unless $|L| = 3$?*[1]

Since all distributive lattices are locally order affine complete, one might think that the same is true in case of modular lattices. The following diagram gives an example showing the failure of such conjecture.

Example 5.3.39 *A finite nonsimple modular lattice which is not order affine complete.*

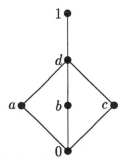

Figure 5.1

Obviously this lattice, let it be $\mathbf{L}$, has two nontrivial congruences: $\rho = \text{Cg}(0, d)$ and $\sigma = \text{Cg}(d, 1)$. Since $\sigma * \rho = L^2$, it follows that the only congruence generated tolerance of **L**, apart from congruences, is $T = \rho * \sigma$ which is determined by $(0, d), (d, 1) \in T$. However, **L** has three more tolerances, for example S, determined by $(0, c), (a, 1) \in S$.

Exercises

1. Show that the least nonmodular lattice $\mathbf{N}_5$ is strictly locally order affine complete (hence also order affine complete) whereas the same lattice with a new greatest element adjoined is not order affine complete.

2. Conclude from the previous exercise and Example 5.3.39 that the only variety of lattices with all members locally order affine complete is that of distributive lattices.

3. Let **L** be a finite lattice with k join-irreducibles. Prove that **L** is order affine complete iff all k-ary compatible order preserving functions on **L** are polynomials. ([Wille, 1977b])

4. Show that a lattice **L** is strictly locally order affine complete iff it satisfies the following condition: for every $a_i, b_i, c, d \in L$ with $a_i \leq b_i$, $c \leq d$, $i = 1, \ldots, m$, if $(c, d) \in \mathrm{Cg}(\mathbf{a}, \mathbf{b})$ then there exists an m-ary polynomial function p on **L** such that $p(\mathbf{a}) = c$ and $p(\mathbf{b}) = d$. ([Kaarli, Täht, 1993])

5.3.4   Order functionally complete lattices

A lattice **L** is said to be **(locally) order functionally complete** if all order preserving functions on **L** are (local) polynomial functions. Finite order functionally complete lattices were studied in [Schweigert, 1974], [Wille, 1977a], and [Kindermann, 1979]. We shall extend their main results to locally order affine complete lattices. We note that up to the present most authors have used the terminology *order polynomially complete* to describe the same concept. Our terminology is, however, more systematic. One can also define strictly locally order functionally complete lattices just omitting the adjective "compatible" in the previous definition. However, the following theorem shows that this does not give anything new.

Theorem 5.3.40 *For a lattice **L**, the following conditions are equivalent:*

(1) **L** *is strictly locally order functionally complete;*

(2) **L** *is locally order functionally complete;*

(3) *all unary order preserving functions on* **L** *are local polynomials;*

(4) $\leq$ *(and then also $\geq$) is a minimal diagonal subuniverse of* $\mathbf{L}^2$;

(5) Δ_L, $\leq$, $\geq$, *and* L^2 *are the only diagonal subuniverses of* $\mathbf{L}^2$;

(6) Δ_I, *and* L^2 *are the only tolerances of* **L**.

PROOF The implications (1) $\Rightarrow$ (2) $\Rightarrow$ (3) and (5) $\Rightarrow$ (6) are obvious. The implication (6) $\Rightarrow$ (1) directly follows from Theorem 5.3.30. Thus we shall be done if we prove the implications (3) $\Rightarrow$ (4) $\Rightarrow$ (5).

(3) $\Rightarrow$ (4). Let D be a diagonal subuniverse of $\mathbf{L}^2$ which lies strictly between Δ_L and $\leq$. We show that there exists a unary order preserving function on **L** which does not preserve D and so cannot be a local polynomial function.

We claim that there exist a triple $a, b, c \in L$ such that $a \leq b \leq c$, $(a, c) \notin D$, but (a, b) or (b, c) is in $D \backslash \Delta_L$. Indeed, by our assumptions we have $s, t, u, v \in L$ such that

$$s \leq t, \quad (s, t) \notin D, \quad (u, v) \in D, \quad u \neq v.$$

Now we may choose a, b, c such that:

- if $(u, v \vee t) \in D$ then $a = u \wedge s, b = u, c = v \vee t$;

- if $(u, v \vee t) \notin D$ then $a = u, b = v, c = v \vee t$.

Note that $(u \wedge s, t \vee v) \notin D$ by Lemma 5.3.16.

Assume that $(b, c) \in D \setminus \Delta_L$ (the case $(a, b) \in D \setminus \Delta_L$ is dual). Now the function

$$g(x) = \begin{cases} c & \text{if } x \geq c \\ a & \text{otherwise} \end{cases}$$

preserves the order relation but not the subuniverse D. This contradicts the condition (3).

(4) $\Rightarrow$ (5). Let D be a diagonal subuniverse of $\mathbf{L}^2$. If D is contained in $\leq$ or $\geq$ then by (4) $D \in \{\Delta_L, \leq, \geq\}$. Suppose D contains the pairs (a, b) and (c, d) with $a < b$, $c > d$. Then by (4) both

$\leq$ and $\geq$ are contained in D. Now if $a, b \in L$ are arbitrary then $(a, a \wedge b), (a \wedge b, b) \in D$; hence $(a, b) = (a, a \wedge b) \vee (a \wedge b, b) \in D$ and $D = L^2$. It remains to notice that by Lemma 5.3.16 the subuniverse $D \neq \Delta_L$ always contains comparable pairs (a, b) with $a \neq b$.   •

Corollary 5.3.41 *A lattice is locally order functionally complete iff it is simple and locally order affine complete.*

Combining Theorem 5.3.40 with Theorem 5.3.32, Theorem 5.3.34, and Proposition 5.3.25, we also have the following corollaries.

Corollary 5.3.42 *Every simple relatively complemented lattice is locally order functionally complete.*

Corollary 5.3.43 *A modular lattice of finite height is locally order functionally complete iff it is relatively complemented.*

Corollary 5.3.44 *For a lattice* **L** *of finite height the following conditions are equivalent:*

(1) **L** *is locally order functionally complete;*

(2) *the only decreasing* $\vee$-*endomorphisms of* **L** *are the identity function and the constant function 0;*

(3) *the only increasing* $\wedge$-*endomorphisms of* **L** *are the identity function and the constant function 1.*

Note that Corollary 5.3.43 does not remain valid for modular lattices of infinite height. A counter example has been pointed out by E. T. Schmidt ([Schmidt, 1979]). The next proposition provides, for lattices of finite height, a reasonably efficient sufficient condition for local order functional competeness.

Proposition 5.3.45 *Let* **L** *be a simple lattice of finite height. If* $1 \in L$ *is a join of atoms of* **L** *or, dually,* $0 \in L$ *is a meet of coatoms of* **L** *then* **L** *is locally order functionally complete.*

PROOF Suppose that $1 \in L$ is a join of atoms of **L** and take a decreasing $\vee$-endomorphism f of **L** which is not the identity function. Since **L** is of finite height, a certain power g of f is idempotent.

Clearly g is a decreasing $\vee$-endomorphism too; so it determines a tolerance T^g. If $(a,b) \in T^g_{\geq}$ then $g(a) \leq b$ and $g(a) \leq g(b)$ because g is idempotent and order preserving. On the other hand, since $a \geq b$ and g is order preserving, we also have $g(a) \geq g(b)$. We showed that $(a,b) \in T^g_{\geq}$ implies $g(a) = g(b)$. Then obviously $T^g_{\leq}$ is transitive, and so T^g is transitive too; hence T_g is a congruence of $\mathbf{L}$. Since $\mathbf{L}$ is simple, $T^g = \Delta_L$, implying that g is constantly 0. Clearly this is only possible if 0 is the only fixed point of f. In particular, $f(a) = 0$ for all atoms of $\mathbf{L}$. Since f is a $\vee$-endomorphism and 1 is a join of atoms, we have $f(1) = 0$. Hence $\mathbf{L}$ is locally order functionally complete by Corollary 5.3.44. $\bullet$

In view of Proposition 5.3.45 it is natural to ask whether the sufficient conditions given there are also necessary. The following example is taken from [Wille, 1977a].

Example 5.3.46 *A finite order functionally complete lattice such that 1 is not a join of atoms, and 0 is not a meet of coatoms.*

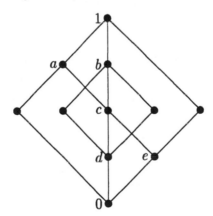

Figure 5.2

We leave the proof of the simplicity of this lattice for the reader. Let f be a decreasing $\vee$-endomorphism of this lattice and assume that f is not the identity map. Then, as proved in Proposition 5.3.45, $f(x) < x$ for every nonzero x. Thus, in particular $f(1) \leq e$, which implies $f(b) \leq e$. Since also $f(b) \leq d$, we have $f(b) \leq b \wedge e = 0$; that is, $f(b) = 0$. The latter together with $f(a) = 0$ yields $f(1) = 0$ and we are done.

So far we have not presented a single result specifically about order functionally complete lattices. The reason is that until very recently the existence problem of infinite order functionally complete lattices remained open. Above, we extended all the known results about finite order functionally complete lattices to infinite locally order functionally complete lattices, in some cases assuming the finiteness of height. It was a long standing conjecture that all order functionally complete lattices are finite. Note that the problem here was considerably more complicated than in case of ordinary functional completeness. Every functionally complete algebra of finite type is finite because $\kappa < 2^\kappa$ for every infinite cardinal κ. It is natural to try to use a simple cardinality argument for the proof of finiteness of order functionally complete lattices. To do that, one has to construct for every infinite simple lattice $\mathbf{L}$ a family of order preserving functions of cardinality greater than $|L|$. It turned out that this is a most nontrivial task and requires heavy machinery of cardinal arithmetic. Since it goes beyond the main focus of this book, we do not present the proofs here but only describe below the main steps of the solution of this important problem.

The first step was made by H. K. Kaiser and N. Sauer. They proved in [Kaiser, Sauer, 1993] that every order functionally complete lattice is bounded and it cannot be countably infinite. It is well known that every infinite lattice $\mathbf{L}$ contains either an infinite chain or an infinite antichain. In both cases Kaiser and Sauer constructed, using the greatest element 1, 2^ω order preserving unary functions on $\mathbf{L}$. Next, M. Haviar and M. Ploščica noticed that for the construction of order preserving functions one can use antichains of $\mathbf{L}^k$ instead of antichains of $\mathbf{L}$. They showed ([Ploščica, Haviar, 1998]) that if $\mathbf{L}^k$, for some natural number k, contains an antichain of cardinality κ, then there exists an n such that $\mathbf{L}^n$ contains an antichain of cardinality 2^κ. Eventually they proved that if there exists an infinite order functionally complete lattice then its cardinality must be greater than each κ_n where $\kappa_0 = \omega$ and $\kappa_{l+1} = 2^{\kappa_l}$, $l = 0, 1, 2, \ldots$.

At the same time M. Goldstern and S. Shelah proved even more. Using the results of [Ploščica, Haviar, 1998] they proved that the cardinality κ of an infinite order functionally complete lattice must be a strong limit cardinal, meaning that for every cardinal α, if $\alpha < \kappa$ then also $2^\alpha < \kappa$. Moreover, using Ramsey type theorems of Erdös and Rado, they proved that κ must be a regular cardinal; that is, it cannot be represented as a sum of less than κ cardinals, each less than κ. Recall that regular strong limit cardinals are referred to as strongly inaccessible cardinals. These results appeared in [Goldstern, Shelah, 1998] in the same issue as the paper [Ploščica, Haviar, 1998]. Finally, in [Goldstern, Shelah, 1999] the problem was completely solved: there are no infinite order functionally complete lattices. It is interesting that Goldstern and Shelah showed that this result holds only in the presence of the axiom of choice: there exist infinite order functionally complete lattices in Zermelo-Fraenkel set theory without the axiom of choice.

5.4 Algebras based on distributive lattices

In this section we consider polynomial completeness properties for two classes of algebras having distributive lattice reducts: Stone algebras and Kleene algebras. There are several important common features of these varieties. First, they are CD and moreover they each have ternary lattice majority term which allows us to invoke condition (5) of Theorem 3.2.2 which asserts that a finite partial function f on an algebra $\mathbf{A}$ has an interpolating polynomial iff all diagonal subuniverses of $\mathbf{A}^2$ are closed under f. Second, each of these varieties is finitely generated, and third, there exist easily described canonical forms for the terms of these varieties.

Stone and Kleene algebras are defined as distributive lattices with an extra unary basic operation. There is a rich general theory of such algebras (see for example [Balbes, Dwinger, 1974] and [Blyth, Varlet, 1994]). Affine completeness of such more general algebras has been studied in [Haviar, 1992] and [Haviar, 1993]. Further, Stone algebras have been generalized in at least two different directions:

1. Distributive lattices $\mathbf{L}$ with two extra unary operations such that one makes $\mathbf{L}$, and the other makes the dual of $\mathbf{L}$, a Stone algebra. These are called double Stone algebras. Affine completeness of double Stone algebras and their generalizations has been studied in [Beazer, 1983], [Haviar, 1995a], and [Haviar, 1996].

2. Not necessarily distributive lattices with an extra unary operation playing the role of pseudocomplementation. These are called p-algebras. Affine completeness problems for them have been considered in [Haviar, 1995b].

5.4.1   Lattices affine complete in a distributive lattice

In what follows we need to handle compatible functions defined on certain sublattices of a given distributive lattice, and we need a criterion for those functions to be restrictions of polynomials of that lattice. For that purpose we introduce the following variation of the affine completeness concept.

Definition Let **B** be a subalgebra of **A**. The algebra **B** is said to be **affine complete in A** if every compatible function of **B** is the restriction to B of some polynomial function of **A**.

The following proposition provides necessary conditions for a lattice to be affine complete in a bounded distributive lattice. Since its proof is very similar to that of the necessity part of Theorem 5.3.12 we leave it for the reader as an exercise. Note that affine completeness of a lattice in a distributive lattice has been studied in detail in [Ploščica, 1996].

Proposition 5.4.1 *Let* **D** *be a sublattice of a bounded distributive lattice* **L**. *If* **D** *is affine complete in* **L** *then the following conditions are satisfied:*

(i) **D** *has no nontrivial Boolean intervals;*

(ii) *if F is an almost principal filter of* **D** *then there exists $a \in L$ such that $F = {\uparrow}a \cap D$;*

(iii) *if I is an almost principal ideal of* **D** *then there exists $a \in L$ such that $I = {\downarrow}a \cap D$.*

When the conditions of Proposition 5.4.1 are satisfied then it is possible to show, modifying the arguments used in the proof of Theorem 5.3.12, that every unary compatible function on **D** is the restriction of some polynomial function of **L** to D. However, as the following example shows, these conditions are not sufficient for **D** to be affine complete in **L**.

Example 5.4.2 ([Ploščica, 1996]) *A bounded distributive lattice* **L** *with a sublattice satisfying the conditions of Proposition 5.4.1 but which is not affine complete in* **L**.

Let $\mathbf{I}$ be the unit interval $[0, 1]$ of real numbers with the usual order,

$$\mathbf{L} = (\mathbf{I} \times \mathbf{I}) \setminus \{(0, 1)\},$$

and

$$\mathbf{D} = \{(x, y) : 0 < x < 1, 0 < y < 1\}.$$

It is not difficult to see that the sublattice $\mathbf{D}$ satisfies the three conditions of Proposition 5.4.1. It obviously has no nontrivial Boolean intervals, and easy calculations show that the other two conditions are fulfilled also. In fact, apart from the principal filters, the lattice $\mathbf{D}$ has almost principal filters $F_a = {\uparrow}(a, 1) \cap D$ and $G_a = {\uparrow}(1, a) \cap D$ where $0 < a < 1$. Almost principal ideals can be described similarly.

However, the lattice $\mathbf{D}$ is not affine complete in $\mathbf{L}$. Consider the binary polynomial function $f(x, y) = y \wedge (x \vee (0, 1))$ of the "closed square" lattice $\mathbf{I} \times \mathbf{I}$. Clearly both the L and D are subuniverses of $\mathbf{I} \times \mathbf{I}$ closed with respect to f. Since distributive lattices enjoy the congruence extension property, the restriction of f to D is a compatible function of the lattice $\mathbf{D}$. However, one easily sees that this function is not a polynomial function of $\mathbf{D}$.

Now we are going to show that in the special case where $\mathbf{D}$ is a filter of $\mathbf{L}$ and $\mathbf{L}$ is bounded, the conditions of Proposition 5.4.1 will be sufficient as well. Certainly in that case the third condition is automatically fulfilled. Our proof is taken from [Haviar, Ploščica, 1995].

Theorem 5.4.3 *Let D be a filter of a bounded distributive lattice $\mathbf{L}$. The lattice $\mathbf{D}$ is affine complete in $\mathbf{L}$ iff the following conditions are satisfied:*

(i) $\mathbf{D}$ *has no nontrivial Boolean intervals;*

(ii) *if F is an almost principal filter of $\mathbf{D}$ then there exists $a \in L$ such that $F = {\uparrow}a \cap D$.*

PROOF Obviously we have to prove only the sufficiency. Let f be an m-ary compatible function on $\mathbf{D}$. Our proof is by induction on m. First let $m = 1$. Since $\mathbf{D}$ has no nontrivial Boolean interval, as in the proof of Theorem 5.3.9, we see that f must preserve the order relation. Consequently Lemma 5.3.11 applies and the set ${\uparrow}f(D)$ is an almost principal filter of $\mathbf{D}$. Hence there exists $a \in L$ such that

$\uparrow f(D) = \uparrow a \cap D$. Now an argument very similar to that used in the proof of Theorem 5.3.12 shows that $f(x)$ is the polynomial function $(a \vee x) \wedge f(1)$.

We proceed with the induction step. Let $m > 1$. For every subset $\alpha \subseteq \underline{m}$ we define a unary function $f_\alpha(x) = f(c_1, \ldots, c_m)$ where

$$c_k = \begin{cases} 1 & \text{if } k \in \alpha, \\ x & \text{if } k \notin \alpha. \end{cases}$$

Obviously every f_α is a compatible function of the lattice **D**. Hence, by the induction hypothesis there exist constants $a_\alpha \in L$ such that

$$f_\alpha(x) = (a_\alpha \vee x) \wedge f(1, \ldots, 1)$$

for every $x \in D$. Note that these constants are not uniquely determined, in general. Any element $a \in L$ such that $\uparrow f_\alpha(D) = \uparrow a \cap D$ can play the role of a_α. In particular, the function $f_{\underline{m}}$ is obviously the constant $f(1, \ldots, 1)$ and we may choose $a_{\underline{m}} = f(1, \ldots, 1)$.

Further, we show that the constants a_α can be chosen so that

$$\alpha \subseteq \beta \quad \Longrightarrow \quad a_\alpha \leq a_\beta. \qquad (5.20)$$

Clearly the condition (5.20) is satisfied for all pairs (α, β) such that $|\alpha| = m$. Suppose it is satisfied for all pairs (α, β) with $|\alpha| > j$. Then it suffices to show that for every α with $|\alpha| = j$, the constant a_α may be replaced by $b_\alpha = \bigwedge_{\alpha \subseteq \beta \subseteq \underline{n}} a_\beta$. Clearly $\uparrow a_\alpha \cap D \subseteq \uparrow b_\alpha \cap D$ and we have to prove the opposite inclusion. Take an arbitrary element $x \in \uparrow b_\alpha \cap D$. Now $x \in D$ implies $x \vee a_\beta \in \uparrow a_\beta \cap D = \uparrow f_\beta(D)$ for every β. Since f is order preserving, $\alpha \subseteq \beta$ implies $\uparrow f_\beta(D) \subseteq \uparrow f_\alpha(D)$. Consequently, $x \vee a_\beta \in \uparrow f_\alpha(D) = \uparrow a_\alpha \cap D$ for every β. Since $\uparrow a_\alpha \cap D$ is a filter of **L**, the meet $\bigwedge_{\alpha \subseteq \beta \subseteq \underline{m}} (x \vee a_\beta)$ is also a member of $\uparrow a_\alpha \cap D$. However, because of $x \geq b_\alpha$ and the distributivity of **L**, this meet is precisely x.

We are going to show that the lattice polynomial

$$p(x_1, \ldots, x_m) = \bigvee_{\beta \subseteq \underline{m}} \left(a_\beta \wedge \bigwedge_{j \in \beta} x_j \right)$$

coincides on D^m with f. As a first step, we show that for every $\alpha \subseteq \underline{m}$, the restriction of p_α to D is precisely f_α. (Here p_α is a unary function which we get if we take p instead of f in the definition of

f_α.) It is obvious that both $p_{\underline{m}}$ and $f_{\underline{m}}$ are the constant functions $a_{\underline{m}} = f(1, \ldots, 1)$. If, however, α is a proper subset of $\underline{m}$ then

$$p_\alpha(x) = (\bigvee_{\beta \subseteq \alpha} a_\beta) \vee (\bigvee_{\gamma \nsubseteq \alpha} (a_\gamma \wedge x)) .$$

In view of condition (5.20), the right hand side of the latter formula is $a_\alpha \vee (a_{\underline{m}} \wedge x) = f_\alpha(x)$.

Now suppose there exists $\mathbf{c} \in D^m$ such that $f(\mathbf{c}) \neq p(\mathbf{c})$. Then there exists a lattice homomorphism $\phi : \mathbf{L} \to \mathbf{D_2}$ which separates $f(\mathbf{c})$ and $p(\mathbf{c})$. Then obviously $\phi(1) = 1$ and $\phi(b) = 0$ for some $b \in D$. Let $\alpha = \{ j \in \underline{m} : \phi(c_j) = 1 \}$ and $\mathbf{b} \in D^m$ have its components defined as follows: $b_j = 1$ if $j \in \alpha$ and $b_j = b$ otherwise. Then $\phi(b_j) = \phi(c_j)$ for every $j \in \underline{m}$; thus the compatibility of f and p yields

$$\phi(f(\mathbf{b})) = \phi(f(\mathbf{c})) \quad \text{and} \quad \phi(p(\mathbf{b})) = \phi(p(\mathbf{c})) .$$

Since $f(\mathbf{b}) = f_\alpha(b) = p_\alpha(b) = p(\mathbf{b})$, we get $\phi(f(\mathbf{c})) = \phi(p(\mathbf{c}))$, a contradiction. ●

5.4.2 Stone algebras

Stone algebras form a variety of algebras which generalize the variety of Boolean algebras. They have a distributive lattice reduct, and a unary operation which can be called a *pseudocomplementation*. More precisely, a **Stone algebra** is defined to be a bounded distributive lattice **L** with a unary operation * such that the following identities are satisfied:

$$x \wedge x^* \approx 0, \quad x \wedge x^{**} \approx x, \quad x^* \vee x^{**} \approx 1 .$$

The distributive lattice reduct of a Stone algebra is called a **Stone lattice**; this is a distinction which parallels that between Boolean algebras and Boolean lattices.

Stone algebras arise from lattice theory and, in fact, a finite distributive lattice **L** becomes a Stone algebra if we take for * the usual pseudocomplementation operation; that is, a^* is the join of all elements $x \in L$ such that $a \wedge x = 0$, and if the equality $x^* \vee x^{**} = 1$ holds for every $x \in L$. It is easy to see that in an arbitrary Stone algebra the same property holds: if $x \wedge y = 0$ then $y \leq x^*$.

Clearly every Boolean algebra is a Stone algebra, the complementation playing the role of *. The simplest nonBoolean Stone algebra is the three element chain $0 < a < 1$ with $0^* = 1$, $a^* = 1^* = 0$. We denote this Stone algebra by $\mathbf{S_3}$. It has a two element subalgebra $\mathbf{B_2}$ with universe $\{0, 1\}$. It is well known that $\mathbf{S_3}$ and $\mathbf{B_2}$ are the only SIs in the variety of Stone algebras. Thus, every Stone algebra can be embedded into a direct power $\mathbf{S_3}^I$ for some index set I. These facts, together with further information, can be found in [Balbes, Dwinger, 1974] and [Blyth, Varlet, 1994].

Let $\mathbf{S}$ be a Stone algebra. An element $b \in S$ is called *closed* if $b = c^*$ for some $c \in S$, and *dense* if $b^* = 0$. The set of all closed elements of $\mathbf{S}$ is a subuniverse of $\mathbf{S}$; in fact it is a Boolean algebra and therefore is denoted by $B(\mathbf{S})$. The set of all dense elements of $\mathbf{S}$ is denoted by $D(\mathbf{S})$. It is a filter in the lattice $\mathbf{S}$.

It is easy to observe that the operation $*$ is a homomorphism of $\mathbf{S}$ onto the dual of $\mathbf{B}(\mathbf{S})$. We shall call the kernel Φ of this homomorphism the **Glivenko congruence** of $\mathbf{S}$. Clearly $D(\mathbf{S})$ is one of the Φ-blocks.

We now list a few further properties of Stone algebras which will be useful in the sequel.

Lemma 5.4.4 *The following identities hold in every Stone algebra:*

$$(x \vee x^*)^* = 0, \qquad (5.21)$$

$$x^{**} \wedge (x \vee x^*) = x. \qquad (5.22)$$

PROOF It suffices to check the identities on $\mathbf{S_3}$. $\quad\bullet$

The formula (5.21) shows that $x \vee x^* \in D(\mathbf{S})$ for every $x \in S$. The formula (5.22) gives an important decomposition law: every element is a meet of two elements, one belonging to $B(\mathbf{S})$ and the other to $D(\mathbf{S})$.

Given a Stone algebra $\mathbf{S}$, if ρ is any congruence of the lattice $\mathbf{D}(\mathbf{S})$ then it can be extended to some congruence σ of the lattice $\mathbf{S}$ (since the variety of distributive lattices has the congruence extension property). Now it is easy to see that $\tau = \Phi \wedge \sigma$ is a congruence of the Stone algebra $\mathbf{S}$, and $\tau|_{D(\mathbf{S})} = \rho$. This implies the following useful fact.

Lemma 5.4.5 *For every Stone algebra $\mathbf{S}$, if f is a compatible function on $\mathbf{S}$ and f preserves $D(\mathbf{S})$ then $f|_{D(\mathbf{S})}$ is a compatible function of the lattice $\mathbf{D}(\mathbf{S})$.*

Now we are going to characterize local polynomial functions on Stone algebras and describe affine complete and locally affine complete Stone algebras. The description of locally affine complete Stone algebras was obtained in [Haviar, 1993] and that of affine complete Stone algebras in [Haviar, Ploščica, 1995]. Earlier, affine complete Stone algebras with a smallest dense element were characterized by R. Beazer ([Beazer, 1982]).

The characterization of local polynomial functions has not appeared in print earlier and this is where we start. Our aim is to obtain an analog of Theorem 5.3.6 for Stone algebras. For that purpose we need to have a relation on any Stone algebra **S** which will play the same role as the order relation in distributive lattices played in Theorem 5.3.6. This relation is the intersection of the Glivenko congruence relation with the lattice order relation on S. It is called the **Glivenko order** of **S** and is denoted by $\leq_G$. Specifically we have

$$b \leq_G c \iff [b^* = c^* \text{ and } b \leq c].$$

In particular, in $\mathbf{S_3}$ the blocks of the Glivenko congruence are $\{0\}$ and $D(\mathbf{S_3}) = \{a, 1\}$. Thus the Glivenko order of $\mathbf{S_3}$ is the subset $\{(0,0), (a,a), (a,1), (1,1)\}$ in $S_3 \times S_3$.

Now we introduce some notation we shall need in the sequel. We consider the set P of all triples $\alpha = (\alpha_0, \alpha_1, \alpha_2)$ where α_0, α_1 and α_2 are disjoint subsets of the set $\underline{m}$. Given any such α, we introduce the Stone term

$$s_\alpha = \left(\bigwedge_{j \in \alpha_0} x_j \right) \wedge \left(\bigwedge_{j \in \alpha_1} x_j^* \right) \wedge \left(\bigwedge_{j \in \alpha_2} x_j^{**} \right).$$

It is easy to see that every m-ary polynomial function on a Stone algebra **S** can be written in the form

$$p(x_1, \ldots, x_m) = \bigvee_{\alpha \in P} (k_\alpha \wedge s_\alpha(x_1, \ldots, x_m))$$

where the k_α are constants from S. Note that we can avoid triples α with intersecting members because of the identities $x \wedge x^* = 0$, $x \wedge x^{**} = x$, $x^* \wedge x^{**} = 0$. Hence every unary polynomial function on **S** has a form:

$$p(x) = k_0 \vee (k_1 \wedge x) \vee (k_2 \wedge x^*) \vee (k_3 \wedge x^{**})$$

with constants $k_0, k_1, k_2, k_3 \in S$.

Lemma 5.4.6 *Let f be a compatible function on a Stone algebra S and assume that f preserves the Glivenko order. If there exists $u \in D(S)$ such that the range of f is contained in $\uparrow u$ then f is a polynomial function of S.*

PROOF Assume that the function f is m-ary. We are going to prove that f coincides with the polynomial function

$$p(\mathbf{x}) = u \vee \bigvee_{\alpha}(k_\alpha \wedge s_\alpha(\mathbf{x}))$$

where α ranges over all members of P with $\alpha_0 \cup \alpha_1 \cup \alpha_2 = \underline{m}$ and the constants are defined as follows: $k_\alpha = f(c_1,\dots,c_m)$ where

$$c_k = \begin{cases} 1 & \text{if } k \in \alpha_0, \\ 0 & \text{if } k \in \alpha_1, \\ u & \text{if } k \in \alpha_2. \end{cases}$$

Thus, we shall prove that $f(\mathbf{b}) = p(\mathbf{b})$ for any $\mathbf{b} \in S^m$. Since f is compatible, it suffices to give the proof just for the case $\mathbf{S} = \mathbf{S_3}$.

If $u = 1$ then both f and p are constantly 1; so they are equal. Since $f(S^m) \subseteq D(\mathbf{S})$, the remaining possibility is $u = a$. Now we distinguish two cases.

1. $f(\mathbf{b}) = 1$. Let $\alpha_0 = \{j : b_j = 1\}$, $\alpha_1 = \{j : b_j = 0\}$, $\alpha_2 = \{j : b_j = a\}$. Then $k_\alpha = f(\mathbf{b}) = 1$ and $s_\alpha(\mathbf{b}) = 1$; hence $p(\mathbf{b}) = 1$.

2. $f(\mathbf{b}) = a$. We have to prove that $k_\alpha \wedge s_\alpha(\mathbf{b}) \leq a$ for every triple α. Suppose the contrary, that is, $k_\alpha = 1 = s_\alpha(\mathbf{b})$ for some α. Then it follows from the definitions of $f(\mathbf{c}) = k_\alpha$ and s_α that $b_j = c_j$ for $j \in \alpha_0 \cup \alpha_1$ and $a = c_j \leq b_j$ if $j \in \alpha_2$. Since f preserves the Glivenko order, this implies $1 = f(\mathbf{c}) \leq f(\mathbf{b}) = a$, a contradiction. ●

Theorem 5.4.7 *A function on a Stone algebra is a local polynomial function iff it preserves the congruences and the Glivenko order.*

PROOF Let f be a local polynomial function on a Stone algebra S. Clearly f preserves the congruences. It is easy to check that the Glivenko order is a diagonal subuniverse in $\mathbf{S} \times \mathbf{S}$; so by Theorem 3.2.2 it is preserved by f.

Now suppose that f is an m-ary compatible function on S which preserves the Glivenko order. We have to prove that f is a local polynomial. By Lemma 5.4.4 f can be decomposed as $f = g \wedge h$ where $g = f^{**}$ and $h = f \vee f^*$. Hence, it suffices to show that g and h are local polynomials.

Since g is compatible and $\mathbf{S}/\Phi$ is Boolean, thus affine complete, there exists a polynomial function q on $\mathbf{S}$ which coincides with g modulo Φ. Then obviously $g = q^{**}$; that is, g is even a (total) polynomial function.

It remains to prove that h is a local polynomial function of $\mathbf{S}$. Let X be a finite subset of S^m. We have to show that h can be interpolated at X by a polynomial function of $\mathbf{S}$. Because we have $h(S^m) \subseteq D(\mathbf{S})$, there exists $u \in D(\mathbf{S})$ such that $h(X) \subseteq \uparrow u$. Taking $h'(\mathbf{x}) = h(\mathbf{x}) \vee u$ we see that the restrictions of h and h' to X coincide. Hence we may assume, without loss of generality, that $h(S^m) \subseteq \uparrow u$. Now our claim follows from Lemma 5.4.6. ●

Having at our disposal a characterization of local polynomial functions, it is not difficult to obtain a description of locally affine complete Stone algebras.

Theorem 5.4.8 *A Stone algebra* **A** *is locally affine complete iff the lattice* **D(S)** *has no nontrivial Boolean intervals.*

PROOF Suppose that the lattice $\mathbf{D(S)}$ has a nontrivial Boolean interval. Then by Proposition 5.3.3 this lattice admits a compatible function g which does not preserve the order. Notice that the function $e(x) = x \vee x^*$ is idempotent and its range is $D(\mathbf{S})$. Thus by Lemma 5.3.1 there exists a compatible function f on $\mathbf{S}$ which extends g. Clearly f does not preserve the Glivenko order; so $\mathbf{S}$ is not locally affine complete.

Now assume that $\mathbf{S}$ has a compatible function f which is not a local polynomial. By Theorem 5.4.7 this function does not preserve the Glivenko order. The argument used in the proof of Theorem 5.3.9 allows us to assume, without loss of generality, that f is unary. By (5.22) $f = g \wedge h$ where $g = f^{**}$ and $h = f \vee f^*$. As in the proof of Theorem 5.4.7, we have that g is a polynomial function. Thus h cannot preserve the Glivenko order, but its restriction to $D(\mathbf{S})$ is a compatible function for the lattice $\mathbf{D(S)}$, by Lemma 5.4.5.

Assume that $\mathbf{S}$ is embedded in some power $\mathbf{S_3}^I$. Now there must exist $i \in I$ such that h_i does not preserve the Glivenko order of the i-th projection $\mathbf{S}_i$ of $\mathbf{S}$. This is only possible if $a \in S_i$ and

$h_i(a) = 1$, $h_i(1) = a$ (because h preserves $D(S)$). Now select $b \in S$ with $b_i = a$ and observe that then also $c_i = a$ where $c = b \vee b^*$. Then obviously $h(c) \not\leq h(1)$, proving that the lattice $D(S)$ is not locally affine complete. Hence, by Theorem 5.3.9, this lattice has a nontrivial Boolean interval.   •

Our next goal is to describe affine complete Stone algebras. We let

$$P_0 = \{\alpha \in P : \alpha_0 = \emptyset, \ \alpha_1 \neq \emptyset, \ \alpha_1 \cup \alpha_2 = \underline{m}\}$$

and for every $\alpha \in P_0$ and an arbitrary m-ary function f on a Stone algebra S, we set $f_\alpha(x_1, \ldots, x_m) = f(y_1, \ldots, y_m)$ where

$$y_k = \begin{cases} 0 & \text{if } k \in \alpha_1, \\ x_k & \text{if } k \in \alpha_2. \end{cases}$$

Thus the function f_α is obtained from f by taking $x_k = 0$ for all $k \in \alpha_1$.

Theorem 5.4.9 *A Stone algebra* **S** *is affine complete iff it satisfies the following two conditions:*

(i) *the lattice* $D(S)$ *does not contain nontrivial Boolean intervals;*

(ii) *if* F *is an almost principal filter of the lattice* $D(S)$ *then there exists* $b \in S$ *such that* $F = \uparrow b \cap D(S)$.

PROOF First assume that **S** is affine complete. Then **S** is locally affine complete and by Theorem 5.4.8 the lattice $D(S)$ does not contain nontrivial Boolean intervals. Let F be an almost principal filter in $D(S)$. By Lemma 5.3.5 the function $f^F(x)$ induced by the filter F is compatible on the lattice $D(S)$. Since $x \vee x^*$ is a compatible unary idempotent function on the Stone algebra **S**, and its range is exactly $D(S)$, by Lemma 5.3.1 the function $f^F(x \vee x^*)$ is a compatible function of **S** as well. Since **S** is affine complete, there exist $k_0, k_1, k_2, k_3 \in S$ such that

$$f^F(x \vee x^*) = k_0 \vee (k_1 \wedge x) \vee (k_2 \wedge x^*) \vee (k_3 \wedge x^{**})$$

for every $x \in S$. In particular, if $x \in D(S)$ then

$$f^F(x) = k_0 \vee (k_1 \wedge x) \vee k_3 = (k_0 \vee k_1 \vee k_3) \wedge (k_0 \vee k_3 \vee x).$$

Since $f^F(1) = 1$, the latter is only possible if $k_0 \vee k_1 \vee k_3 = 1$ and thus $f^F(x) = b \vee x$ for every $x \in D(\mathbf{S})$ where $b = k_0 \vee k_3$. Now, in view of $F = \{x : f^F(x) = x\}$ it is easy to conclude that $F = {\uparrow}b \cap D(\mathbf{S})$.

In order to prove the sufficiency of the two conditions assume that $\mathbf{S}$ is a Stone algebra satisfying these conditions and f is an m-ary compatible function on $\mathbf{S}$. As in the proof of Theorem 5.4.7, the general case reduces to the one with $f(S^m) \subseteq D(\mathbf{S})$. By Lemma 5.4.5 the restriction of f to $D(\mathbf{S})$ is a compatible function of the lattice $\mathbf{D}(\mathbf{S})$. Since by Theorem 5.4.3 the lattice $\mathbf{D}(\mathbf{S})$ is affine complete in the lattice $\mathbf{S}$, there exists a lattice polynomial p of $\mathbf{S}$ which coincides with f on $D(\mathbf{S})$. We continue with induction on m. Clearly f is a polynomial if $m = 0$. Thus assume that $m > 0$ and that all $(m-1)$-ary compatible functions on $\mathbf{S}$ are polynomials. In particular all the functions f_α, $\alpha \in P_0$, are polynomials. Now the proof will be complete as soon as we prove the identity

$$f(\mathbf{x}) = (x_1^{**} \wedge \cdots \wedge x_m^{**} \wedge p(\mathbf{x})) \vee \bigwedge_{\alpha \in P_0} (s_\alpha(\mathbf{x}) \wedge f_\alpha(\mathbf{x})). \qquad (5.23)$$

If $\mathbf{x} \in D(\mathbf{S})^m$ then $x_k^{**} = 1$, $k = 1, \ldots, m$, and all $s_\alpha(\mathbf{x})$ are zero because of $\alpha_1 \neq \emptyset$. Hence the right hand side of the formula (5.23) is $p(\mathbf{x})$ which is equal to $f(\mathbf{x})$ in view of the choice of p. If $\mathbf{x} \notin D(\mathbf{S})^m$ then we have to use the embedding $\mathbf{S} < \mathbf{S}_3^I$; in fact this embedding allows us to just replace an arbitrary algebra $\mathbf{S}$ by $\mathbf{S}_3$. Let $\beta_1 = \{k : x_k = 0\}$ and $\beta_2 = \{k : x_k \neq 0\}$. If $\beta_2 = \underline{m}$ then $\mathbf{x} \in D(\mathbf{S})^m$. Therefore we may assume that $\beta_2 \neq \underline{m}$ which easily implies that both sides of the equality (5.23) are $f_\beta(\mathbf{x})$.   •

Corollary 5.4.10 ([Beazer, 1982]) *Let $\mathbf{S}$ be a Stone algebra such that the lattice $\mathbf{D}(\mathbf{S})$ has a smallest element. Then $\mathbf{S}$ is affine complete iff $\mathbf{D}(\mathbf{S})$ has no nontrivial Boolean intervals.*

Having characterized locally affine complete and affine complete Stone algebras we finally observe that the only strictly locally affine complete Stone algebras are the Boolean algebras. To see this note that if $\mathbf{S}$ is not Boolean then the set $D(\mathbf{S})$ of dense elements (which is also a block of the Glivenko congruence) contains at least two elements, for otherwise the decomposition 5.22 implies that $x \vee x^* = 1$ which implies that $\mathbf{S}$ is Boolean. We may also suppose that this pair of elements is comparable. Then the function which interchanges them is a finite partial compatible function which cannot be interpolated by any Stone polynomial.

We conclude with two examples.

Example 5.4.11 ([Haviar, Ploščica, 1995]) *A Stone algebra which is affine complete and whose lattice of dense elements is not affine complete.*

Let $\mathbf{I}$ be a unit interval of real numbers with the pseudocomplementation defined by $x^* = 0$ for all $x > 0$. Let $\mathbf{S} = \mathbf{I}^2$. Clearly $D(\mathbf{S}) = \{(x,y) : x > 0, y > 0\}$ has no nontrivial Boolean intervals. It is not difficult to check that all almost principal filters of $\mathbf{D}(\mathbf{S})$ have the form $\uparrow a \cap D(\mathbf{S})$ for some $a \in S$; so $\mathbf{S}$ is affine complete. However, for example $\uparrow(0,1) \cap D(\mathbf{S})$ is an almost principal filter of $\mathbf{D}(\mathbf{S})$ which is not a principal filter. Thus the lattice $\mathbf{D}(\mathbf{S})$ is not affine complete.

Example 5.4.12 ([Haviar, Ploščica, 1995]) *A locally affine complete Stone algebra which is not affine complete.*

Let $\mathbf{R}$ be a real line and $\mathbf{L} = \{0\} \cup \mathbf{R} \times \mathbf{R} \cup \{1\}$, with the natural order relation. Let $\mathbf{S}$ be the Stone algebra whose lattice reduct is $\mathbf{L}$ and * be defined so that $S = \{0\} \cup D(\mathbf{S})$. Then obviously $\mathbf{D}(\mathbf{S})$ has no nontrivial Boolean intervals; so $\mathbf{S}$ is locally affine complete. However, the subset $F = \{(x,y) : x \geq 0\} \cup \{1\}$ is an almost principal filter of $\mathbf{D}(\mathbf{S})$ which cannot be represented in the form $\uparrow a \cap D(\mathbf{S})$ for some $a \in S$. Thus $\mathbf{S}$ is not affine complete.

5.4.3   Kleene algebras

Kleene algebras form another variety of algebras generalizing Boolean algebras and whose origin lies in logic. We recall the definition and some terminology for Kleene algebras. For more information see [Balbes, Dwinger, 1974]. By definition a **Kleene algebra** is a bounded distributive lattice with an additional unary operation (complementation) denoted by $'$ and satisfying the identities:

$$(x \vee y)' = x' \wedge y' , (x \wedge x') \vee (y \vee y') = y \vee y' , 0' = 1 , x'' = x$$

and their duals. Clearly every Boolean algebra is a Kleene algebra. A special Kleene algebra which is not Boolean is

$$\mathbf{K_3} = \langle \{0, a, 1\}; \vee, \wedge, ', 0, 1 \rangle$$

with $0 < a < 1$ and $a' = a$. It is known that the variety of Kleene algebras is generated by $\mathbf{K_3}$. Moreover, $\mathbf{K_3}$ and its subalgebra $\mathbf{B_2}$ with universe $\{0, 1\}$ are the only subdirectly irreducible Kleene algebras. Hence every Kleene algebra $\mathbf{K}$ can be embedded into a suitable power $\mathbf{K_3}^I$.

In what follows we need canonical forms for Kleene polynomials, that is, polynomial functions on Kleene algebras. Let $\alpha = (\alpha_0, \alpha_1)$ where $\alpha_0, \alpha_1 \subseteq \underline{m}$. We assign to every such pair α the m-ary Kleene term

$$e_\alpha(x_1, \ldots, x_m) = (\bigvee_{i \in \alpha_0} x_i) \vee (\bigvee_{j \in \alpha_1} x_j').$$

It follows easily from the axioms of Kleene algebras that every m-ary Kleene polynomial on $\mathbf{K}$ can be represented as a meet of *elementary polynomials* $k_\alpha \vee e_\alpha(x_1, \ldots, x_m)$ where k_α are constants from K. In particular, if $m = 1$ then, identifying $\emptyset = 0$, $\underline{1} = 1$, and writing ij instead of (i, j), we have the following canonical form of a unary Kleene polynomial:

$$p(x) = k_{00} \wedge (k_{10} \vee x) \wedge (k_{01} \vee x') \wedge (k_{11} \vee x \vee x').$$

For every Kleene algebra $\mathbf{K}$ we let $K^\vee = \{x \vee x' | x \in K\}$. The subset $K^\wedge$ is defined dually. It is easy to check that $K^\vee$ and $K^\wedge$ are a filter and an ideal of the distributive lattice reduct of $\mathbf{K}$, respectively. Obviously the complementation operation induces an antiisomorphism between them. Note that the union $K^\vee \cup K^\wedge$ is always a subuniverse of $\mathbf{K}$ and every congruence of the lattice $K^\vee$ extends in a natural way to a congruence of this Kleene subalgebra. It is known ([Berman, 1977]) that the variety of Kleene algebras has the congruence extension property. Hence we have the following important lemma.

Lemma 5.4.13 *For every Kleene algebra $\mathbf{K}$, if f is a compatible function on $\mathbf{K}$ and f preserves $K^\vee$ then $f|_{K^\vee}$ is a compatible function of the lattice $\mathbf{K}^\vee$.*

Now we define a binary relation on an arbitrary Kleene algebra **K** which turns out to be of great importance. First, for every $k \in K$ we let

$$x \sqsubseteq_k y \quad \Longleftrightarrow \quad x \wedge k \leq y \leq x \vee k'.$$

It is easy to check that all these relations are reflexive and transitive.

The **uncertainty order** of the Kleene algebra **K** is the join of all of the relations $\sqsubseteq_s$ where $s \in K^\vee$. It will be denoted by the symbol $\sqsubseteq$. Thus

$$x \sqsubseteq y \quad \Longleftrightarrow \quad x \wedge s \leq y \leq x \vee s' \quad \text{for some} \quad s \in K^\vee.$$

Note that in case $\mathbf{K} = \mathbf{K_3}$ the uncertainty order is the following relation:

$$\{(0,0),(a,a),(1,1),(0,a),(1,a)\}.$$

If **K** is Boolean then $K^\vee = \{1\}$ and $K^\wedge = \{0\}$. Hence, in the case of Boolean algebras the uncertainty order becomes the equality relation.

As was done in the case of Stone algebras, we shall characterize local polynomial functions, and the classes of affine complete and locally affine complete Kleene algebras. These results originally appeared in [Haviar, Kaarli, Ploščica, 1997]. For a slightly different approach see [Haviar, Ploščica, 1997]. Some earlier partial results appeared in [Haviar, 1993].

Our first goal is to establish an analog of Theorem 5.3.6 for Kleene algebras where the usual order relation is replaced by the uncertainty order. For that purpose we need some lemmas. Given a Kleene algebra **K** and $s \in K^\vee$, we denote

$$K^s = \{x \in K \,|\, x \vee x' \geq s\}.$$

Lemma 5.4.14 *For every $s \in K^\vee$, if two m-ary compatible functions of* **K** *coincide on* $\{0,s,1\}^m$ *then they coincide on all of* $(K^s)^m$.

PROOF Let f and g be m-ary compatible functions on a Kleene algebra **K** and assume that their restrictions to $\{0,s,1\}^m$ coincide. Suppose we have an embedding $\mathbf{K} < \mathbf{K_3}^I$ and denote the projection map from $\mathbf{K_3}^I$ onto the i-th direct factor by p_i. Clearly it suffices to show that the functions f_i and g_i, induced by f and g, respectively, on $p_i(K)$ coincide on $p_i(K^s)$ for every $i \in I$. The latter directly follows from our assumption as soon as we see that $p_i(K^s) = p_i(\{0,s,1\})$. This equality is obvious if $p_i(K^s) = \{0,1\}$. If, however, $a \in p_i(K^s)$ then it is easy to see that $p_i(s) = a$ and we are done.   ●

Lemma 5.4.15 *Let* **K** *be embedded in* $\mathbf{K_3}^I$. *Then for every pair of elements* $u, v \in K$,

$$u \sqsubseteq v \quad \Longleftrightarrow \quad (\forall i \in I) \; [u_i \sqsubseteq v_i].$$

PROOF If $u \sqsubseteq v$ then trivially $u_i \sqsubseteq v_i$. Now suppose that $u_i \sqsubseteq v_i$ holds for every $i \in I$. Then put $s = (u \vee u') \wedge (v \vee v')$ and observe that $s \in K^\vee$ and $u \wedge s \leq v \leq u \vee s'$. •

The importance of Lemma 5.4.15 is that many questions concerning compatible functions which preserve the uncertainty order are reduced to the case of the algebra $\mathbf{K_3}$. In this respect the following lemma will be useful.

Lemma 5.4.16 *Let* $f : K_3^m \to K_3$ *be a function which preserves the uncertainty order. If* $f(a, \ldots, a) \in \{0, 1\}$ *then* f *is constant.*

PROOF This is true because

$$x \sqsubseteq 0 \quad \Longrightarrow \quad x = 0, \qquad x \sqsubseteq 1 \quad \Longrightarrow \quad x = 1,$$

for each $x \in K_3$. •

Lemma 5.4.17 *Let* f *be an m-ary compatible function on a Kleene algebra* **K**. *The function* f *preserves the uncertainty order iff it preserves all relations* $\sqsubseteq_k$ *where* $k \in K$.

PROOF The sufficiency is obvious, so assume that f preserves the uncertainty order. Since all the relations $\sqsubseteq_k$ are transitive, the same argument as used in the proof of Theorem 5.3.9 allows us to reduce the problem to the case of unary function.

Let $b, c \in K$, $f(b) = u$, $f(c) = v$, and $b \sqsubseteq_k c$. Suppose **K** is embedded in $\mathbf{K_3}^I$. Then $k_i \wedge b_i \leq c_i \leq b_i \vee k_i'$ for every $i \in I$. Now we consider the cases $k_i = 1, 0, a$.

If $k_i = 1$ then $b_i = c_i$ and because of the compatibility of f, $u_i = f_i(b_i) = f_i(c_i) = v_i$, hence $k_i \wedge u_i \leq v_i \leq u_i \vee k_i'$. If $k_i = 0$ then obviously $k_i \wedge u_i \leq v_i \leq u_i \vee k_i'$.

The case $k_i = a$ is more complicated and only here do we need the assumption that f preserves the uncertainty order. Since $b \sqsubseteq_k c$, we have $b_i = c_i$ or $c_i = a$. In the first case $u_i = v_i$ and we are done. If $(b_i, c_i) = (0, a)$ then put $d = c \wedge c'$ and observe that $0 \sqsubseteq d$. Hence

$f(0) \sqsubseteq f(d)$ which implies $u_i = f_i(b_i) \sqsubseteq f_i(c_i) = v_i$. The latter, however, is equivalent to $u_i \sqsubseteq_a v_i$. Finally, if $(b_i, c_i) = (1, a)$ then take $d = c \vee c'$, observe that $1 \sqsubseteq d$, and again get $u_i \sqsubseteq_a v_i$.   •

Theorem 5.4.18 *A function on a Kleene algebra is a local polynomial function iff it preserves the congruences and the uncertainty order.*

PROOF The congruences of a Kleene algebra **K** together with the uncertainty order are the subuniverses of $\mathbf{K}^2$ containing the diagonal. Hence by Theorem 3.2.2 they are preserved by all local polynomial functions of **K**.

For the converse, suppose that f is an m-ary compatible function of **K** which preserves the uncertainty order. We have to show that, given any finite subset $X \subseteq K^m$, there exists an m-ary polynomial p of **K** such that $f|_X = p|_X$. Since every finite subset of K is contained in some K^s, $s \in K^\vee$, by Lemma 5.4.14 we may restrict to the case $X = (\{0, s, 1\})^m$. Moreover, since **K** has a ternary lattice majority term, by part (3) of Theorem 3.2.2 it suffices to find interpolating polynomials separately for every two element subset $\{\mathbf{b}, \mathbf{c}\} \subseteq (\{0, s, 1\})^m$.

To do this first observe that the polynomial function

$$p(x) = (x \vee v) \wedge (v \vee u) \wedge (u \vee x') \tag{5.24}$$

maps 1 and 0 to arbitrarily chosen elements u and v of K. Hence the interpolating polynomial exists whenever 1 and 0 occur as the j-th coordinates of the vectors **b** and **c**. This allows us to assume in the sequel that if $b_j \neq c_j$ then $\{b_j, c_j\} = \{0, s\}$ or $\{b_j, c_j\} = \{1, s\}$. We let $f(\mathbf{b}) = u$, $f(\mathbf{c}) = v$ and, for the beginning, consider the following two cases:

1. For every j, either $b_j = c_j$ or $(b_j, c_j) = (1, s)$, and there exists j such that $(b_j, c_j) = (1, s)$. Now $\mathbf{b} \sqsubseteq_s \mathbf{c}$ and, by Lemma 5.4.17, $u \sqsubseteq_s v$. Straightforward calculations show that the polynomial (5.24) works again, that is, $p(1) = u$ and $p(s) = v$.

2. For every j, if $b_j \neq c_j$, then $(b_j, c_j) = (0, s)$ or $(b_j, c_j) = (1, s)$, and there exists j such that $(b_j, c_j) = (0, s)$. Then $\mathbf{b} \sqsubseteq_{s'} \mathbf{c}$ and by Lemma 5.4.17 $u \sqsubseteq_{s'} v$. Now one can check that the polynomial $q(x) = p(x')$ works, that is, $q(0) = u$ and $q(s) = v$.

The remaining cases are more complicated. We go through in detail the specific case where for every j, we have either $b_j = c_j$ or $(b_j, c_j) \in \{(1, s), (0, s), (s, 1)\}$, and $(0, s)$ and $(s, 1)$ actually do occur as the coordinate pairs of **b** and **c**. The other cases can be handled similarly.

Let us set $\mathbf{d} = (d_1, \ldots, d_m)$, where

$$d_j = \left\{ \begin{array}{ll} b_j & \text{if } b_j = c_j \text{ or } c_j = s, \\ c_j & \text{otherwise,} \end{array} \right.$$

and let $f(\mathbf{d}) = w$. Then $\mathbf{d} \sqsubseteq_s \mathbf{b}$ and $\mathbf{d} \sqsubseteq_{s'} \mathbf{c}$; thus by Lemma 5.4.17 $w \sqsubseteq_s u$ and $w \sqsubseteq_{s'} v$. Clearly it suffices to construct a binary polynomial q such that $q(0, s) = u$ and $q(s, 1) = v$. We define

$$\begin{array}{lll} q_1(x, y) & = & (x \vee u \vee w) \wedge (y \vee u) \wedge v, \\ q_2(x, y) & = & (y' \vee v \vee w) \wedge (x' \vee v) \wedge u \end{array}$$

and then put $q = q_1 \vee q_2$. It is easy to check that q is the required polynomial.

Handling the other cases similarly completes the proof. •

Next we describe locally affine complete Kleene algebras.

Theorem 5.4.19 *A Kleene algebra* **K** *is locally affine complete iff the lattice* $\mathbf{K}^{\vee}$ *does not contain nontrivial Boolean intervals.*

Remark Of course, in view of Theorem 5.3.9 and the antiisomorphism between the lattices $\mathbf{K}^{\vee}$ and $\mathbf{K}^{\wedge}$, the condition of the theorem is equivalent to any of the following three: (1) the lattice $\mathbf{K}^{\vee}$ is locally affine complete; (2) the lattice $\mathbf{K}^{\wedge}$ does not contain nontrivial Boolean intervals; (3) the lattice $\mathbf{K}^{\wedge}$ is locally affine complete.

PROOF Suppose $\mathbf{K}^{\vee}$ contains a nontrivial Boolean interval. Then by Theorem 5.3.9 the lattice $\mathbf{K}^{\vee}$ has a compatible function $g(x)$ which does not preserve the order relation. Let $f : K \rightarrow K$ be defined via $f(x) = g(x \vee x')$. Obviously the restriction of f to $K^{\vee}$ coincides with g, and it easily follows from Lemma 5.3.1 that f is a compatible function of **K**. Since the restriction of the uncertainty order to $K^{\vee}$ coincides with the usual order relation of the lattice $\mathbf{K}^{\vee}$, we see that f does not preserve the uncertainty order. Thus, by Theorem 5.4.18, f is not a local polynomial function.

For the converse suppose that $\mathbf{K}$ has a compatible function f which is not a local polynomial. This means, by Theorem 5.4.18, that f does not preserve the uncertainty order. We are going to prove that then there exists a compatible function on the lattice $\mathbf{K}^\vee$ which does not preserve the order relation. In view of Theorem 5.3.9 this implies that $\mathbf{K}^\vee$ has a nontrivial Boolean interval.

As in the proof of Theorem 5.3.9, we may assume that f is unary. We also assume that $\mathbf{K} < \mathbf{K_3}^I$ for some index set I. By Lemma 5.4.15, if $b, c \in K$ are such that $b \sqsubseteq c$ but $f(b) \not\sqsubseteq f(c)$ then there must exist $i \in I$ such that $b_i \sqsubseteq c_i$ but $f_i(b_i) \not\sqsubseteq f_i(c_i)$. This is possible only if

$$b_i \in \{0, 1\}, \quad c_i = a, \quad f_i(a) \in \{0, 1\}.$$

We may assume that $f_i(a) = 1$ because otherwise we could replace the function $f(x)$ by $f(x)'$. Since f_i cannot be constant, either $f_i(1) \neq 1$ or $f_i(0) \neq 1$. We may assume $f_i(1) \neq 1$ because otherwise we could replace $f(x)$ by $f(x')$.

Let $s = c \vee c'$ and consider the function $g(x) = f(x) \vee s$. Clearly g is a compatible function of the Kleene algebra $\mathbf{K}$ and it preserves $K^\vee$. By Lemma 5.4.13 the restriction of g to $K^\vee$ is compatible for the lattice $\mathbf{K}^\vee$. However, the function $h = g|_{K^\vee}$ is not order preserving. Indeed,

$$h_i(1) = f_i(1) \vee s_i = f_i(1) \vee a = a$$

and

$$h_i(a) = f_i(a) \vee a = 1 \vee a = 1.$$

Consequently $h(1) \not\leq h(s)$ and we are done.   ●

Now we describe affine complete Kleene algebras.

Theorem 5.4.20 *A Kleene algebra $\mathbf{K}$ is affine complete iff it satisfies the following two conditions:*

(i) *the lattice $\mathbf{K}^\vee$ has no nontrivial Boolean intervals;*

(ii) *for every almost principal filter D in the lattice $\mathbf{K}^\vee$, there exists $b \in K$ such that $D = {\uparrow}b \cap K^\vee$.*

PROOF If $\mathbf{K}$ is affine complete then it is locally affine complete and by Theorem 5.4.19 $\mathbf{K}^\vee$ has no nontrivial Boolean intervals. Let D be an almost principal filter in $\mathbf{K}^\vee$ and, referring to the definition preceding Lemma 5.3.5, consider the function $f^D : K^\vee \to K^\vee$. We know, by Lemma 5.3.5, that f^D is a compatible function on the lattice $\mathbf{K}^\vee$. Hence it follows from Lemma 5.3.1 that the function $g(x) = f^D(x \vee x')$ is a compatible function on $\mathbf{K}$. If $\mathbf{K}$ is affine complete then there must exist constants $k_{ij} \in K$ such that

$$g(x) = k_{00} \wedge (k_{10} \vee x) \wedge (k_{01} \vee x') \wedge (k_{11} \vee x \vee x')$$

for every $x \in K$. Choosing $x = 1$ we have $1 = k_{00} \wedge k_{01}$, implying $k_{00} = k_{01} = 1$. Hence

$$g(x) = (k_{10} \vee x) \wedge (k_{11} \vee x \vee x')$$

for every $x \in K$. If $x \in K^\vee$ then $x' \leq x$ and $g(x) = f^D(x)$. Therefore

$$f^D(x) = (k_{10} \vee x) \wedge (k_{11} \vee x) = (k_{10} \wedge k_{11}) \vee x$$

for every $x \in K^\vee$, implying $D = {\uparrow}b \cap K^\vee$ where $b = k_{10} \wedge k_{11}$. This proves the necessity part of the theorem.

Now we prove the sufficiency of the two conditions. We assume that $\mathbf{K}$ is embedded in $\mathbf{K_3}^I$ and denote the i-th projection mapping by π_i, $i \in I$. Then every $\pi_i(\mathbf{K})$ is either $\mathbf{K_3}$ or $\mathbf{B_2}$. Let f be an m-ary compatible function on $\mathbf{K}$. Our aim is to find a finite set P_2 of polynomial functions on $\mathbf{K}$ such that for every two m-tuples $\mathbf{k} = (k^1, \ldots, k^m)$, $\mathbf{l} = (l^1, \ldots, l^m)$ in $K_3{}^m$ there exists $p \in P_2$ such that $p_i(\mathbf{k}) = f_i(\mathbf{k})$ and $p_i(\mathbf{l}) = f_i(\mathbf{l})$ for all $i \in I$ for which it makes sense. The latter means that if a occurs among coordinates of $\mathbf{k}$ or $\mathbf{l}$ then we must have $\pi_i(\mathbf{K}) = \mathbf{K_3}$. In fact, if we have such a set P_2 then the rest is easy. Using the majority term $u(x,y,z)$ we shall define

$$P_{k+1} = \{u(p_1,p_2,p_3) : p_1, p_2, p_3 \in P_k\}, \quad k \geq 2.$$

Then every set of polynomials P_k has the property that given an arbitrary k-element subset X of $K_3{}^m$, there exists $p \in P_k$ such that $p_i(\mathbf{x}) = f_i(\mathbf{x})$ for every $\mathbf{x} \in X$ and every $i \in I$ for which it makes sense. Since $|K_3{}^m| = 3^m$, the set P_{3^m} consists of a single polynomial function which must coincide with f.

We start the construction of the polynomials which form the set P_2, as we shall see later. Let $\alpha = (\alpha_1, \alpha_2, \alpha_3)$ be a triple of disjoint subsets of $\underline{m}$ whose union is $\underline{m}$. We allow some of α_i to be empty; therefore we cannot call α a partition of $\underline{m}$. With every such triple

α we associate a subset $S_\alpha \subseteq K^m$ consisting of all of the m-tuples $\mathbf{b} = (b^1, \ldots, b^m)$ such that

$$b^j \in \{0, 1\} \quad \text{if} \quad j \in \alpha_1,$$
$$b^j \in K^\vee \quad \text{if} \quad j \in \alpha_2,$$
$$b^j \in K^\wedge \quad \text{if} \quad j \in \alpha_3.$$

We are going to find, for every α, a polynomial p_α such that $f|_{S_\alpha} = p_\alpha|_{S_\alpha}$. First consider the case $\alpha_1 = \emptyset$. Now, because the complementation $'$ is an antiisomorphism between the lattices $K^\vee$ and $K^\wedge$, we may assume without loss of generality that $\alpha_3 = \emptyset$, that is, $S_\alpha = (K^\vee)^m$. We show that a suitable polynomial will be the following:

$$
\begin{aligned}
p(x_1, \ldots, x_m) \;=\; & (f(1, \ldots, 1) \wedge (f(x_1, \ldots, x_m) \vee x_1 \vee \cdots \vee x_m)) \\
\vee \;& (f(x_1', x_2, \ldots, x_m) \wedge x_1') \\
& \cdots\cdots\cdots\cdots\cdots \\
\vee \;& (f(x_1, \ldots, x_{m-1}, x_m') \wedge x_m').
\end{aligned}
$$

That p is a polynomial follows from Theorem 5.4.3. Indeed, the function

$$f(x_1, \ldots, x_m) \vee x_1 \vee \cdots \vee x_m$$

preserves the filter $K^\vee$ and its restriction to $K^\vee$ is compatible. Since, by Theorem 5.4.3, the lattice $K^\vee$ is affine complete in the lattice reduct of K, the restriction of this function to $K^\vee$ must be a lattice polynomial of K. Similarly, the restrictions of all functions

$$g_j(x_1, \ldots, x_m) = (f(x_1, \ldots, x_{j-1}, x_j', x_{j+1}, \ldots, x_m) \wedge x_j')',$$

$j = 1, \ldots, m$, are lattice polynomials of K and then the same applies to the functions

$$g_j(x_1, \ldots, x_m)' = f(x_1, \ldots, x_{j-1}, x_j', x_{j+1}, \ldots, x_m) \wedge x_j'.$$

We have to show that $f(\mathbf{b}) = p(\mathbf{b})$ whenever $\mathbf{b} \in (K^\vee)^m$. Keeping in mind the embedding $\mathbf{K} < \mathbf{K_3}^I$, it suffices to consider the case $\mathbf{K} = \mathbf{K_3}$ and $b^j \in \{a, 1\}$, $j = 1, \ldots, m$. Now the equality $f(\mathbf{b}) = p(\mathbf{b})$ is obvious if $\mathbf{b} = (1, \ldots, 1)$. If both a and 1 occur among b^j, we have

$$p(\mathbf{b}) = (f(1, \ldots, 1) \wedge f(\mathbf{b})) \vee (f(\mathbf{b}) \wedge a).$$

If $f(\mathbf{b}) \in \{0, 1\}$ then, by Lemma 5.4.16, $f(1, \ldots, 1) = f(\mathbf{b})$, yielding $p(\mathbf{b}) = f(\mathbf{b}) \vee (f(\mathbf{b}) \wedge a) = f(\mathbf{b})$. If however, $f(\mathbf{b}) = a$, then

$$p(\mathbf{b}) = (f(1, \ldots, 1) \wedge a) \vee a = a = f(\mathbf{b}) .$$

The remaining case $\mathbf{b} = (a, \ldots, a)$ is similar to the preceding one. Again it is necessary to distinguish the cases $f(\mathbf{b}) \neq a$ and $f(\mathbf{b}) = a$.

Finally consider the subsets S_α corresponding to the triples α with $\alpha_1 \neq \emptyset$. Now p_α is constructed by induction on the size of α_1. Suppose, without loss of generality, that $m \in \alpha_1$ and take polynomials q_0 and q_1 such that

$$q_k(\mathbf{b}) = f(\mathbf{b}) \quad \text{if} \quad \mathbf{b} \in S_\alpha \text{ and } b^m = k .$$

Note that the polynomials q_0 and q_1 exist by the induction hypothesis. Now define

$$q(\mathbf{x}) = (q_0(\mathbf{x}) \wedge x'_m) \vee (q_1(\mathbf{x}) \wedge x_m) .$$

It is easy to see that $q(\mathbf{b}) = f(\mathbf{b})$ for every $\mathbf{b} \in S_\alpha$.

It remains to observe that the set of all polynomials p_α has the property that P_2 must satisfy. We take $\mathbf{k}, \mathbf{l} \in K_3{}^m$ and define a triple α as follows:

$$j \in \alpha_1 \quad \text{if} \quad k^j, l^j \in \{0, 1\},$$
$$j \in \alpha_2 \quad \text{if} \quad a \in \{k^j, l^j\} \text{ but } 0 \notin \{k^j, l^j\},$$
$$j \in \alpha_3 \quad \text{if} \quad \{k^j, l^j\} = \{0, a\}.$$

Next, take $\mathbf{b}, \mathbf{c} \in K^m$ such that $\mathbf{b}_i = \mathbf{k}$ and $\mathbf{c}_i = \mathbf{l}$. Note that such m-tuples $\mathbf{b}$ and $\mathbf{c}$ exist in S_α. Indeed, for $j \in \alpha_1$ we simply take b^j and c^j in $\{0, 1\}$ so that they agree with k^j and l^j, respectively. For $j \in \alpha_2$ take some b^j and c^j with their i-th projections k^j and l^j respectively, and then replace them by $b^j \vee (b^j)'$ and $c^j \vee (c^j)'$, respectively. The case $j \in \alpha_3$ is dual to $j \in \alpha_2$. Now clearly $(p_\alpha)_i(\mathbf{k}) = f_i(\mathbf{k})$ and $(p_\alpha)_i(\mathbf{l}) = f_i(\mathbf{l})$. This proves the theorem. ●

Now we have characterized both the locally affine complete and affine complete Kleene algebras. An argument like that for Stone algebras shows that the only strictly locally affine complete Kleene algebras are Boolean. We leave this as an exercise.

Exercise

Show that the only strictly locally affine complete Kleene algebras are the Boolean algebras.

5.5   Semilattices

In this section we completely describe affine complete and locally affine complete semilattices. These results deserve attention because the variety of semilattices is the only nontrivial one for which such results are known and which has the property that no nontrivial lattice equation holds in the congruence lattices throughout the variety. The trivial example is the variety of sets. (See [Hobby, McKenzie, 1988] where it is, however, shown that a nontrivial congruence equation, i.e., involving ∘ as well as ∨ and ∧, does hold in all congruence lattices of semilattices.) The paper [Kaarli, Márki, Schmidt 1983] is the source of most of the material contained in this section.

5.5.1   Compatible functions on semilattices

Throughout the section meet semilattices are considered. We recall some elementary concepts of semilattice theory. An *ideal of a semi-lattice* **S** is a nonempty subset $I \subseteq S$ such that for all $x \in I$ and $y \in S$, $y \leq x$ implies $y \in I$. A *filter of a semilattice* is defined exactly as in the case of lattices. A subset A of semilattice **S** is said to be *inductive* if any two elements of A have a common upper bound in A. *Principal ideals* and *almost principal ideals* of semilattices are defined exactly as in lattice theory. Also, as in the lattice case, an almost principal ideal I of a semilattice **S** defines a function $f_I : S \to S$ such that $\downarrow f(a) = \downarrow a \cap I$ for every $a \in S$.

     Clearly, a function f on a semilattice **S** is a polynomial of **S** iff it has the form
$$f(x_1, \ldots, x_m) = a \wedge x_1 \wedge \cdots \wedge x_m$$
where a is either the empty symbol or an element of S and the set of variables may also be empty. In the following lemma we list basic properties of local polynomial functions of semilattices; all are easy consequences of this description.

Lemma 5.5.1 *Let f be an m-ary local polynomial function on a semilattice* **S**. *Then:*

  (i) *f is order preserving;*

  (ii) *f is a **contraction**, that is, $f(a_1, \ldots, a_m) \leq a_i$ for every $a_1, \ldots, a_m \in S$ whenever f depends on x_i;*

(iii) *if $m = 1$ then f is idempotent;*

(iv) *if f is not constant then $f(S^m)$ is an ideal of* **S**.

It is easy to observe that $\text{Cg}(a, b) = \text{Cg}(a \wedge b, a) \vee \text{Cg}(a \wedge b, b)$ for any elements a and b of a semilattice S. This implies that a unary function f on S is compatible with Con **S** iff it is compatible with all principal congruences $\text{Cg}(a, b)$ of **S** such that $a < b$. The following lemma describes these congruences. Its straightforward proof is left to the reader.

Lemma 5.5.2 *Let $a < b$ in a semilattice* **S**. *Then*

$$\text{Cg}(a, b) = \{(c, d) \in S^2 : c = d \text{ or } c, d \leq b \text{ and } a \wedge c = a \wedge d\}.$$

Now we show that every almost principal ideal of a semilattice gives rise to a unary compatible function.

Lemma 5.5.3 *Let I be an almost principal ideal of a semilattice* **S**. *Then:*

(i) *f_I is an idempotent, order preserving, compatible contraction on* **S** *and I is the set of all fixed points of f_I;*

(ii) *f_I is a local polynomial function iff $I = S$ or I is inductive;*

(iii) *f_I is a polynomial function iff $I = S$ or I is a principal ideal.*

PROOF (i) Clearly f_I is an order preserving contraction and I is the set of its fixed points. It is easy to show, using Lemma 5.5.2, that f_I is a compatible function of **S**.

(ii) If $I = S$ then obviously f_I is the identity function. Now let I be a proper ideal of **S** such that f_I is a local polynomial function. Take a finite subset $A \subseteq I$ and $b \in S \setminus I$. Then there exists a unary polynomial p of **S** which interpolates f at $A \cup \{b\}$. Since $f(b) \neq b$ and $f(a) = a$ for every $a \in A$, the polynomial p must have the form $s \wedge x$ for some $s \in S$. Now we see that s is an upper bound of A in **S**. Then, since f_I is idempotent and order preserving with range I, $f(s)$ is an upper bound of A in I. Conversely, if I is inductive and $s \in I$ is an upper bound of a finite subset $A \subseteq I$, then the polynomial $s \wedge x$ interpolates f_I at A.

(iii) The proof is similar to that of the preceding claim. ●

Next we describe unary compatible functions of semilattices.

Proposition 5.5.4 *A unary function f on a semilattice $\mathbf{S}$ is compatible iff it has one of the following forms:*

1. *$f = f_I$ for some almost principal ideal I of $\mathbf{S}$;*

2. *there exists an element $0 \neq a \in S$ and an almost principal ideal I of the subsemilattice $\uparrow a$ of $\mathbf{S}$ such that the restriction of f to $\uparrow a$ is f_I and $f(x) = a$ for $x \not\geq a$; moreover, the ideal I has the property: if $u \in I$ and $u > a$ then $\downarrow u = \downarrow a \cup [a, u]$. (In this case f is order preserving but not a contraction.)*

3. *there are elements $a, b \in S$ such that a covers b, $c \wedge a \leq b$ for all $c \not\geq a$, and $f(x) = b$ for $x \geq a$ and $f(x) = a$ otherwise. (In this case f is neither order preserving nor a contraction.)*

PROOF Using Lemma 5.5.2 it is easy to check that the functions of all three types are compatible. We show that there are no more unary compatible functions.

Let f be an arbitrary unary compatible function on $\mathbf{S}$. We first show that the subset $D = \{x \in S : f(x) \leq x\}$ is a filter of $\mathbf{S}$. If $x \in D$ and $y \in S$, $x \leq y$, then by Lemma 5.5.2 also $y \in D$. Next observe that if $x \in S \setminus D$ and $z \in \downarrow x$ then $f(z) = f(x)$ because of $f(z) \equiv f(x)$ $(\mathrm{Cg}(z, x))$. Consequently, if $x, y \in S \setminus D$ then we have $f(x) = f(x \wedge y) = f(y)$. Now, if $y = f(x)$ then $f(y) = y$, a contradiction with $y \notin D$. This proves that, given an arbitrary element $x \in S$, either x or $f(x)$ is contained in D, in particular, $D \neq \emptyset$. Finally, if $x \in D$ and $z \in \downarrow x$ then $f(z) \equiv f(x)$ $(\mathrm{Cg}(z, x))$ implies $f(z) \leq x$. Hence, if $x, y \in D$ then $f(x \wedge y) \leq x \wedge y$ and we are done.

Now observe that D is a principal filter of $\mathbf{S}$ unless $D = S$. We just showed that $f(S \setminus D)$ is a one element set whenever $D \neq S$. Moreover, if $\{a\} = f(S \setminus D)$ then $a \in D$. Suppose there exists $b \in D$, $b < a$, and let c be an arbitrary element in $S \setminus D$. Then $b \wedge c \notin D$ and therefore $f(b \wedge c) = a$. Hence

$$(f(b), a) = (f(b), f(b \wedge c)) \in \mathrm{Cg}(b, b \wedge c),$$

which is possible only if $f(b) = a$ (because $a \not\leq b$). This, however, contradicts $b \in D$. Thus, $D = \uparrow a$.

Let J denote the ideal of $\mathbf{S}$ generated by the subset $f(D)$. Obviously $\downarrow f(x) \subseteq \downarrow x \cap J$ for every $x \in D$. If the latter inclusion were strict, there would exist $z \in S$ and $y \in D$ such that $z \leq x \wedge f(y)$ but $z \not\leq f(x)$. Let ρ be the smallest congruence of S which collapses the principal filter $\uparrow z$. It is easy to check that $z/\rho = \uparrow z$; hence $f(x) \not\equiv z \; (\rho)$. On the other hand, $x \equiv y \; (\rho)$ which, by the compatibility of f, implies $f(x) \equiv f(y) \; (\rho)$. Since also $f(y) \equiv z \; (\rho)$, we have $f(x) \equiv z \; (\rho)$, a contradiction. This proves the equality

$$\downarrow f(x) = \downarrow x \cap J \tag{5.25}$$

for every $x \in D$.

Next we show that the intersection $I = J \cap D$ is precisely the set of all fixed points of f. If $f(x) = x \in S$ then clearly $x \in I$. For the converse, take an arbitrary element $x \in I$. By $x \in J$ there is a $y \in D$ such that $x \leq f(y)$. Since $x, y \in D$, we have $f(x) \leq x \leq f(y) \leq y$. Therefore $f(x) \equiv f(y) \; (\text{Cg}(x, y))$ implies $f(x) \wedge x = f(y) \wedge x = x$; hence $f(x) \geq x$. Again taking into account $x \in D$, we have $f(x) = x$.

Now we distinguish three cases:

1. $D = S$. Then f has the form 1.

2. $D \neq S$ and $I \neq \emptyset$. Since $D \neq S$, the filter D is principal. Let $x \in D = \uparrow a$. Since $I \neq \emptyset$, we can find $y \in I$ such that $y \leq x$. Then $(y, f(x)) = (f(y), f(x)) \in \text{Cg}(y, x)$ which, due to $f(x) \leq x$, implies $y \leq f(x)$; thus $f(x) \in D$. Consequently formula (5.25) yields the equality $\downarrow f(x) = \downarrow x \cap J$ in D. Since $f(S \setminus D) = \{a\}$, the function f has the form 2.

If $I = \{a\}$ then f turns out to be the constant function a. Otherwise take $u \in I$, $u > a$, and $z \in \downarrow u$ such that $z \notin D$. Then

$$(a, u) = (f(z), f(u)) \in \text{Cg}(z, u),$$

which is possible only if $a \wedge z = u \wedge z = z$. Hence $z \leq a$ and we see that every $z \in \downarrow u$ is comparable with a; in other words $\downarrow u = \downarrow a \cup [a, u]$.

3. $D \neq S$ and $I = \emptyset$. Let $b = f(a)$. By $a \in D$ and $I = \emptyset$ we have $b < a$. Suppose there is $c \in S$ such that $b < c < a$. Then $c \notin D$; hence $f(c) = a$ and $(b, a) = (f(a), f(c)) \in \text{Cg}(a, c)$ which is impossible. This contradiction proves that a covers b. For arbitrary $c \not\geq a$ we have $c \wedge a < a$; thus $(a, b) = (f(c \wedge a), f(a)) \in \text{Cg}(c \wedge a, a)$ which implies $c \wedge a = b \wedge c \wedge a = b \wedge c \leq b$.

Now let $d \geq a$. Then $(b, f(d)) = (f(a), f(d)) \in \text{Cg}(a, d)$ implying $f(d) \wedge a = b \wedge a = b$. Further, since $f(d) \notin D$, we have $f(f(d)) = a$; hence by compatibility of the function $f(x) \wedge f(d)$, we have

$$(f(d), b) = (f(d) \wedge f(d), f(f(d)) \wedge f(d)) \in \text{Cg}(d, f(d)),$$

which implies $f(d) = b \wedge f(d) \leq b$. Together with $f(d) \wedge a = b$ this yields $f(d) = b$. Hence the function f has the form 3.   •

Clearly, no local polynomial can be of the forms 2 or 3. In view of Lemma 5.5.3 we obtain therefore the following description of unary (local) polynomial functions of semilattices.

Corollary 5.5.5 *Let f be a unary nonconstant nonidentity function on a semilattice* **S**. *The function f is a local polynomial of* **S** *iff $f = f_I$ for some inductive almost principal ideal I of* **S**. *It is a polynomial iff the ideal I is principal.*

Our next aim is to prove three lemmas which reduce the solutions of the main problems of the present section to the case of unary functions. While the first two are easy, the proof of the third is rather tedious.

Lemma 5.5.6 *Let f be an essentially m-ary compatible contraction on a semilattice* **S**. *Then*

$$f(x_1, \ldots, x_m) = f(y, \ldots, y) \quad \text{where} \quad y = x_1 \wedge \cdots \wedge x_m .$$

PROOF More generally, we prove the implication

$$a_1 \wedge \cdots \wedge a_m = b_1 \wedge \cdots \wedge b_m \quad \Longrightarrow$$
$$f(a_1, \ldots, a_m) = f(b_1, \ldots, b_m) .$$

Obviously it is sufficient to show that $a_1 \wedge \cdots \wedge a_m = b_1 \wedge \cdots \wedge b_m$ implies

$$f(b_1, \ldots, b_{i-1}, a_i, \ldots, a_m) = f(b_1, \ldots, b_i, a_{i+1}, \ldots, a_m)$$

for every i. However, the latter easily follows from the fact that f is a compatible contraction depending on x_i. One simply has to observe that by Lemma 5.5.2 $(c, d) \in \mathrm{Cg}(a_i, b_i)$ with $c, d \leq a_i \wedge b_i$ is only possible if $c = d$.   •

Given an m-ary function f on a set A we obtain a series of unary functions by identifying some of the variables and giving to the remaining ones constant values in A. We shall refer to all such functions as *unary functions induced by f*. For example, if $m = 4$ then $f(x, a, x, b)$, $f(a, x, b, c)$, $f(x, x, a, x)$ where a, b, c are constants in A are some of the unary functions induced by f.

Lemma 5.5.7 *Let* **S** *be a semilattice and* $f : S^m \to S$ *be a function such that all unary induced functions of* f *are local polynomials of* **S**. *Then*

(i) $f(a_1, \ldots, a_m) = u$ *implies* $f(b_1, \ldots, b_m) = u$ *for every choice of* b_i *in* $\{a_i, u\}$, $i = 1, \ldots, m$; *in particular,* $f(u, \ldots, u) = u$;

(ii) *if there exist* $a_1, \ldots, a_m \in S$ *such that* $u = f(a_1, \ldots, a_m) \not\leq a_i$ *for every* i *then* f *is a constant function.*

PROOF (i) It is enough to show that $f(u, a_2, \ldots, a_m) = u$. The latter, however, follows from the fact that by Lemma 5.5.1 the local polynomial function $f(x, a_2, \ldots, a_m)$ is idempotent.

(ii) We use induction on m. The claim holds if $m = 1$. By the induction hypothesis the function $f(a_1, x_2, \ldots, x_m)$ is the constant u. Then $f(a_1, b_2, \ldots, b_m) = u \not\leq a_1$ for every $b_2, \ldots, b_m \in S$ which implies that the local polynomial function $f(x, b_2, \ldots, b_m)$ is constant. Since the latter holds for all $(m-1)$-tuples $(b_2, \ldots, b_m)$, we conclude that f is constant. ●

Lemma 5.5.8 *Let* **S** *be a semilattice without atoms and let* f *be a function on* S *such that all induced unary functions of* f *are (local) polynomials of* **S**. *Then* f *itself is a (local) polynomial function.*

PROOF We prove only the more general statement about local polynomials. The same argument also proves the statement about polynomial functions. Note that the compatibility of all unary induced functions implies the compatibility of f. Suppose the claim of the lemma is not valid and choose a counter-example f of the smallest arity m. Then clearly $m \geq 2$, f depends on all m variables and all $(m-1)$-ary induced functions are local polynomials. Since f is essentially m-ary and not a local polynomial function, Lemma 5.5.6 implies that f is not a contraction. Hence there exists an m-tuple $(a_1, \ldots, a_m)$ in S^m such that $f(a_1, \ldots, a_m) = u \not\leq a_i$ for some i. We may choose this m-tuple so that the cardinality

$$|\{j : f(a_1, \ldots, a_m) \leq a_j\}| = l$$

is minimal. Moreover, without loss of generality, let $u \leq a_j$ for $j = 1, \ldots, l$.

By Lemma 5.5.7 $l \geq 1$. We first show that l and m can be only 1 and 2, respectively. Consider the function $f(x, \ldots, x, x_{l+1}, \ldots, x_m)$. Since by Lemma 5.5.7 $f(u, \ldots, u, a_{l+1}, \ldots, a_m) = u \not\leq a_m$, this function is not a contraction. We show that it is essentially $(m-l+1)$-ary; then $l = 1$ follows from our minimality assumption about m. Suppose first that $f(x, \ldots, x, x_{l+1}, \ldots, x_m)$ does not depend on x. Then in particular the function $f(x, \ldots, x, a_{l+1}, \ldots, a_m)$ is the constant u; thus $f(a_m, \ldots, a_m, a_{l+1}, \ldots, a_m) = u$ implying, by Lemma 5.5.7, that f is a constant function, a contradiction. Since f depends on x_m, there exist $b_1, \ldots, b_m, c_m \in S$ such that

$$v = f(b_1, \ldots, b_{m-1}, b_m) \neq f(b_1, \ldots, b_{m-1}, c_m) = w.$$

By Lemma 5.5.7 then also

$$v = f(v, \ldots, v, b_m) \neq f(w, \ldots, w, c_m) = w$$

showing that $f(x, \ldots, x, x_{l+1}, \ldots, x_m)$ depends on x_m. Similar arguments prove that this function depends on $x_{l+1}, \ldots, x_{m-1}$.

Now, having $l = 1$, consider the function $g(x, y) = f(x, y, \ldots, y)$. Note that the function $f(u, x_2, \ldots, x_m)$ is the constant u; therefore $u = g(u, a_m) \not\leq a_m$ and g cannot be a contraction. We show that g is essentially binary; then $m = 2$ follows from the minimality assumption about m. Since f is essentially m-ary, there exists $a \in S$ such that the $(m - 1)$-ary function $f(a, x_2, \ldots, x_m)$ is not constant. Since this function is a local polynomial, the function $f(a, x, \ldots, x)$ is not constant either. This proves that $g(x, y)$ depends on y. Assuming that $g(x, y)$ does not depend on x and taking into account that $f(u, x_2, \ldots, x_m)$ is constantly u, we see that $g(x, y)$ is constantly u as well. This contradicts the fact that $g(x, y)$ depends on y.

Thus, f is a binary function and there are $a_1, a_2 \in S$ such that $f(a_1, a_2) = f(u, a_2) = u \leq a_1$ and $u \not\leq a_2$. We are going to prove that f does not depend on x_2. Let A be the set of all such $a \in S$ that $f(a, y)$ is a constant function. We need several steps in order to show that $A = S$.

1. $a \in A \Rightarrow f(a, y) \leq a$. Otherwise we would have $f(a, a) \not\leq a$ which would imply by Lemma 5.5.7 that f is constant.

2. If $a \in A$ then $0 \neq w \leq v = f(a, y)$ implies $w \in A$. Since S has no atoms there exists $z \in S$ such that $z < w$. Suppose that $w \notin A$. Then, in particular, $f(w, z) \leq z$. On the other hand we have

$f(w, z) \equiv f(a, z)$ $(\mathrm{Cg}(w, a))$ by the compatibility of f. Since, by step 1, $w < a$, Lemma 5.5.2 implies

$$f(w, z) \wedge w = f(a, z) \wedge w = v \wedge w = w,$$

and hence $f(w, z) \geq w > z$, a contradiction.

3. $f(x, y) \leq x$ for every $x, y \in S$. Suppose $v = f(b_1, b_2) \not\leq b_1$ for some $b_1, b_2 \in S$. Then $f(x, b_2)$ is a nonzero constant function. In particular, if $a \in A$ then $f(a, b_2) = v$. Hence $v \leq a$ by step 1 and step 2 implies that v is an atom of $\mathbf{S}$ which is impossible.

4. A is an ideal of $\mathbf{S}$. Let $a \in A$ and $b < a$. Since $a \in A$, the local polynomial function $f(a, x)$ is constant. Because $f(b, x) \leq b$ by step 3, compatibility of f yields $f(a, x) \wedge b = f(b, x)$. Hence, $f(b, x)$ is a constant function also.

5. If $b \notin A$, $0 \neq c \leq b$ and $c \in f(S^2)$ then $c \notin A$. Since $b \notin A$, the local polynomial function $f(b, x)$ depends on x. Assume that $c \in A$ and take $d \in S$, $d < c$. Then by Lemma 5.5.7 $f(c, d) = f(c, c) = c$. Now the compatibility of f implies

$$c = c \wedge f(c, d) = c \wedge f(b, d);$$

hence $c \leq f(b, d)$. On the other hand, since the function $f(b, x)$ depends on x, we have $f(b, d) \leq d < c$, a contradiction.

6. $A = S$. Suppose there exists $b \in S \setminus A$. Then the local polynomial function $f(b, x)$ depends on x. Hence there exists $d \in S$ such that $f(b, d) = c \neq 0$. By step 3 we know that $c \leq b$. Since c cannot be an atom and by Lemma 5.5.1 the range of $f(b, x)$ is an ideal of $\mathbf{S}$, we may assume $c < b$. Note that by step 5, $c \notin A$.

Now consider the unary local polynomial function $g(x) = f(x, c)$. We know that $g(c) = c$, $g(u) = f(u, c) = f(u, u) = u$ and that $g(b) \leq c < b$. Hence $g(x)$ is neither a constant function ($u \in A$, $c \notin A$) nor the identity function. Consequently there exists $t \in S$ such that $g(c) = t \wedge c$ and $g(u) = t \wedge u$, or equivalently, $c \leq t$ and $u \leq t$. By step 4 the first of these inequalities implies $t \notin A$ but then by step 5, $u \notin A$, a contradiction. $\bullet$

Remark In Lemma 5.5.8 the requirement that the semilattice have no atoms is essential. For example the binary function f defined on the two element semilattice by

$$f(0, 0) = 0, \quad f(0, 1) = f(1, 0) = f(1, 1) = 1$$

is not a polynomial though all of its induced unary functions are.

5.5.2　Completeness theorems

Having all of the necessary technical results, it is now relatively easy to describe the classes of semilattices having various completeness properties. We know that all local polynomial functions on a semilattice are order preserving, compatible contractions. Hence it is reasonable to characterize the class of semilattices in which all order preserving, compatible contractions are (local) polynomials. This will be done by the next theorem.

Theorem 5.5.9 *For an arbitrary semilattice* **S** *the following are equivalent:*

(1) *every compatible contraction on* **S** *is a (local) polynomial function of* **S***;*

(2) *every order preserving, compatible contraction on* **S** *is a (local) polynomial function of* **S***;*

(3) *every proper almost principal ideal of* **S** *is principal (inductive).*

PROOF We give the proof only for polynomial functions. The case of local polynomial functions is similar.

The implication (1) $\Rightarrow$ (2) is obvious.

(2) $\Rightarrow$ (3). If I is an almost principal ideal of **S** then the function f_I is an order preserving, compatible contraction by Lemma 5.5.3. Hence in view of condition (2) the same lemma implies that I is inductive.

(3) $\Rightarrow$ (1). We have to show that under condition (3) every compatible contraction on **S** is a polynomial function. By Lemma 5.5.6 it suffices to consider only unary functions. However, by Proposition 5.5.4 every such function has the form f_I for some almost principal ideal I of **S** and by Lemma 5.5.3 this function is a polynomial provided $I = S$ or I is a principal ideal of **S**.

This completes the proof.　　　　　　　　　　　　　　　　　　●

Corollary 5.5.10 *Let* **S** *be a semilattice with a greatest element and* f *a function on* S. *Then the following are equivalent:*

(1) f *is a polynomial function of* **S***;*

(2) f *is a local polynomial function of* **S***;*

(3) *f is a compatible contraction on* S.

PROOF It suffices to notice that all almost principal ideals of a semilattice with a greatest element are principal. •

Now we characterize (locally) affine complete semilattices. For that purpose we need the following notion. An element a of a semilattice S is said to be *thick* if there is no $b < a$, $b \neq 0$, such that $\downarrow b \cup [b, a] = \downarrow a$.

Theorem 5.5.11 *A semilattice S is (locally) affine complete iff it satisfies the following conditions:*

(i) *every proper almost principal ideal of* S *is principal (inductive);*

(ii) *every element of* S *is thick;*

(iii) S *has no atoms.*

PROOF We prove only half of the theorem: the statement about affine completeness. The other half about local affine completeness can be proved similarly.

Assume first that S is affine complete. Then condition (i) is satisfied by Theorem 5.5.9. Suppose S contains an element u which is not thick. Then there exists $0 \neq a \in \downarrow u$ such that $a \neq u$ but $\downarrow u = \downarrow a \cup [a, u]$. Now the function of form 2 in Proposition 5.5.4 defined by the element a and the ideal $I = \uparrow a \cap \downarrow u$ is a compatible function which is not a polynomial. This proves (ii). If b is an atom of S then the function of form 3 in Proposition 5.5.4 defined by b and $a = 0$ is a compatible function which is not a polynomial. This proves (iii).

Assume now that the conditions (i)–(iii) are satisfied and let f be an arbitrary compatible function on S. Due to Lemma 5.5.8 it suffices to consider only unary functions. Suppose that f is not a contraction. Then it must be of one of the forms 2 or 3 of Proposition 5.5.4. In the former case we see that no $a \neq u \in I$ is thick. Thus it follows from condition (ii) that $I = \{a\}$; that is, the function f is constant. If f is of form 3 then it is easy to observe that a is either an atom or a nonthick element. Hence conditions (ii) and (iii) imply that f cannot be of form 3. Thus f is a contraction and so by Theorem 5.5.9 it is a polynomial function. •

A natural idea which arises when comparing Theorems 5.5.9 and 5.5.11 is that there should be a third theorem characterizing (locally) order affine complete semilattices, that is, semilattices on which all order preserving compatible functions are (local) polynomials. Clearly such semilattices must satisfy the first two conditions of Theorem 5.5.11 and these conditions ensure that all unary order preserving compatible functions are (local) polynomials. However, the example of the two element semilattice shows that something has to be added. Indeed, the function considered in the remark after the proof of Lemma 5.5.8 is order preserving and compatible but not a polynomial function. So we have the following problem.

Problem 5.5.12 *Characterize (locally) order affine complete semilattices.*

An obvious consequence of Theorem 5.5.11 is that no finite semilattice and no chain is (locally) affine complete. We conclude this section with some examples and exercises. The first example shows that the existence of a greatest element is not necessary for the affine completeness of a semilattice.

Example 5.5.13 *An affine complete semilattice without a greatest element.*

Let **A** be the semilattice of all integers ≤ 0 with the natural order relation, and put $S = A^2 \setminus \{(0,0)\}$. Clearly S is a subuniverse of $\mathbf{A}^2$. Obviously **S** has no atoms and it is very easy to observe that all elements of **S** are thick. Note that **S** has two maximal elements: $(0,-1)$ and $(-1,0)$. Let I be a proper ideal of **S** and assume that I is not principal. Now it is not difficult to observe that I must necessarily contain an element (a,b) such that $(a,b+1),(a+1,b) \in I$ but $(c,d) = (a+1,b+1) \in S\setminus I$. Then $I \cap \downarrow(c,d)$ is not a principal ideal of **S**. This proves that all almost principal ideals of **S** are principal.

Example 5.5.14 *A locally affine complete semilattice which is not affine complete.*

Let **A** be the semilattice of all integers with the natural order relation, and put $\mathbf{S} = \mathbf{A}^2$. Obviously **S** has no atoms and all elements of **S** are thick. Let I be an almost principal ideal of **S**. Then it is easy to observe that I contains with every two elements also their join in **S**. Hence there are three possibilities. First, if none of the coordinates of $(a, b) \in I$ is bounded then $I = S$. Secondly, if both of the coordinates of $(a, b) \in I$ are bounded then I is a principal ideal of **S**. Finally, if only one, say the first, coordinate is bounded, then there exists an integer a such that I consists of all pairs (a, x) where x is an integer. Obviously then I is inductive but not principal.

Exercises

1. Find necessary and sufficient conditions for a direct product of semilattices to be affine complete.

2. Show that a free semilattice is affine complete iff it has an infinite set of free generators.

5.6 Miscellaneous results

5.6.1 p-rings and subalgebra primal algebras

In the introduction of Chapter 4 we indicated that early in the study of affine completeness A. Iskander characterized affine complete p-rings. Recall that p-rings (associative rings satisfying the equation $x^p \approx x$) form the variety generated by the Galois field $GF(p)$ with operations $+$ and $\times$ (without a distinguished identity element). By the discussion of Section 1.2.1 and Theorem 3.4.1, this algebra is subalgebra primal with one proper subalgebra, consisting of the element 0. Therefore, by Corollary 4.2.6, it follows that there are p-rings which are not affine complete although, since the variety is arithmetical, all finite p-rings are affine complete. We also know, of course, that the variety of p-rings consists (up to isomorphism) of subdirect powers of $GF(p)$. Using relatively elementary ring theory A. Iskander established the following precise characterization.

Theorem 5.6.1 *Let a p-ring* **R** *be embedded in* **S** $= GF(p)^I$. *The ring* **R** *is affine complete iff* $\{s \in S : Rs \subseteq R\} = \mathrm{Sg}^{\mathbf{R}}(R, 1)$.

Rather than prove this theorem directly, we shall first establish a particular case of the description of all affine complete members of the variety generated by a subalgebra primal algebra which has a single proper subalgebra, and which is due to D. Clark and H. Werner ([Clark, Werner, 1981]). (Our particular case is of course that in which the subalgebra primal algebra has a single subalgebra consisting of just one element.) This will provide a glimpse of the topological methods which they employed and from their topological result we shall then easily infer Iskander's algebraic result.

Let $\mathbf{S}$ be a subalgebra primal algebra, $\{0\}$ the only subuniverse of $\mathbf{S}$ and $V = V(\mathbf{S})$. It follows from Theorems 3.4.1 and 3.4.4 that $\mathbf{S}$ is quasiprimal, hence it admits a ternary discriminator term $t(x, y, z)$. By [Keimel, Werner, 1974] we know that every $\mathbf{A} \in V$ has a Boolean representation that we are going to describe now. Let $\mathbf{I}$ be a Boolean topological space with a distinguished element which we also denote by 0. Such spaces (we shall call them *pointed Boolean spaces*) form a category $\mathcal{B}_0$ with zero preserving continuous maps in the role of morphisms. In particular, the algebra $\mathbf{S}$ can be considered as a member of that category when endowed with the discrete topology and taking the zero of the algebra in the role of the zero of the Boolean space. Let $\mathbf{I} \in \mathcal{B}_0$ and $A = \mathcal{B}_0(\mathbf{I}, \mathbf{S})$ be the set of all morphisms from $\mathbf{I}$ to $\mathbf{S}$ in the category $\mathcal{B}_0$. Clearly the set A becomes an algebra of the variety V if we define operations on it pointwise. Now, it is an important fact that every $\mathbf{A} \in V$ has such a representation. In fact, V is even categorically dually equivalent to the category $\mathcal{B}_0$. Thus if we want to describe affine complete members of V, it is reasonable to do it in topological terms. This is exactly what we are going to do now. We shall characterize pointed Boolean spaces $\mathbf{I}$ such that the algebra $\mathcal{B}_0(\mathbf{I}, \mathbf{S})$ is affine complete.

We start with two technical lemmas. Clearly $\mathbf{A} = \mathcal{B}_0(\mathbf{I}, \mathbf{S})$ can be considered as a subalgebra of $\mathbf{S}^I$. Then the i-th projection of $\mathbf{A}$ consists of all $a(i)$ where a is a continuous zero preserving mapping from $\mathbf{I}$ to $\mathbf{S}$. We denote this projection by $\mathbf{A}_i$. Obviously $A_0 = \{0\}$ and $A_i = S$ if $i \neq 0$. Consequently $0 \in I$ is superfluous from a purely algebraic point of view. However, because of the topological nature of our representation, 0 is the most significant member of I.

Due to the embedding $\mathbf{A} < \mathbf{S}^I$, every compatible function f on $\mathbf{A}$ can be represented as a family $(f_i)_{i \in I}$ where f_i is a function on A_i, $i \in I$. We first establish which families $(f_i)_{i \in I}$ preserve A.

Lemma 5.6.2 *Let $\mathbf{A} = \mathcal{B}_0(\mathbf{I}, \mathbf{S})$ and $f = (f_i)_{i \in I}$ be a family of m-ary functions as described above. The function f preserves A iff, for* *every m-ary function g on S, the set $I_g = \{i \in I : f_i = g\}$ is closed with respect to the topology of* I. *(Here $f_0 = g$ means $g(0, \ldots, 0) = 0$).*

PROOF Assume that f preserves A and prove that then all sets $I \setminus I_g = \{i \in I : f_i \neq g\}$ are open. Let $j \in I$ be such that $f_j \neq g$. Then there exist $a_1, \ldots, a_m \in A$ such that

$$f_j(a_1(j), \ldots, a_m(j)) \neq g(a_1(j), \ldots, a_m(j)).$$

Since the functions $a_1, \ldots, a_m$ are continuous, there exists a clopen neighborhood N of j such that all these functions are constant on N. Hence

$$f_j(a_1(i), \ldots, a_m(i)) \neq g(a_1(i), \ldots, a_m(i))$$

for every $i \in N$ which proves that $N \subseteq \{i \in I : f_i \neq g\}$.

Now assume that all subsets I_g are closed and prove that f preserves A. Let $a_1, \ldots, a_m$ be arbitrary elements of A. Obviously $f(a_1, \ldots, a_m)$ preserves zero; so we only have to show that $f(a_1, \ldots, a_m)$ is a continuous function from **I** to **S**. Because the topology of **I** is discrete, it suffices to find a clopen neighborhood for every $j \in I$, on which the function $f(a_1, \ldots, a_m)$ is constant. We first put $J = \bigcup\{I_g : g \neq f_j\}$ if $j \neq 0$ and $J = \bigcup\{I_g : g(0, \ldots, 0) \neq 0\}$ if $j = 0$. In both cases J is closed because there are only finitely many functions g on S. Hence $I \setminus J$ is an open neighborhood of j and therefore there exists a clopen neighborhood N_1 of j contained in $I \setminus J$. Next choose a clopen neighborhood N_2 of j such that $a_1, \ldots, a_m$ are constant on N_2. This can be done because these functions are continuous. Obviously $N = N_1 \cap N_2$ is also a clopen neighborhood of j and it is easy to check that $f(a_1, \ldots, a_m)$ is constant on that N. •

Lemma 5.6.3 *Let* $\mathbf{A} = \mathcal{B}_0(\mathbf{I}, \mathbf{S})$ *and* $f = (f_i)_{i \in I}$ *be a compatible function on* **A**. *The function* f *is a polynomial of* **A** *iff there exists a function g on S and a clopen neighborhood N of 0 such that $f_i = g$ for every $i \in N$.*

PROOF Let f have the form $u(x_1, \ldots, x_m, a_1, \ldots, a_n)$ where u is a term and $a_1, \ldots, a_n \in A$. Since $a_1, \ldots, a_n$ are continuous functions from **I** to **S**, there exists a clopen neighborhood N of 0 on which they all are constant. Then $u(x_1, \ldots, x_m, a_1(i), \ldots, a_n(i))$ where $i \in I$ is a suitable function g. This proves the necessity.

To prove the sufficiency, we first observe that A contains for every $s \in S$ the function a^s such that $a^s(i) = 0$ if $i \in N$ and $a^s(i) = s$ otherwise. Since the variety V is arithmetical, by Theorem 3.3.3 there exists a polynomial function r of $\mathbf{A}$ which interpolates f at the finite set $X = \{a^s : s \in S\}$. However, the set X projects onto every A_i, $i \notin N$, implying that $f_i = r_i$ for every $i \notin N$.

On the other hand, $f_0 = g$ implies $g(0, \ldots, 0) = 0$. Since $\mathbf{S}$ is subalgebra primal with a unique subuniverse $\{0\}$, this means that g is a term function. Let u be a term such that $u^{\mathbf{S}} = g$.

Now choose a nonzero $s \in S$ and consider the polynomial function $p(\mathbf{x}) = t(t(a^0, a^s, u(\mathbf{x})), t(a^0, a^s, r(\mathbf{x})), r(\mathbf{x}))$ where $t(x, y, z)$ is a discriminator term. It is easy to check that $p_i = u_i = g$ if $i \in N$ and $p_i = r_i$ if $i \notin N$. Consequently, $f = p$. $\bullet$

Now we are ready to establish an easy, purely topological criterion for $\mathbf{A}$ to be affine complete.

Theorem 5.6.4 *The algebra* $\mathbf{A} = \mathcal{B}_0(\mathbf{I}, \mathbf{S})$ *is affine complete if and only if, for any given pair of closed subsets* $I_1, I_2 \subseteq I$ *such that* $I_1 \cup I_2 = I$ *and* $I_1 \cap I_2 = \{0\}$, *either* I_1 *or* I_2 *is clopen.*

PROOF Assume that $\mathbf{A}$ is affine complete and take two subsets $I_1, I_2 \subseteq I$ such that $I_1 \cup I_2 = I$ and $I_1 \cap I_2 = \{0\}$. Consider the unary function $f = (f_i)_{i \in I}$ on S^I where f_i is the identity function on S if $i \in I_1$ and f_i is constantly 0 for $i \in I_2$. By Lemma 5.6.2 this function preserves A. Moreover, since $\mathbf{A}$ has no skew congruences, the function f is compatible; so by affine completeness of $\mathbf{A}$ it is a polynomial function of $\mathbf{A}$. Now Lemma 5.6.3 implies the existence of a clopen neighborhood N of 0 such that f_i is the same function on S for all $i \in N$. Obviously this is only possible if $N \subseteq I_1$ or $N \subseteq I_2$. If $N \subseteq I_1$ then $I_2 \setminus N = I_2 \setminus \{0\}$ is closed; hence $I_1 = I \setminus (I_2 \setminus \{0\})$ is closed. This shows that I_1 is clopen and proves the necessity.

For the sufficiency, we first notice that our topological condition is equivalent to the following stronger one: if $I_1, \ldots, I_n$ are closed subsets of $\mathbf{I}$ such that $I_1 \cup \cdots \cup I_n$ is clopen and $I_j \cap I_k = \{0\}$ for every two distinct j and k, then at least one of the sets I_j is clopen. This can be proved by an easy induction argument.

Take an arbitrary compatible function $f = (f_i)_{i \in I}$ on $\mathbf{A}$ and let $g_1, \ldots, g_n$ be the list of all zero preserving functions on S. Then the union of the subsets I_{g_j}, $j = 1, \ldots, n$, is clopen and all of their pairwise intersections are $\{0\}$. Hence some of these subsets I_{g_j} must be clopen. By Lemma 5.6.3 this implies that f is a polynomial function. •

We show now that in case of p-rings Iskander's condition is equivalent to the topological condition established in Theorem 5.6.4. Let a p-ring $\mathbf{R}$ be subdirect in $GF(p)^J$. We know ([Keimel, Werner, 1974]) that there is a canonical way to embed J into a pointed Boolean space $\mathbf{I}$ so that $\mathbf{R}$ is isomorphic to $\mathbf{R}' = \mathcal{B}_0(\mathbf{I}, GF(p))$ and this isomorphism assigns to every $r \in R$ considered as a function $J \rightarrow GF(p)$ an extension $r' : I \rightarrow GF(p)$. In particular we can embed $\mathbf{R}$ subdirectly into $GF(p)^I$ and it is easy to observe that Iskander's condition is satisfied for the inclusion $\mathbf{R} < GF(p)^J$ iff it is satisfied for the inclusion $\mathbf{R} < GF(p)^I$.

We assume first that Iskander's condition is satisfied and let I_1 and I_2 be closed subsets of I with $I = I_1 \cup I_2$ and $I_1 \cap I_2 = \{0\}$. Take an element $a = (a_i)_{i \in I} \in GF(p)^I$ where $a_i = 0$ if $i \in I_1$ and $a_i = 1$ otherwise. It follows from Lemma 5.6.2 that $Ra \subseteq R$; hence by Iskander's condition $a = r + n \cdot 1$ where $r \in R$ and n is an integer. Since the function r is continuous, there is a partition of I with finitely many clopen blocks such that r is constant on each block. Clearly then also a is constant on each block. This implies that every block is contained either in I_1 or I_2. Consequently that one of the latter which contains the block containing zero must be clopen.

Now assume that the Boolean space $\mathbf{I}$ satisfies the condition from Theorem 5.6.4. Take $a = (a_i)_{i \in I} \in GF(p)^{I \setminus \{0\}}$ such that $Ra \subseteq R$ and prove that $a \in \mathrm{Sg}(R, 1)$. We shall do this by induction on the number of different components of a. If this number is 1 then clearly a is an integral multiple of 1 and we are done. Since $Ra \subseteq R$, Lemma 5.6.2 yields that all sets $I_s = \{i \in I : a_i = s\} \cup \{0\}$ where $s \in GF(p)$ are closed. Hence I is a union of nonempty closed subsets I_s, $s \in GF(p)$. Thus by our assumption one of those subsets, say I_{s_0}, must be clopen. Let I_{s_1} be another closed nonempty subset. Obviously the element $r = (r_i)_{i \in I}$ with

$$r_i = \begin{cases} 0 & \text{if } i \in I_{s_0}, \\ s_1 - s_0 & \text{otherwise,} \end{cases}$$

is an element of R. It is easy to see that $b = a - s_0 \cdot 1 - r$ has fewer different components than a and $Rb \subseteq R$. By the induction hypothesis b is an element of $\mathrm{Sg}^{\mathbf{R}}(R, 1)$ but then so is a.

We conclude the subsection with two examples. Note that all finite p-rings are affine complete as finite members of an arithmetical variety. Also, Iskander's theorem immediately yields that all p-rings with identity element are affine complete.

Example 5.6.5 *A p-ring which is not affine complete.*

Let $\mathbf{R} = GF(p)^{(I)}$ be a weak direct power of $GF(p)$ where the set I is infinite. Recall that $\mathbf{R}$ is a subalgebra of $GF(p)$ consisting of all $a = (a_i)_{i \in I}$ with only finitely many nonzero components a_i. Let I be the join of two infinite disjoint subsets I_1 and I_2 and take the element $a \in GF(p)$ such that $a_i = 1$ if $i \in I_1$, $a_i = 0$ if $i \in I_2$. Then obviously $aR \subseteq R$ but $a \notin \mathrm{Sg}(R, 1)$.

Example 5.6.6 *An affine complete p-ring which does not have an identity element.*

Let I be an infinite set and $\mathbf{R}$ be a maximal ideal of the ring $GF(p)^I$ such that $GF(p)^{(I)} \subseteq R$. Then clearly

$$\{a \in GF(p)^I : aR \subseteq R\} = GF(p)^I = \{a + n \cdot 1 : a \in R, \ n \text{ an integer}\}.$$

Thus $\mathbf{R}$ is affine complete by Iskander's criterion.

Exercise

Find Boolean representations for the two p-rings from the preceding examples. Use Theorem 5.6.4 to show that one of them is affine complete and the other is not.

5.6.2   Median algebras

Every lattice $\mathbf{L}$ gives rise to a new algebraic structure $\langle L; u \rangle$ where $u(x, y, z) = (x \vee y) \wedge (y \vee z) \wedge (z \vee x)$ is one of the two dual median operations. In what follows we denote this structure by $\mathbf{M_L}$. We restrict our attention to the most thoroughly studied case in which $\mathbf{L}$ is a distributive lattice. In this case the dual median operations are the same.

Definition A **median algebra** is any subalgebra of any $\mathbf{M_L}$ where $\mathbf{L}$ is a distributive lattice.

Median algebras form a subclass of a more general class of algebras which have been introduced for the abstract study of the betweenness relation (see [Isbell, 1980]). There does not seem to be a good general reference for median algebras. The reader will find the necessary background for our purposes in [Bandelt, 1993], [Ploščica, 1995], and in the papers cited there.

In the sequel, as is customary in the theory of median algebras, we shall simply write (x, y, z) instead of $u(x, y, z)$. It can be proved that the class of all median algebras forms a variety which can be defined by the following equations:

$$(x, x, y) \approx x,$$
$$(x, y, z) \approx (x, z, y) \approx (y, x, z) \approx (y, z, x) \approx (z, x, y) \approx (z, y, x),$$
$$(x, u, v) \approx (x, (x, u, v), (z, u, v)).$$

The variety of median algebras is the only known variety of type consisting of a single ternary operation symbol for which the locally affine complete and affine complete members have been described. Since median algebras are close to distributive lattices, we shall present only results without proofs.

It is known that $\mathbf{M_{D_2}}$ is the only SI median algebra, up to isomorphism. This fact implies that median algebras also satisfy the following stronger equation

$$((x, y, z), u, v) \approx (x, (y, u, v), (z, u, v)). \tag{5.26}$$

One can easily check that $\mathbf{M_L}$ where $\mathbf{L}$ is the three element chain is the only three element median algebra, up to isomorphism. However, there exists a four element median algebra which is not associated with any distributive lattice.

A *split congruence* of a median algebra is one with only two blocks. Clearly every congruence of a median algebra is a meet of split congruences. Moreover, one can easily prove that all split congruences of the median algebra $\mathbf{M_L}$ where $\mathbf{L}$ is any distributive lattice are actually the congruences of the lattice $\mathbf{L}$. This immediately implies the equality $\operatorname{Con} \mathbf{L} = \operatorname{Con} \mathbf{M_L}$.

Given an arbitrary median algebra $\mathbf{M}$ and $a \in M$, we define the binary relation $\leq_a$ by

$$x \leq_a y \iff (a, x, y) = x.$$

It is easy to check that the relation $\leq_a$ is a semilattice order on the set M, and is a diagonal subuniverse of $\mathbf{M}^2$. Note that the transitivity of $\leq_a$ follows from equation (5.26). The relations $\leq_a$ are called *principal semilattice orders* of the median algebra $\mathbf{M}$.

If a, b, c are elements of a median algebra $\mathbf{M}$ and $(a, b, c) = a$ then we say that a *lies between* b *and* c. The set of all elements between b and c is called a *segment between* b *and* c. It is easy to see that this segment consists of all elements (x, b, c) where $x \in M$. If $\mathbf{M} = \mathbf{M_L}$ and $a, b \in L$ then the segment between a and b is precisely the interval $[a \wedge b, a \vee b]$ in the lattice $\mathbf{L}$.

If $b = c$ then the segment between b and c consists of a single element b. Such segments are called *trivial*. A subset $C \subseteq M$ is called *convex* if, for every $b, c \in C$, the whole segment between b and c is contained in C. Obviously the intersection of every system of convex subsets is convex. This allows us to speak of the convex subset generated by an arbitrary subset $S \subseteq M$. We shall denote this convex subset by $\mathrm{Conv}\, S$. It is easy to observe that the segment between b and c is precisely $\mathrm{Conv}\{b, c\}$. Clearly every convex subset of a median algebra is a subuniverse. A segment $\mathrm{Conv}\{b, c\}$ of a median algebra is said to be *Boolean*, if it is a Boolean lattice with respect to the principal semilattice order $\leq_b$. (Note that for arbitrary $x, y \in \mathrm{Conv}\{b, c\}$, $x \leq_b y$ iff $y \leq_c x$.)

A convex subset C of $\mathbf{M}$ is called a *halfspace* or a *prime convex set* if $C \neq M$, $C \neq \emptyset$, and the complement $M \setminus C$ is convex, too. It is not difficult to prove, using Zorn's lemma, that every proper convex subset is contained in a halfspace. Also it is an easy exercise to check that every halfspace C of $\mathbf{M}$ determines the split congruence of $\mathbf{M}$ with blocks C and $M \setminus C$. On the other hand, the two blocks of any split congruence of $\mathbf{M}$ are halfspaces.

Now we are able to state the main results of [Bandelt, 1993] which characterize local polynomial functions of median algebras and describe locally affine complete median algebras.

Theorem 5.6.7 *Let* $\mathbf{M}$ *be a median algebra, and let* $\leq$ *be any semilattice order on* M *which is a subuniverse of* $\mathbf{M}^2$. *Then the following statements are equivalent for a function* $f : M^m \to M$:

(1) f *is a local polynomial function;*

(2) f *preserves* $\leq$ *and all split congruences of* $\mathbf{M}$;

(3) *f preserves all principal semilattice orders of* **M**.

Theorem 5.6.8 *A median algebra is locally affine complete iff it has no nontrivial Boolean segments.*

In [Ploščica, 1995] affine complete median algebras are described. The result is given in terms of Chebyshev sets. A subset C of a median algebra **M** is said to be a *Chebyshev set* if for every $y \in M$ there exists $x \in C$ such that $(x, y, z) = x$ for every $z \in C$. It is known that all finitely generated convex subsets are Chebyshev sets. They are usually referred to as *finitely bounded* Chebyshev sets.

Theorem 5.6.9 *A median algebra* **M** *is affine complete iff the following two conditions are satisfied:*

(i) **M** *has no nontrivial Boolean segments;*

(ii) *every proper Chebyshev set of* **M** *is finitely bounded.*

We conclude the subsection with three easy observations and an example. First, it is easy to observe that all finite median algebras have nontrivial Boolean segments; so none of them can be locally affine complete. Second, a median algebra $\mathbf{M_L}$ is locally affine complete iff the underlying distributive lattice **L** is already locally affine complete. Third, since $\mathrm{Con}\, \mathbf{L} = \mathrm{Con}\, \mathbf{M_L}$, the median algebra $\mathbf{M_L}$ can be affine complete only if the distributive lattice **L** is affine complete. That the converse of the latter is not true, and that there are locally affine complete median algebras, which are not affine complete, is verified by the following example.

Example 5.6.10 *An affine complete distributive lattice* **L** *such that the median algebra* $\mathbf{M_L}$ *is locally affine complete but not affine complete.*

Take any affine complete distributive lattice **L** without 0 (see Example 5.3.14). Then clearly $\mathbf{M_L}$ is locally affine complete. However, if we take any $1 \neq a \in L$ then it is easy to check that the subset $\downarrow a$ of L is a proper Chebyshev set of $\mathbf{M_L}$ which is not finitely bounded. Thus Theorem 5.6.9 implies that $\mathbf{M_L}$ is not affine complete.

5.6.3   Commutative inverse semigroups

Recall that a semigroup $\langle S; + \rangle$ is called *inverse* if for every $x \in S$ there exists a unique element $y \in S$ such that $x + y + x = x$ and $y + x + y = y$. Such an element y is called the *inverse* of x and in the sequel it will be denoted by $-x$. It is convenient to consider inverse semigroups as algebras of type $\{+, -\}$ where $+$ and $-$ are binary and unary operation symbols, respectively. Then the class of all commutative inverse semigroups (briefly CISs) is precisely the variety defined by the equations

$$x+y \approx y+x, \quad x+(y+z) \approx (x+y)+z, \quad x+(-x)+x \approx x, \quad -(-x) \approx x.$$

This variety contains two important subvarieties defined by additional equations $x + (-x) \approx y + (-y)$ and $x + x \approx x$, respectively. Clearly these are the varieties of abelian groups and semilattices. Moreover, these two subvarieties generate the variety of CISs.

Any CIS $\mathbf{S}$ has an idempotent endomorphism $\epsilon(x) = x + (-x)$. The range of ϵ is the semilattice of all idempotents of $\mathbf{S}$. This semilattice $\mathbf{E}$ is called the *structure semilattice* of $\mathbf{S}$. We shall denote the restriction of the addition operation of $\mathbf{S}$ to E by $\wedge$. If $\alpha \in E$ then the set $A_\alpha = \epsilon^{-1}(\alpha)$ is a subuniverse of $\mathbf{S}$. It is easy to see that $\mathbf{A}_\alpha$ is an abelian group with α in the role of zero, and $\mathbf{S}$ is the disjoint union of the $\mathbf{A}_\alpha$, $\alpha \in E$. The latter are called the *(abelian) group components* of $\mathbf{S}$.

Suppose $\alpha, \beta \in E$ and $\beta \leq \alpha$. Then the mapping $\phi_{\alpha\beta} : A_\alpha \to A_\beta$ defined by $\phi_{\alpha\beta}(x) = x + \beta$ is a homomorphism of abelian groups. The $\phi_{\alpha\beta}$ are called the *connecting homomorphisms* of $\mathbf{S}$. Straightforward calculations show that

$$\phi_{\alpha\alpha} = 1_{A_\alpha} \quad \text{and} \quad \phi_{\alpha\gamma} = \phi_{\beta\gamma}\phi_{\alpha\beta} \tag{5.27}$$

for every $\alpha, \beta, \gamma \in E$ with $\gamma \leq \beta \leq \alpha$. Also, it is easy to check that actually the group components together with the connecting homomorphisms entirely determine the structure of $\mathbf{S}$:

$$x + y = \phi_{\alpha\gamma}(x) + \phi_{\beta\gamma}(y) \tag{5.28}$$

where $x \in A_\alpha$, $y \in A_\beta$ and $\gamma = \alpha \wedge \beta$.

On the other hand, suppose we have a meet semilattice $\mathbf{E}$, an abelian group $\mathbf{A}_\alpha$ for every $\alpha \in E$, and a system of group homomorphisms $\phi_{\alpha\beta} : \mathbf{A}_\alpha \to \mathbf{A}_\beta$, $\alpha, \beta \in E$, $\beta \leq \alpha$, satisfying conditions

(5.27). Then the disjoint union of the A_α, $\alpha \in E$, becomes a CIS if we define addition by formula (5.28). Note that if $a \in A_\alpha$ then $-a$ will be exactly the abelian group inverse of a in $\mathbf{A}_\alpha$. Recall that in the usual semigroup terminology the above construction is called a strong semilattice of the abelian groups.

The variety of CISs is generated by two varieties (abelian groups and semilattices) for which the affine completeness problem was completely solved. Also the polynomial functions of such semigroups have a simple canonical form and their principal congruences are easily described. All of this suggests that it is possible to describe affine complete CISs in terms of the affine completeness of the structure semilattice and group components. An attempt to do this was made in [Kaarli, Márki, 1995]. It turned out that actually very much depends on the connecting homomorphisms. For example, if all connecting homomorphisms are zero, there is practically no connection between the different group components and it is very easy to find nonpolynomial compatible functions.

Here we present, without proofs, some results characterizing affine complete members in certain classes of CISs. The proofs can be found in [Kaarli, Márki, 1995]. We first describe three relatively simple properties which are necessary for affine completeness of a CIS **S**.

Proposition 5.6.11 *Let* **S** *be an affine complete CIS. Then the following conditions are satisfied:*

(i) *The structure semilattice* **E** *is affine complete.*

(ii) *If* **E** *does not have a greatest element then the only family* $(a_\alpha)_{\alpha \in E}$ *such that* $a_\alpha \in A_\alpha$ *for every* $\alpha \in E$ *and* $\phi_{\alpha\beta}(a_\alpha) = a_\beta$ *for every* $\alpha, \beta \in E$ *with* $\beta \leq \alpha$, *consists of zeros of the group components. This is called the* **string condition**.

(iii) *For every* $\alpha \in E$, *the intersection of kernels of all* $\phi_{\alpha\beta}$ *where* $\beta \leq \alpha$ *is zero. This is called the* **kernel condition**.

Note that the affine completeness of **E** easily follows from the fact that it is the range of a unary idempotent polynomial function.

Suppose now that all connecting homomorphisms are injective (in this case **S** is called **E-unitary**). Then the kernel condition is trivially satisfied and the family of group components together with connecting homomorphisms forms a direct system. Hence there exists a direct limit $\mathbf{A} = \varinjlim \mathbf{A}_\alpha$.

Theorem 5.6.12 *Let* **S** *be a CIS with structure semilattice* **E** *and assume that all connecting homomorphisms* $\phi_{\alpha\beta}$ *of* **S** *are injective. Then* **S** *is affine complete iff the following conditions are satisfied:*

(i) *the semilattice* **E** *is affine complete;*

(ii) *the string condition is satisfied;*

(iii) *the direct limit of the direct system of group components of* **S** *is affine complete.*

The case when all connecting homomorphisms are surjective has a satisfactory solution also.

Theorem 5.6.13 *Let* **S** *be a CIS with structure semilattice* **E** *and assume that all connecting homomorphisms* $\phi_{\alpha\beta}$ *of* **S** *are surjective. Then* **S** *is affine complete iff the following conditions are satisfied:*

(i) *the semilattice* **E** *is affine complete and has a greatest element unless* $S = E$;

(ii) *the kernel condition is satisfied;*

(iii) *every group component of* **S** *is affine complete.*

There are some other classes of CISs for which affine complete members have been characterized. Such characterizations involve, as a rule, certain "connectedness" conditions for the structure semilattice. Connectedness means here, roughly speaking, that the connecting homomorphisms are sufficiently far from being zero. As an example, we present here the result for the torsion case.

We say that a CIS is *torsion* (*primary* for some prime p) if all of its group components are torsion (*primary* for some prime p). One can easily prove, as in the abelian group case, that every affine complete torsion CIS is bounded. A bounded CIS **S** can be represented as a direct product of finitely many primary CISs. The latter are called the primary components of **S**. All of them have the same structure semilattice as **S**. It turns out that a bounded CIS is affine complete iff all of its primary components are affine complete.

Let **S** be a p-primary CIS and let **E** be its structure semilattice. We need to impose certain connectedness conditions on all principal ideals of **E** and on all of **E**. Therefore we introduce a set E^1 such that $E^1 = E$ if **E** has a greatest element and $E^1 = E \cup \{1\}$ otherwise. Now $\downarrow 1 = E$ and we can handle E as a principal ideal. For every $\alpha \in E^1$ and positive integer k we introduce a graph $G^k(\alpha)$ with the elements of $\downarrow \alpha$ in the role of vertices and with an edge between β and γ if they are comparable, say $\gamma \leq \beta$, and the exponent of the range of the connecting homomorphism $\phi_{\beta\gamma}$ is at least p^k.

Theorem 5.6.14 *Let* **S** *be a commutative inverse semigroup of exponent p^n. Then* **S** *is affine complete iff it satisfies the three conditions given in Proposition 5.6.11 and the following two conditions:*

(iv) *for every $\alpha \in E$ with $A_\alpha \neq 0$, there exists $\beta \leq \alpha$ such that A_β is an affine complete abelian group of exponent p^n;*

(v) *for every $\alpha \in E^1$ and $k \in \underline{n}$, the set $\{\beta \in \downarrow \alpha : \exp A_\beta \geq p^k\}$ is a connected component of the graph $G^k(\alpha)$.*

We conclude with examples which demonstrate that the conditions of Proposition 5.6.11 are independent, and that the affine completeness of a CIS does not depend on the affine completeness of its group components.

Example 5.6.15 *A CIS which is not affine complete though both its structure semilattice and group components are affine complete and all connecting homomorphisms are bijective.*

Choose an arbitrary affine complete semilattice **E** which does not have a greatest element (see Example 5.5.13), and an arbitrary nontrivial affine complete abelian group **A**. All group components A_α are chosen to be copies of **A** and all connecting homomorphisms to be the identity maps. Clearly then the kernel condition is satisfied but the string condition is not. Hence the CIS we get in this way is not affine complete.

Example 5.6.16 *A CIS which is not affine complete though it satisfies the string and the kernel conditions and all connecting homomorphisms are bijective.*

This is almost the same as the preceding example. The only change is that $\mathbf{E}$ has to be chosen with a greatest element and not affine complete. Any finite lattice with at least two elements will do.

Example 5.6.17 *An affine complete CIS with all group components not affine complete.*

Let $\mathbf{I}$ be the semilattice of negative integers, with the natural order, and $\mathbf{E} = \mathbf{I}^2$. For a prime p define $\mathbf{B}_n = \mathbf{Z} \oplus \mathbf{Z}_{p^n}$. For any $\alpha = (i,j) \in E$, put $\mathbf{A}_\alpha = \mathbf{B}_{-(i+j)}$ and for $\alpha \geq \beta$ define the connecting homomorphism $\phi_{\alpha\beta}$ as the canonical embedding. Let $\mathbf{S}$ be the CIS with structure semilattice $\mathbf{E}$, group components $\mathbf{A}_\alpha$, and connecting homomorphisms $\phi_{\alpha\beta}$. The semilattice $\mathbf{E}$ is affine complete by Theorem 5.5.11 and the string condition is satisfied because $\mathbf{E}$ has a greatest element. So the CIS will be affine complete if affine completeness is verified for the direct limit $\mathbf{A} = \varinjlim \mathbf{A}_\alpha$. It remains to observe that $\mathbf{A}$ is isomorphic to $\mathbf{Z} \oplus \mathbf{Z}_{p^\infty}$ and the latter is affine complete by Theorem 5.2.21. However, the group components $\mathbf{A}_\alpha$ are not affine complete by Theorem 5.2.22.

Example 5.6.18 *A CIS which is not affine complete though the structure semilattice is affine complete, the string condition is satisfied, all group components are affine complete, and all connecting homomorphisms are injective.*

The construction of this example is similar to the preceding one. The difference is that now $\mathbf{B}_n = \mathbf{Z}_{p^n} \oplus \mathbf{Z}_{p^n}$. It is easy to see that now $\varinjlim \mathbf{A}_\alpha$ is isomorphic to $\mathbf{Z}_{p^\infty} \oplus \mathbf{Z}_{p^\infty}$ which is not affine complete by Lemma 5.2.23. However, all group components $\mathbf{A}_\alpha$ are affine complete by Proposition 5.2.19.

5.6.4   Lattice ordered groups

Let a group and a lattice have the same universe A and assume that the group multiplication preserves the lattice order. The algebra obtained in this way is called a *lattice ordered group* (briefly, an *l*-group). The condition that multiplication preserves the order can be given by one of the following equations:

$$x(y \vee z)w \approx xyw \vee xzw, \qquad x(y \wedge z)w \approx xyw \wedge xzw .$$

Clearly the class of all l-groups forms a variety. We recommend [Darnel, 1995] as a good reference book for l-groups. The variety of l-groups is arithmetical because its defining equations include both the group and lattice axioms. Consequently all l-groups are strictly locally affine complete.

Now we recall some well known facts about l-groups which will be needed in the sequel. Every l-group is a torsion free group and a distributive lattice. The underlying lattice of a nontrivial l-group is never complete because it never has a greatest and a smallest element. However it is possible that a nonempty subset X of an l-group $\mathbf{L}$ has $\bigvee X$ whenever X has an upper bound. If this occurs, the l-group $\mathbf{L}$ is said to be *complete*. The most common examples of complete l-groups are the group of integers and the group of real numbers, with the natural order relation. It is known that all complete l-groups have abelian group reducts. Let $\mathbf{L}$ be an l-group and $\rho \in \mathrm{Con}\, \mathbf{L}$. Since $\mathbf{L}$ has a group reduct, there must exist a normal subgroup $\mathbf{H}$ of the group reduct of $\mathbf{L}$ such that $a \equiv_\rho b$ iff $ab^{-1} \in H$. It is easy to see that H must be a convex subset of the lattice reduct of $\mathbf{L}$. On the other hand, it is known that every convex normal subgroup of an l-group determines a congruence of that l-group. In the theory of l-groups such subgroups are usually referred to as *l-ideals*.

Given an element a of an l-group $\mathbf{L}$, we define $|a| = a \vee a^{-1}$. The element $|a|$ is called the *modulus* of a. If X is any subset in $\mathbf{L}$ then the set $X^{\perp} = \{y \in L : (\forall x \in X)\ |x| \wedge |y| = 1\}$ is an l-ideal of $\mathbf{L}$. An l-group $\mathbf{L}$ is called *projectable* if $\mathbf{L}$ is a direct product of l-ideals $a^{\perp}$ and $a^{\perp\perp}$, for every $a \in L$. It is known that all projectable l-groups are complete and have abelian group reducts.

The affine completeness problem for abelian l-groups has been studied by J. Jakubík and M. Csontóová. In [Jakubík, 1995] it was proved that every projectable abelian l-group is either a chain or not affine complete. In [Jakubík, Csontóová, 1998] the authors proved that no nontrivial abelian l-group which is a chain can be affine complete. In particular, every nontrivial abelian projectable l-group is not affine complete. So far no nontrivial affine complete l-group is known. Therefore we turn our attention to the following problem.

Problem 5.6.19 *Does there exist any nontrivial affine complete l-group?*

If the answer to this problem is negative, it would solve our last open problem; an appropriate conclusion to this chapter and this book.

Problem 5.6.20 *Does there exist any variety which contains no nontrivial affine complete members?*

The arguments of J. Jakubík and M. Csontóová were based on the explicit construction of nonpolynomial compatible functions. We shall show below that it is possible to disprove affine completeness of a large class of l-groups by using only cardinality arguments. We shall call an l-ideal H of an l-group $\mathbf{L}$ *essential* if, for every l-ideal G of $\mathbf{L}$,

$$|G| > 1 \quad \Longrightarrow \quad |G \cap H| > 1.$$

Theorem 5.6.21 *If H is an l-ideal of an affine complete l-group $\mathbf{L}$ such that $\mathbf{L}/H$ is subdirectly irreducible then H is an essential l-ideal of $\mathbf{L}$.*

PROOF Suppose, on the contrary, that there exists an l-ideal G of $\mathbf{L}$ such that $|G| \neq 1$ but $|G \cap H| = 1$. Then $\mathbf{L}$ can be identified with a subdirect product of the SI l-group $\mathbf{L}_1 = \mathbf{L}/H$ and the l-group $\mathbf{L}_2 = \mathbf{L}/G$. Since the variety of l-groups is congruence permutable, by Theorem 1.2.13 this subdirect product is a rectangular subuniverse in $\mathbf{L}_1 \times \mathbf{L}_2$. Hence, there exist l-ideals I_1 and I_2 of $\mathbf{L}_1$ and $\mathbf{L}_2$, respectively, and an isomorphism $\phi : \mathbf{L}_1/I_1 \to \mathbf{L}_2/I_2$ such that $(x, y) \in L$ iff $\phi(xI_1) = yI_2$. Then, in particular $I_1 \times I_2 \subseteq L$. Note that because of $G \neq \{1\}$ we have $I_1 \neq \{1\}$. Let M be the monolith l-ideal of $\mathbf{L}_1$. Then every function from L_1 to M is compatible for the l-group $\mathbf{L}_1$. Since L_1 is infinite and l-groups are algebras of finite type, the cardinality of M^{L_1} is bigger than the cardinality of the set of polynomial functions on $\mathbf{L}_1$. Therefore there exists a compatible nonpolynomial function $f : L_1 \to M$. Since $M \subseteq I_1$ and $I_1 \times I_2 \subseteq L$, the function $g(x, y) = (f(x), 1)$ where $x \in L_1$, $y \in L_2$, preserves L. Moreover, since the variety of l-groups is CD, the function g is compatible. However, the function g cannot be a polynomial because it induces a nonpolynomial function on the subdirect factor $\mathbf{L}_1$.   ●

Corollary 5.6.22 *A direct product of a subdirectly irreducible l-group and any l-group is never affine complete.*

Bibliography

[Aichinger, 2000]   Aichinger, E. *On Hagemann's and Herrmann's characterization of strictly affine complete algebras*, Algebra Universalis, to appear.

[Artin, 1957]   Artin, E. *Geometric Algebra*, Interscience Publishers, New York.

[Baker, Pixley, 1975]   Baker, K. A., Pixley, A. F., *Polynomial interpolation and the Chinese remainder theorem for algebraic systems*, Math. Z. **143**, 165-174.

[Balbes, Dwinger, 1974]   Balbes, R., Dwinger, P., *Distributive Lattices*, Univ. Miss. Press.

[Bandelt, 1993]   Bandelt, H.-J., *Diagonal subalgebras and local polynomial functions of median algebras*, Algebra Universalis **30**, 20-26.

[Beazer, 1982]   Beazer, R., *Affine complete Stone algebras*, Acta Math. Acad. Sci. Hungar. **39**, 169-174.

[Beazer, 1983]   Beazer, R., *Affine complete double Stone algebras with bounded core*, Algebra Universalis **16**, 237-244.

[C. Bergman, 1998]   Bergman, C., *Categorical equiv-alence of algebras with a majority term*, Algebra Universalis **40**, 149-175.

[Bergman, Berman, 1996]   Bergman, C., Berman, J., *Morita Equivalence of almost-primal clones*, J. Pure Appl. Algebra **108**, 175-201.

[G. Bergman, 1977]   Bergman, G. M., *On the existence of subalgebras of direct products with prescribed d-fold projections*, Algebra Universalis **7**, 341-356.

[G. Bergman, 1988]   Bergman, G. M., *Embedding arbitrary algebras in groups*, Algebra Universalis **25**, 107-120.

[Berman, 1977]   Berman, J., *Distributive lattices with an additional unary operation*, Aequationes Math. **16**, 165-171.

[Blyth, Varlet, 1994]   Blyth, T., Varlet, J. C., *Ockham Algebras*, Oxford Univ. Press.

[Burris, Sankappanavar, 1981]   Burris, S., Sankappanavar, H. P., *A Course in Universal Algebra*, Graduate Texts in Mathematics, Springer, New York.

[Clark, Davey, 1998]   Clark, D., Davey, B., *Natural Dualities for the Working Algebraist*, Cambridge Univ. Press, Cambridge, New York, Melbourne.

[Clark, Krauss, 1969]   Clark, D., Krauss, P., *Paraprimal algebras*, Algebra Universalis **6**, 165-192.

[Clark, Werner, 1981]   Clark, D., Werner, H., *Affine completeness in semiprimal varieties*, Colloq. Math. Soc. J. Bolyai **28** (Finite algebra and multiple valued logic), North-Holland, Amsterdam, 809-823.

[Csákány, 1994]   Csákány, B., *Functional Completeness of Algebras*, TEMPUS JEP-06015-93 Lecture Notes, Rome, 1994, 1-14.

[Darnel, 1995]   Darnel, M., *Theory of lattice ordered groups*, Marcel Dekker, New York.

[Davey, Werner, 1983]   Davey, B., Werner, H., *Dualities and equivalences for varieties of algebras*, Colloq. Math. Soc. J. Bolyai **33** (Contributions to lattice theory), North-Holland, Amsterdam, 101-275.

[Denecke, Lüders, 2000]   Denecke, K., Lüders, O., *Category equivalence of varieties and invariant relations*, Algebra Universalis, to appear.

[Dorninger, 1980]   Dorninger, D., *A note on local polynomial functions over lattices*, Algebra Universalis **11**, 135-138.

[Dorninger, Eigenthaler, 1982]   Dorninger, D., Eigenthaler, D., *On compatible and order preserving functions on lattices,*

Banach Center Publ. **9**, Semester 1978, Universal Algebra and Applications, 97-104.

[Dorninger, Nöbauer, 1979]   Dorninger, D., Nöbauer, W., *Local polynomial functions on lattices and universal algebras*, Colloq. Math. **42**, 83-93.

[Faith, 1973]   Faith, C., *Algebra: Rings, Modules, Categories I*, Springer, New York-Heidelberg-Berlin.

[Fleischer, 1955]   Fleischer, I., *A note on subdirect products*, Acta Math. Acad. Sci. Hungar. **6**, 463-465.

[Foster, 1953]   Foster, A. L., *Generalized "Boolean" theory of universal algebras*,
Part I: Subdirect sums and normal representation theorem, Math. Z. **58**, 306-336.
Part II: Identities and subdirect sums in functionally complete algebras, Math. Z. **59**, 191-199.

[Foster, Pixley, 1964]   Foster, A. L., Pixley, A. F., *Semi-categorical algebras II*, Math. Z. **85**, 169-184.

[Fraser, Horn, 1970]   Fraser, G., Horn, A., *Congruence relations in direct products*, Proc. Amer. Math. Soc. **26**, 390-394.

[Fried, Kiss, 1983]   Fried, E., Kiss, E. W., *Connection between congruence lattices and polynomial properties*, Algebra Universalis **17**, 227-262.

[Fried, Pixley, 1979]   Fried, E., Pixley, A. F., *The dual discriminator function in universal algebra*, Acta Sci. Math. (Szeged) **41**, 83-100.

[Fried, Pixley, 1994]   Fried, E., Pixley, A. F., *Compatible functions and the Chinese Remainder Theorem*, Algebra Univeralis **32**, 478-492.

[Fuchs, 1949]   Fuchs, L., *Über die Ideale arithmetischer Ringe*, Comment. Math. Helv. **23**, 334-341.

[Fuchs, 1970, 1973]   Fuchs, L., *Infinite Abelian Groups*, Academic Press, Pure and Appl. Math., vol. 36, New York, London.

[Geiger, 1968]   Geiger, D., *Closed systems of functions and predicates*, Pacific J. Math. **27**, 95-100.

[Gierz, 1994]   Gierz, G., *Morita equivalence of quasiprimal algebras and sheaves*, Algebra Universalis **35**, 570-576.

[Goldstern, Shelah, 1998]   Goldstern, M., Shelah, S., *Order polynomially complete lattices must be large*, Algebra Universalis **39**, 197-209.

[Goldstern, Shelah, 1999]   Goldstern, M., Shelah, S., *There are no infinite order polynomially complete lattices, after all*, Algebra Universalis **42**, 49-57.

[Grätzer, 1962]   Grätzer, G., *On Boolean functions (Notes on lattice theory II)*, Rev. Math. Pures Appl. (Bucarest) **7**, 693-697.

[Grätzer, 1964]   Grätzer, G., *Boolean functions on distributive lattices*, Acta Math. Acad. Sci. Hung. **15**, 195-201.

[Grätzer, 1979]   Grätzer, G., *Universal Algebra*, 2nd edition, Springer, New York.

[Gumm, 1978]   Gumm, H. P., *Is there a Mal'cev theory for single algebras?*, Algebra Universalis **8**, 320-329.

[Gumm, 1979]   Gumm, H. P., *Algebras in permutable varieties: Geometrical properties of affine algebras*, Algebra Universalis **9**, 8-34.

[Hagemann, Herrmann, 1982]   Hagemann, J., Herrmann, C., *Arithmetical locally equational classes and representation of partial functions*, Math. Soc. J. Bolyai **29** (Universal algebra), North Holland, Amsterdam, 345-360.

[Haviar, 1992]   Haviar, M., *On affine completeness of distributive p-algebras*, Glasgow Math. J. **34**, 365-368.

[Haviar, 1993]   Haviar, M., *Affine complete algebras abstracting Kleene and Stone algebras*, Acta Math. Univ. Comenian. **2**, 179-190.

[Haviar, 1995a]   Haviar, M., *The study of affine completeness for quasi-modular double p-algebras*, Acta Univ. M. Belii. Series Math. **3**, 17-30.

[Haviar, 1995b]   Haviar, M., *Construction and affine completeness of princi-*

pal p-algebras, Tatra Mt. Math. Publ. **5**, 217-228.

[Haviar, 1996] Haviar, M., *Affine complete algebras abstracting double Stone and Kleene algebras*, Acta Univ. M. Belii. Series Math. 4, 39-52.

[Haviar, Kaarli, Ploščica, 1997] Haviar, M., Kaarli, K., Ploščica, M., *Affine completeness of Kleene algebras*, Algebra Universalis **37**, 477-490.

[Haviar, Ploščica, 1995] Haviar, M., Ploščica, M., *Affine complete Stone Algebras*, Algebra Universalis **34**, 355-365.

[Haviar, Ploščica, 1997] Haviar, M., Ploščica, M., *Affine completeness of Kleene algebras II*, Acta Univ. M. Belii **5**, 51–61.

[Herrmann, 1973] Herrmann, C., *S-verklebte Summen von Verbänden*, Math. Z. **130**, 255-274.

[Hobby, McKenzie, 1988] Hobby, D., McKenzie, R., *The Structure of Finite Algebras*, Amer. Math. Soc. Contemporary Mathematics Volume 76, Providence.

[Hu, 1969] Hu, T. K., *Stone duality for primal algebra theory*, Math. Z. **110**, 180-198.

[Hu, 1971] Hu, T. K., *Characterization of algebraic functions in equational classes generated by independent primal algebras*, Algebra Universalis **1**, 187-191.

[Huhn, 1972]   Huhn, A. P., *Weakly distributive lattices*, preprint, József Attila University, Szeged.

[Isbell, 1980]   Isbell, J. R., *Median Algebras*, Trans. Amer. Math. Soc. **260**, 319-362.

[Iskander, 1972]   Iskander, A., *Algebraic functions on p-rings*, Colloq. Math. **25**, 37-42.

[Jakubík, 1995]   Jakubík, J., *Affine completeness of complete lattice ordered groups*, Czechoslovak Math. J. **45**, 571-576.

[Jakubík, Csontóová, 1998]   Jakubík, J., Csontóová, M., *Affine completeness of projectable lattice ordered groups*, Czechoslovak Math. J. **48**, 359-363.

[Jónsson, 1967]   Jónsson, B., *Algebras whose congruence lattices are distributive*, Math. Scand. **21**, 110-121.

[Jónsson, 1972]   Jónsson, B., *Topics in universal algebra*, Lecture Notes in Math., vol 250, Springer, Berlin-Heidelberg-New York.

[Kaarli, 1982]   Kaarli, K., *Affine complete Abelian groups*, Math. Nachr. **107**, 235-239.

[Kaarli, 1983]   Kaarli, K. *Compatible function extension property*, Algebra Universalis **17**, 200-207.

[Kaarli, 1985]   Kaarli, K., *Locally affine complete Abelian groups* Tartu Ülik. Toimetised **700**, 11-16.

[Kaarli, 1990]   Kaarli, K., *On affine complete varieties generated by hemiprimal algebras with Boolean congruence lattices*, Tartu Ülik. Toimetised **878**, 23-38.

[Kaarli, 1992]   Kaarli, K., *On varieties generated by functionally complete algebras*, Algebra Universalis **29**, 495-502.

[Kaarli, 1993]   Kaarli, K., *On certain classes of algebras generating arithmetical affine complete varieties*, General algebra and applications, Research and Exposition in Mathematics **20**, Heldermann, Berlin, 152-161.

[Kaarli, 1997]   Kaarli, K., *Locally finite affine complete varieties*, J. Austral. Math. Soc. (Series A) **62**, 141-159.

[Kaarli, Márki, 1995]   Kaarli, K., Márki, L., *Affine complete commutative inverse semigroups*, Algebra Colloquium **2**, 51-78.

[Kaarli, Márki, Schmidt 1983]   Kaarli, K., Márki, L., Schmidt, E. T., *Affine complete semilattices*, Monatsh. Math. **99** (1985), 297-309.

[Kaarli, McKenzie, 1997]   Kaarli, K., McKenzie, R., *Affine complete varieties are congruence distributive*, Algebra Universalis **38**, 329-354.

[Kaarli, Pixley, 1987]   Kaarli, K., Pixley, A. F., *Affine complete varieties*, Algebra Universalis **24**, 74-90.

[Kaarli, Pixley, 1994]   Kaarli, K., Pixley, A. F., *Finite arithmetical affine complete algebras having no proper subalgebras*, Algebra Universalis **31**, 557-579.

[Kaarli, Täht, 1993]   Kaarli, K., Täht, K., *Strictly locally order affine complete lattices*, Order **10**, 261-270.

[Kaiser, Sauer, 1993]   Kaiser, H. K., Sauer, N., *On order polynomially complete lattices*, Algebra Universalis **30**, 171-176.

[Kasch, 1982]   Kasch, F., *Modules and rings*, London Math. Soc. Monograph No. 17, Academic Press Inc., London.

[Keimel, Werner, 1974]   Keimel, K., Werner, H., *Stone duality for varieties generated by quasi-primal algebras*, Mem. Amer. Math. Soc. **148**, 59-85.

[Kindermann, 1979]   Kindermann, M., *Über die Äquivalenz von Ordnungspolynomvollständigkeit und Toleranzeinfachheit endlicher Verbände*, Contributions to General Algebra (Proc. Klagenfurt Conf. 1978), Verlag J. Heyn, Klagenfurt, 145-149.

[Kiss, Márki, Pröhle, Tholen, 1983]   Kiss, E. W., Márki, L., Pröhle, P., Tholen, W., *Categorical algebraical properties. A compendium of amalgamation, congruence extension, epimorphisms, residual smallness and injectivity*, Studia Sci. Math. Hungar. **18**, 79-141.

[Köhler, Pigozzi 1980]   Köhler, P., Pigozzi, D., *Varieties with equationally definable principal congruences*, Algebra Universalis **11**, 213-219.

[Korec, 1978]   Korec, I., *A ternary function for distributivity and permutability of an equivalence lattice*, Proc. Amer. Math. Soc. **69**, 8-10.

[Korec, 1981]   Korec, I., *Concrete representation of some equivalence lattices*, Math. Slovaca **31**, 13-22.

[MacLane, 1997]   MacLane, S., *Categories for the Working Mathematician*, Springer, New York.

[Mal'cev, 1954]   Mal'cev, A. I., *On the general theory of algebraic systems* (Russian), Mat. Sb. **35** (77), 3-20.

[Maurer, Rhodes, 1965]   Maurer, W. D., Rhodes, J. L., *A property of finite simple non-abelian groups*, Proc. Amer. Math. Soc. **16**, 552-554.

[McKenzie, 1975]   McKenzie, R., *On spectra, and the negative solution of the decision problem for identities having a finite nontrivial model*, J. Symbolic Logic **40**, 186-196.

[McKenzie, 1976]   McKenzie, R., *On minimal locally finite varieties with permutable congruence relations*, preprint.

[McKenzie, 1978]   McKenzie, R., *Para primal varieties: A study of finite axiomatizability and definable principal congruences in locally finite varieties*, Algebra Universalis **8**, 336-348.

[McKenzie, McNulty, Taylor, 1987] McKenzie, R., McNulty, G., Taylor, W., *Algebras, Lattices, Varieties*, Wadsworth and Brooks/Cole, Monterey, California.

[Michler, Wille, 1970] Michler, G., Wille, R., *Die primitiven Klassen arithmetischer Ringe*, Math. Z. **113**, 369-372.

[Mitschke, 1978] Mitschke, A., *Near unanimity identities and congruence distributivity in equational classes*, Algebra Universalis **8**, 29-32.

[Murskiĭ, 1975] Murskiĭ, V. L., *The existence of finite bases of identities, and other properties of "almost all" finite algebras* (Russian), Problemy Kibernet, 30, 43-56.

[Nöbauer, 1976] Nöbauer, W., *Über die affinvollständigen, endlich erzeugbaren Moduln*, Monatsh. Math. **82**, 187-198.

[Nöbauer, 1978] Nöbauer, W., *Affinvollständige Moduln*, Math. Nachr. **86**, 85-96.

[Pálfy, Szabó, Szendrei, 1982] Pálfy, P. P., Szabó, L., Szendrei, Á., *Automorphism groups and functional completeness*, Algebra Universalis **15**, 385-400.

[Pippenger, 1997] Pippenger, N., *Theories of Computability*, Cambridge Univ. Press, Cambridge, New York, Melbourne.

[Pixley, 1961]   Pixley, A. F., *Clusters of algebras: identities and structure lattices*, Ph.D. Dissertation, U.C. Berkeley.

[Pixley, 1963]   Pixley, A. F., *Distributivity and permutability of congruence relations in equational classes of algebras*, Proc. Amer. Math. Soc. **14**, 105-109.

[Pixley, 1971]   Pixley, A. F., *The ternary discriminator function in universal algebra*, Math. Ann. **191**, 167-180.

[Pixley, 1972a]   Pixley, A. F., *Completeness in arithmetical algebras*, Algebra Universalis **2**, 179-196.

[Pixley, 1972b]   Pixley, A. F., *Local Weak Independence and Primal Algebra Theory*, Boll. Un. Mat. Ital. (4) **5**, 381-399.

[Pixley, 1979]   Pixley, A. F., *Characterizations of arithmetical varieties*, Algebra Universalis **9**, 87-98.

[Pixley, 1982]   Pixley, A. F., *A survey of interpolation in universal algebra*, Colloq. Math. Soc. J. Bolyai **29**, (Universal algebra), North-Holland, Amsterdam, 583-608.

[Pixley, 1984]   Pixley, A. F., *Principal congruence formulas in arithmetical varieties*, Universal Algebra and Lattice Theory, Lecture Notes in Math. **1149**, Springer, New York, 238-254.

[Ploščica, 1994]   Ploščica, M., *Affine complete distributive lattices*, Order **11**, 385-390.

[Ploščica, 1995]   Ploščica, M., *Affine complete median algebras*, Contributions to General Algebra **9**, 261-270.

[Ploščica, 1996]   Ploščica, M., *Affine completions of distributive lattices*, Order **13**, 295-311.

[Ploščica, Haviar, 1998]   Ploščica, M., Haviar, M., *On order-polynomial completeness of lattices*, Algebra Universalis **39**, 217-219.

[Pöschel, 1979]   Pöschel, R., *Concrete representation of algebraic structures and a general Galois theory*, Contributions to General Algebra (Proc. Klagenfurt Conf. 1978), Verlag J. Heyn, Klagenfurt, 249-972.

[Pöschel, Kalužnin, 1979]   Pöschel, R., Kalužnin, L., *Funktionen- und Relationen-algebren*, Birkhäuser, Basel.

[Post, 1941]   Post, E. L., *The two-valued iterative systems of mathematical logic*, Ann. of Math. Studies, 5, Princeton University Press, N. J.

[Quackenbush, 1971]   Quackenbush, R. W., *Demi-semiprimal algebras and Mal'-cev-type conditions*, Math. Z. **122**, 166-176.

[Quackenbush, 1974]   Quackenbush, R. W., *Some classes of idempotent functions*

	and their compositions, Colloq. Math., **29**, 71-81.
[Quackenbush, 1979]	Quackenbush, R. W., *A new proof of Rosenberg's primal algebra characterization theorem*, Colloq. Math. Soc. J. Bolyai **28** (Finite algebra and multiple valued logic), North-Holland, Amsterdam, 603-634.
[Quackenbush, Wolk, 1971]	Quackenbush, R. W., Wolk, B., *Strong representation of congruence lattices*, Algebra Universalis **1**, 165-166.
[Rosenberg, 1970]	Rosenberg, I., *Über die funktionale Vollständigkeit in den mehrwertigen Logiken*, Rozpravy Československe Akad. Věd., Ser. Math. Nat. Sci. **80**, 3-93.
[Rosenberg, 1979]	Rosenberg, I., *On a Galois connection between algebras and relations and its applications* (Proc. Klagenfurt Conf. 1978), Verlag J. Heyn, Klagenfurt, 273-289.
[Rosenberg, Szabó, 1984]	Rosenberg, I., Szabó, L., *Local Completeness I*, Algebra Universalis **18**, 308-326.
[Rosenbloom, 1942]	Rosenbloom, P. C., *Post algebras I, Postulates and general theory*, Amer. J. Math. **64**, 167-188.
[Rousseau, 1967]	Rousseau, G., *Completeness in finite algebras with a single op-*

eration, Proc. Amer. Math. Soc. **18**, 1009-1013.

[Saks, 1985a]

Saks, A., *On affine completeness of modules* (Russian), Tartu Ülik. Toimetised **700**, 71-79.

[Saks, 1985b]

Saks, A., *On affine completeness of decomposable modules* (Russian), Tartu Ülik. Toimetised **764**, 123-135.

[Schmidt, 1979]

Schmidt, E. T., *Remark on compatible and order preserving functions on lattices*, Studia Sci. Math. Hung. **14**, 139-144.

[Schmidt, 1987]

Schmidt, E. T., *On locally order-polynomially complete lattices*, Acta Math. Hungar. **49**, 481-486.

[Schweigert, 1974]

Schweigert, D., *Über die endliche, ordnungspolynomvollständige Verbände*, Monatsh. Math. **78**, 68-76.

[Sierpiński, 1945]

Sierpiński, W., *Sur les fonctions de plusieurs variables*, Fund. Math. **33**, 169-173.

[Snow, 1998]

Snow, J. W., *Relations on Algebras and Varieties as Categories*, Ph D. Dissertation, Vanderbilt Univ., Nashville, Tennessee.

[Stone, 1972]

Stone, M. G., *Subalgebra and automorphism structure in universal algebras: a concrete characterization*, Acta Sci. Math. (Szeged) **33**, 45-48.

[Szendrei, 1981] Szendrei, Á., *A new proof of the McKenzie-Gumm theorem*, Algebra Universalis **13**, 133-135.

[Szendrei, 1992] Szendrei, Á., *The primal algebra characterization theorem revisited*, Algebra Universalis **29**, 41-60.

[Taylor, 1973] Taylor, W., *Characterizing Mal'cev conditions*, Algebra Universalis **3**, 351-397.

[Wade, 1945] Wade, L. I., *Post algebras and rings*, Duke Math. J. **12**, 389-395.

[Werner, 1970] Werner, H., *Eine Charakterisierung funktional vollständiger Algebren*, Arch. Math. (Basel) **21**, 381-385.

[Werner, 1971] Werner, H., *Produkte von Kongruenzklassengeometrien universeller Algebren*, Math. Zeit. **121**, 111-140.

[Werner, 1978] Werner, H., *Discriminator Algebras*, Studien zur Algebra und ihre Anwendungen, vol. 6, Academie-Verlag, Berlin.

[Wille, 1970] Wille, R., *Kongruenzklassengeometrien Universeller Algebren*, Lecture Notes in Math. 113, Springer, Berlin-Heidelberg-New York.

[Wille, 1977a] Wille, R., *Eine Characterisierung endlicher, ordnungspolynomvollständiger Verbände*, Arch. Math. (Basel) **28**, 557-560.

[Wille, 1977b]   Wille, R., *Über endliche, ord-*
                              nungaffinvollständige Verbände,
                              Math. Z. **155**, 103-107.

[Zariski, Samuel, 1958]   Zariski, O., Samuel, P., *Com-*
                              mutative Algebra, vol 1, Van
                              Nostrand, Princeton.

Index

Milton Keynes UK
Ingram Content Group UK Ltd.
UKHW021637071024
449327UK00020BA/1333

9 780367 398330